Y0-ARC-684

Postglacial Vegetation of Canada

Postglacial Vegetation of Canada

J. C. Ritchie

University of Toronto
Scarborough College
Toronto, Canada

Cambridge University Press

Cambridge

New York New Rochelle

Melbourne Sydney

Published by the Press Syndicate of the University of Cambridge
The Pitt Building, Trumpington Street, Cambridge CB2 1RP
32 East 57th Street, New York, NY 10022, USA
10 Stamford Road, Oakleigh, Melbourne 3166, Australia

First published 1987

Printed in the United States of America

Library of Congress Cataloging-in-Publication Data
Ritchie, J. C. (James Cunningham), 1929-
Postglacial vegetation of Canada.
Bibliography: p.
Includes index.
1. Paleobotany – Quaternary. 2. Paleobotany – Canada.
I. Title.
QE931.R573 1987 561′.1971 87-15072

British Library Cataloguing-in-Publication Data
Ritchie, J.C.
Postglacial vegetation of Canada.
1. Paleobotony – Canada
I. Title
561′.1971 QE938.A1
ISBN 0 521 30868 2

For Samantha

ARCTIC OCEAN
PACIFIC OCEAN
BEAUFORT SEA
Queen Elizabeth Islands
Ellef Ringnes
Amund Ringnes
Prince Patrick Island
Bathurst I.
Melville I.
Banks Island
Viscount Melville Sound
Victoria Island
Tuktoyaktuk Pen.
Melville Hills
Brooks Range
ALASKA
FAIRBANKS
Tanana River
OLD CROW
INUVIK
Anderson River
Horton River
YUKON TERRITORY
Mackenzie River
Great Bear Lake
Coppermine R.
NORTHWEST TERRITORIES
DISTRICT OF MACKENZIE
DISTRICT OF KEEWATIN
WHITEHORSE
FORT SIMPSON
YELLOWKNIFE
Liard River
Great Slave Lake
Queen Charlotte Islands
BRITISH COLUMBIA
Peace River
Lake Athabasca
Athabasca River
CHURCHILL
Churchill River
ALBERTA
Fraser River
Vancouver Island
North Saskatchewan River
SASKATCHEWAN
MANITOBA
Lake Winnipeg
Olympic Pen.
Kootenay R.
South Saskatchewan River
WASHINGTON
Columbia River
Missouri River
Lake-of-the-Woods
OREGON
MONTANA
NORTH DAKOTA
MINNESOTA
IDAHO
SOUTH DAKOTA
CALIFORNIA
Mississippi
WYOMING
NEVADA
UTAH
NEBRASKA
IOWA
0 200 km
120
110
100
MCR 78

Ellesmere Island
Devon Island
Baffin Island
Clyde Inlet
Cumberland Sound
Southampton Island
GREENLAND
ARCTIC CIRCLE
ATLANTIC OCEAN
HUDSON BAY
Ungava Bay
NOUVEAU QUEBEC
Kaniapiscau River
Hamilton Inlet
NEWFOUNDLAND
LABRADOR
NEWFOUNDLAND ISLAND
Avalon Peninsula
Anticosti
Gulf of St. Lawrence
Gaspé Pen.
PRINCE EDWARD
NEW BRUNSWICK
NOVA SCOTIA
QUEBEC
ONTARIO
Lake Abitibi
Lake Temiscaming
QUEBEC
MONTREAL
OTTAWA
St. Lawrence River
MAINE
Manitoulin I.
Lake Superior
MICHIGAN
WISCONSIN
Lake Michigan
Lake Huron
TORONTO
L. Ontario
Lake Erie
NEW YORK
PENNSYLVANIA
VERMONT
NEW HAMPSHIRE
MASS.
CONN.
RHODE ISLAND
NEW JERSEY
60
50
40
90
80
70
60
PRODUCED BY THE SURVEYS AND MAPPING BRANCH. DEPARTMENT OF ENERGY, MINES & RESOURCES

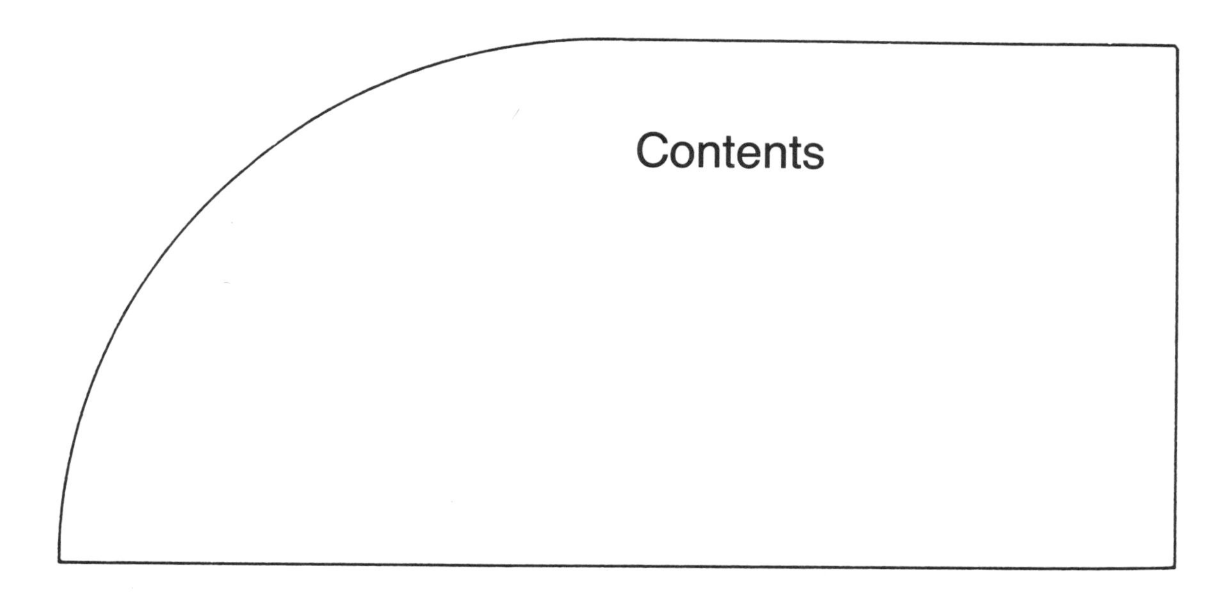

Contents

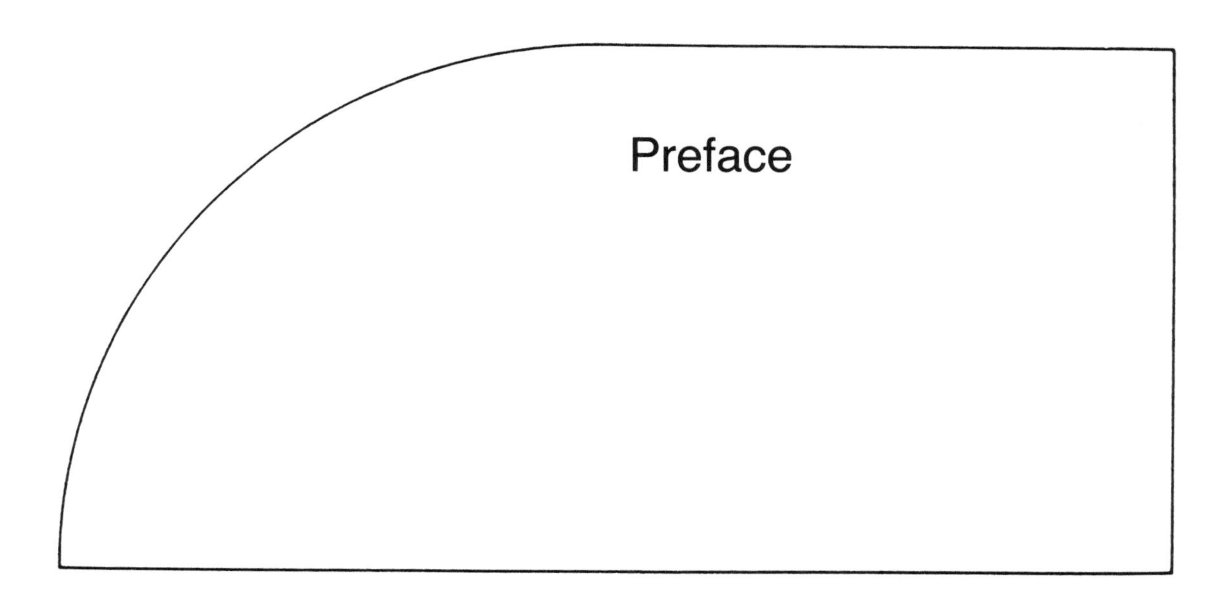

Preface

The aim of the book is threefold: first, to assemble the published information on the history of the plant cover of Canada for the latest part of the Quaternary – during and following the most recent major ice age; second, to search for patterns of vegetation change and to evaluate such alternative explanations of these changes as climatic factors, varied rates of species spread from Pleistocene refugia, biological factors that might have controlled the spread and abundance of species, soil and geomorphological factors, chance events, and other influences; and third, to expose interesting, useful, and challenging problems in palaeoecology that might be solved by imaginatively designed searches for new data or novel analyses of the existing record.

Why do I suppose that such a conspectus might be useful, or even feasible, as the landmass involved is so immense, diversified, and patchily represented by modern investigations? Others have written perceptively on the rationale for studies of the Quaternary and its biota (Wright 1976, 1984). Let me add simply that few thoughtful observers can doubt that the rate of culturally induced environmental change is accelerating faster than our ability to predict consequences. While one need not subscribe to the views of those who foresee an early apocalyptic outcome, or even those who call for a new social order to grapple with the disarray of the times, if such it be, it is clear that an improved understanding of past changes in plant cover and its relations with environment will be helpful if the future is to be predicted accurately. Recent concerns have propelled palaeoecology to front stage, for example, in the evaluations of the possible effects of changes in global climate that might result from increases in atmospheric CO_2; in the attempts to disentangle longer-term climatic controls from short-term cultural effects on arid region ecosystems; and even in the predictions of the responses of vegetation to various versions of the "nuclear winter" effect. The trite notion that the past has much to offer in our attempts to look forward is well founded, though not always rigorously exploited.

In addition to these "useful" attributes, I hope that this compilation will contribute in some way to the ongoing development of new concepts in plant ecology. In contrast with Europe, the Canadian landscape has a short, though locally brutal history of human interference. So, for many regions north of our fringe of human settlement, the latest products of vegetational history can still be examined at first hand. We can also usefully study such topics, essential for the understanding of past vegetation, as the relationships between pollen deposition and modern vegetation.

Vegetation can be an object of scientific study from several points of view: composition, structure, function, environment, distribution, taxonomy, and history, no better elaborated than by Egler (1977). This book deals primarily with the last and is restricted to only the most recent sliver of geological time — the latest roughly 18,000 years. Canada offers the largest single landmass anywhere to have been almost completely denuded of plant cover and then almost completely revegetated in the geologically very short time span of less than 10,000 years.

Is the sprawl and variety of the country so great, its history of palaeoecology so short, and are the political boundaries so unnatural that the proposed task is unmanageable? The answer to all three parts of the question is "probably yes," and a treatment of all of North America would have seemed more logical. However, there are at hand very recent, detailed, modern treatises on the vegetational history and related palaeoecology of the United States (Wright 1983), whereas Canada remains without any synthesis volume. Certainly there are huge gaps, partly because the pool of investigators is small, partly because distances are great – Canada is second only to the USSR in area, with almost ten million square kilometers. There are more sites (about 500) with pollen data on vegetational history in Switzerland (Lang 1985) with a land surface of 41,277 km^2, three-quarters the size of Nova Scotia, than in all of Canada.

However, science is much more than data collection. In spite of the gaps and thin areas, some patterns are emerging from the recent flurry of modern investigations. An attempt to pull these threads together should help to identify and clarify the areas for fruitful future research; to identify new problems in palaeoecology; to pose the appropriate questions, answerable within the time and scope of current research; and, equally important, to suggest regions or topics where the continued accumulation of data is not particularly useful or necessary. A dense network of fossil sites, with detailed pollen and macrofossil diagrams, is not of itself important. Until the data are interpreted rigorously in terms of vegetational change and palaeoenvironments, they remain only as monuments to someone's enormous labour in the laboratory.

The lack of coincidence of our international borders with any natural, geographic boundary might create an unfortunate restriction on the scope of this book. I deal with it, partly at least, by drawing in whatever portions of the record from the adjacent United States that seem necessary to yield a balanced treatment. So the appropriate records from New England, the southern Great Lakes Basin, the northern Great Plains, Washington, and Alaska are included.

The book was written for a heterogeneous readership of fairly broad background. As I have written, I have had in mind those colleagues and senior students who, either because they live and work outside North America or because their special field of scholarship is peripheral though related to Quaternary plant ecology, require a balanced, up-to-date synthesis of a large body of literature that probably lies just beyond the range of their intimate familiarity. For that reason, I have assembled and condensed information that is already variously familiar to the small band of active palaeoecologists whose investigations are centred in North America. Of course, I hope they too will find something of interest, but my chief aim is to interest ecologists, physical geographers, geologists, foresters, archaeologists, soil scientists, and historians, and therefore to provide a coherent document that requires for comprehension the very minimum of specialized or technical knowledge.

In reviewing the relatively large body of relevant literature, I have been assailed by the reality of what has, of course, been known for some time – first, that we are making slow progress in developing rigorously quantitative methods of linking fossil, chiefly pollen data, to palaeoclimate, and, second, that the vast majority of pollen sites in the literature register past vegetation at regional scales with very little information at the level of plant communities. As a result, reconstruction has often been imprecise, discursive, even anecdotal, and commonly riddled with circular reasoning. To counter this tendency, I have tried to separate the factual material from the interpretation and to present enough information on the modern ecology and pollen representation of the important taxa so that the interested reader can develop interpretations of the fossil data independent of the original authors.

I view the book, hopefully, as one of the last of a particular genre – a conventional, descriptive, narrative account of the past record of vegetation, capped (or soiled!) with a final chapter that rests on my subjective, sometimes blinkered, but I hope rarely idiosyncratic or arrogant assessments of what it all means in palaeoecological terms. If current attempts in the more active laboratories of palynology succeed, Quaternary plant ecology might make the transition that Birks (1985) describes from the descriptive and narrative to the analytical and rigorously quantitative – hot on the heels of the parent subject, ecology (Harper 1977). One major impediment however, as all palynologists and indeed all investigators who record microfossils have long appreciated, is that the accumulation of data requires laborious, relatively slow work at the microscope. Dense networks of fossil sites are essential, each with detailed, carefully registered inventories of microfossils, if the statistical precepts of quantitative analysis are to be observed. In Canada, only southern Quebec is close to meeting these requirements, thanks to the excellent team directed by Professor Pierre Richard. Indeed, I have been influenced in my choice of format for this book by the major compilation of the first modern pollen records for Quebec by Richard (1977a). In that superb document, overlooked by our linguistically parochial American colleagues I regret to say, he sets out the modern framework of environment, flora and vegetation, the modern pollen rain, and then the fossil record, followed by a perceptive interpretation that touches on many of the central themes addressed here.

Acknowledgements

The preparation of this book was the first phase of a continuing project, to document the late-Quaternary pollen records of Canada, supported by a Special Research Grant from the University of Toronto. In addition, my specific field and laboratory investigations have been generously supported by the National Science and Engineering Research Council of Canada. Permission to reproduce certain photographs and figures was granted by: the Canadian National Institute for Forest Research, the Canadian Journal of Earth Sciences (C. W. Barnosky 1984), Quaternary Research (M. B. Davis, R. W. Spear and L.C.K. Shane 1980), Canadian Journal of Forest Research (S. Payette 1985), and by R. G. Baker, K. D. Bennett, L. C. Cwynar, R. B. Davis, S. A. Edlund, D. G. Green, G. A. King, K-B. Liu, G. M. MacDonald, C. E. Schweger, and S. C. Zoltai. I acknowledge gratefully the prompt and helpful response of these colleagues, and the following, to my request for advanced information and comment on their investigations: T. W. Anderson, P. A. and H. R. Delcourt, E. C. Grimm, R. J. Hebda, G. L. Jacobson, S. T. Jackson, J. K. Jeglum, M. S. Kearney, B. H. Luckman, R. N. Mack, J. B. MacPherson, R. W. Mathewes, N. G. Miller, R. J. Mott, P. J. Mudie, R. W. Spear, J. Terasmae, B. G. Warner, D. R. Whitehead, and H. E. Wright, Jr.

Constructive reviews of early versions of individual chapters were provided by F. K. Hare and W. A. Watts, while the thorough critiques of the first draft of the entire manuscript by H. J. B. Birks, L. C. Cwynar and K. Gajewski, and of a later draft by K. D. Bennett, were enormously helpful. I acknowledge the great value of being able to participate in two international cooperative projects – Project 158b of the International Geological Correlation Programme (IGCP), and the Cooperative Holocene Mapping Project (COHMAP). The Graphics Department of Scarborough College provided excellent illustrative (Diane Gradowski, Tony Westbrook) and photographic (David Harford) work. I greatly appreciate the meticulous and constructive efforts of Christine Rogers and Kate Hadden in preparing and checking the illustrative materials, reading the text, and in providing helpful suggestions on all aspects of the endeavour. Sue West played a key role by her skillful and careful handling of word processing and index preparation, and by her efficient control of most of the exchanges and discussions with technical and editorial personnel at the Press. Penny Goddard, with skill and equanimity, assumed responsibility for entering the final version and assisting with the compilation of the index. I have dedicated the book to Samantha, who supervised most of the writing, and to June Hope who inspired me to launch the project and who provided the essential support and understanding to enable me to realize it.

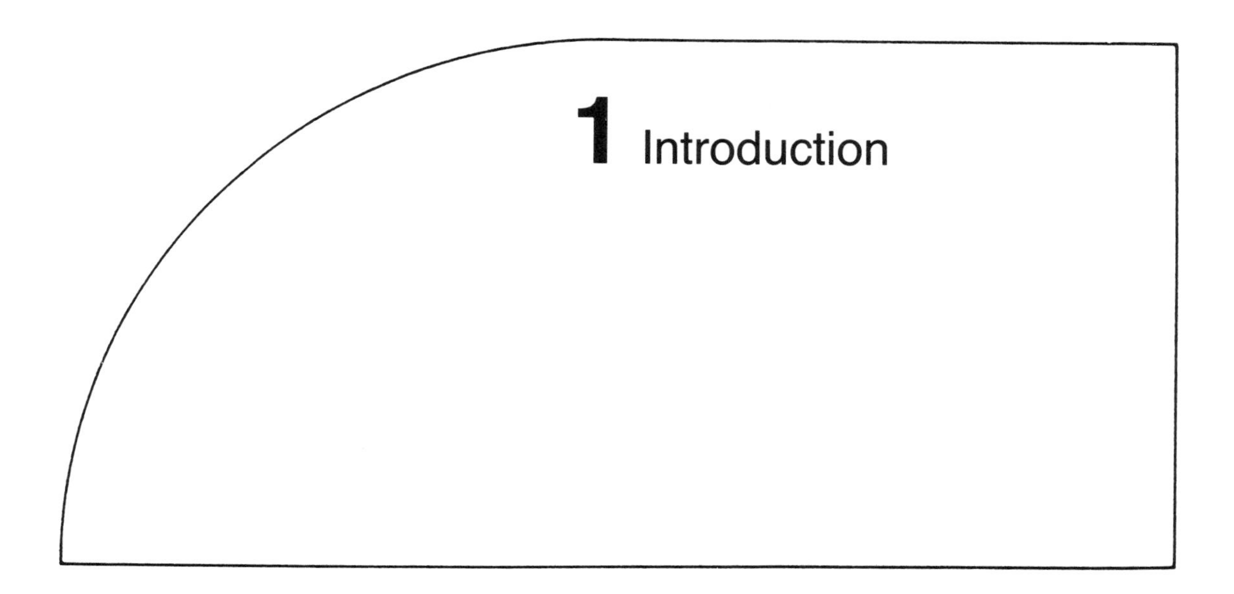

1 Introduction

The vast and varied landscapes of Canada are mantled by a tapestry of vegetation whose diversity and richness stimulate our senses, tantalize our intellects, and underpin our national economy. This green cover ranges from the soaring rain forests of the Pacific coasts, through the transcontinental monotony of the taiga, to the transient seasonal brilliance of the northern tundra and the plains grasslands. Few countries have such a range of vegetation diversity set out on such a massive scale.

Furthermore, in both its present landscape and its immediate past, Canada has been dominated by glaciations on a vast scale. On the one hand, the periodic inundation of Canada by Quaternary continental and montane ice sheets resulted in the almost complete removal and destruction of soft, fossil-bearing sediments deposited during past interglacial periods. On the other hand, there is the remarkable opportunity to trace the details of the process of revegetation of buried and denuded landscapes across the full width of the continent. Indeed, the central theme of this book will be that there were several different patterns of revegetating the country as it emerged from the waning ice sheets of the latest glacial period.

The plan of the book is to provide a concise account of whatever background information is essential to an understanding of vegetational history – the physical setting, consisting of climate, geology, and soils; the modern flora and vegetation as expressed in the pattern of landform–vegetation regions; and a summary of the autecology of the main species of the past and present vegetation and of their representation in modern pollen spectra. It is important to stress that in the chapter on the physical setting, I am presenting only those aspects that I consider to have direct relevance to the later chapters. The reader will find nothing on tectonics and pre-Quaternary geological history; synoptic climatology is kept to a bare minimum, and I have combined the description of regional climates with concise accounts of the modern vegetation. Aspects of the physical environment are emphasized that appear to have a direct role in influencing the ecological relations of a regional vegetation complex – physiography, surface roughness and materials, and climatic factors – that determine the thermal and moisture characteristics of sites and regions. Then follows the core of the book, an account of the record of vegetational history since the latest glaciation for each of the major regions, based chiefly on data from pollen analysis. Finally, attempts are made to collate these data in the form of regional accounts of vegetational history and to relate these trends and events to climatic and other environmental changes. In the presentation of the fossil data, I have exercised selection, attempting to use results from sites that I consider to be both representative of large areas and of optimum quality in terms of the details of pollen identification, control of the chronology by isotope and other dating methods, and interpretive rigour. No attempt is made to provide a catalogue of every site in Canada from which some palaeobotanical data have been recorded – such listings can be found elsewhere (Harington and Rice 1984; Bryant and Holloway 1985).

In its broad, philosophical outlines, the book follows roughly an ecological schema that is certainly not novel or original, but whose roots go back in both the development of ecological ideas and in the molding of my own complexion of biases, predilections, and insights. It is based on the notion that the most useful unit of study is the landscape region defined loosely as an area with similar topography or landforms, macroclimate, and zonal vegetation. This theme has its origins in the landshaft–vegetationen groupings of Ludi and other European plant ecologists (Sukachev 1960), and is expressed in such modern approaches as the IGCP Project 158b investigation of mires and lakes (Berglund and Digerfeldt 1976), and various biophysical and ecosystem classifications in North America (Hills 1960; Thie and Ironside 1976). I used the approach in a regional summary of part of the Western Interior of Canada (Ritchie 1976), and interested readers can find a lucid statement of the concept in an essay by Rowe (1984).

A critical question in the use of such an approach is that of scale. In regions with a dense network of sites, each with detailed palaeobotanical records, the ecological regions can be small, perhaps approaching the recommended size of a few thousand square kilometres (Berglund 1986). However, in Canada, which combines vast land area with relatively few investigated sites, an appropriate size of an ecological region would be 50,000 to 150,000 km^2 in the plains region and smaller in the western montane belt.

It is pertinent here to clarify the general question of the spatial scale of the pollen record used to reconstruct the past vegetation of Canada, so that the limitations of the method can be appreciated fully. The problem of spatial scale has been recognized for many years and discussed fully, and somewhat repetitively, in the recent literature. It is that "paleoecologists are frequently frustrated by the difficulty of defining precisely the geographic area represented by pollen data derived from the sediments of a given lake" (Jacobson and Bradshaw 1981, p. 83). These authors provide a useful summary of the literature and show that in forested regions, closed-drainage lakes less than 1 ha receive 75 to 100 percent of their pollen from sources within 20 m of the site. That is, changes in forest stands can be expected to be registered in such very small sites. On the other hand, lakes larger than 5 ha receive pollen chiefly from a regional source area, defined loosely as an area with a radius of several kilometres surrounding the site. A recent analysis by Bradshaw and Webb (1985) of the areal representation of pollen spectra from lakes of different sizes in Wisconsin and Michigan provides an excellent illustration of the value of detailed, dense networks of pollen sites and reliable quantitative estimates of forest cover. This signal contribution addresses the central problem of the relationship between the proportions of a taxon in the pollen sum and in the surrounding landscape, as it is influenced by both the size of the sedimentary repository and the area of the contributing vegetation. I will return to their findings in Chapter 3, where the pollen productivity and representation of individual taxa are examined. Similarly, a detailed investigation by Janssen (1984) of the pollen spectra in surface samples from a large mire complex in Minnesota demonstrates convincingly that the different scales of vegetation pattern are represented by pollen types of different source areas. The interested reader can find excellent reviews of these matters in Birks and Birks (1980, Chapter II); Birks and Gordon (1985, Chapter 6), and Prentice (1983, 1985), and an exemplary practical demonstration by Heide (1984) of the different spatial resolution of pollen records from a lake, and from a small hollow in the adjacent forest.

The general conclusion of these and other studies in the forested regions of the temperate zone, excluding montane areas, is that pollen spectra in lakes of 5 to 75 ha are derived primarily from regional sources. Forest regions are defined loosely as areas "in which the same vegetation succession will occur on the same physiographic site, providing the type and degree of disturbance is the same" (Hills 1960, p. 410). In practical terms, only minor attention need be paid to the problem of spatial scale – the reality is that even in the area of Canada and the adjacent United States with the most dense network of sites, the pollen record is predominantly regional because lake size is almost invariably greater than 5 ha and the level of detailed pollen identification and enumeration is rarely adequate to decipher local vegetation changes at the stand level.

Generalizations about spatial scale should be accepted cautiously. Most that have appeared recently (Davis 1969; Webb, Yeracaris, and Richard 1978; Delcourt and Delcourt 1979; Delcourt, Delcourt, and Webb 1983) have relevance only to forested plains regions, as Berglund (1986) stresses. Montane and arctic areas, which make up roughly one-half of the land surface of Canada, present quite different problems of pollen representation. Spear (1981), for example, has documented the difficulty of pollen analysis in mountainous regions – first, montane vegetation types are relatively unproductive of pollen so that "only a small percentage of the total pollen assemblage is definitely produced by high-elevation species" (p. 52); and second, as earlier investigators have concluded, "Surface-sample data show that a large amount of pollen from low-elevation trees is carried upward far beyond elevations where these trees occur" (p. 71). These difficulties are confronted but not resolved in Chapter 7 on Pacific and Cordilleran vegetational history.

Arctic regions are equally intractable in the interpretation of the pollen record, and several investigations from different sites in the circumpolar tundra zone have shown that the regional, treeless vegetation, particularly in the higher latitudes, produces a small proportion (20 to 50%) of the pollen recorded in lake sediments (summarized by Birks 1973).

The reader may quickly comprehend the central thrust of the book by examining Figures 1.1 to 1.3. They set out the rudiments of our undertaking.

Figure 1.1. The approximate extent of glaciers in North America at 18,000 yr BP, after Prest (1984), to illustrate that all of Canada, except for areas in the far northwest, was covered by ice. The broken line marks the approximate boundary between the Keewatin and Labradorean Sectors.

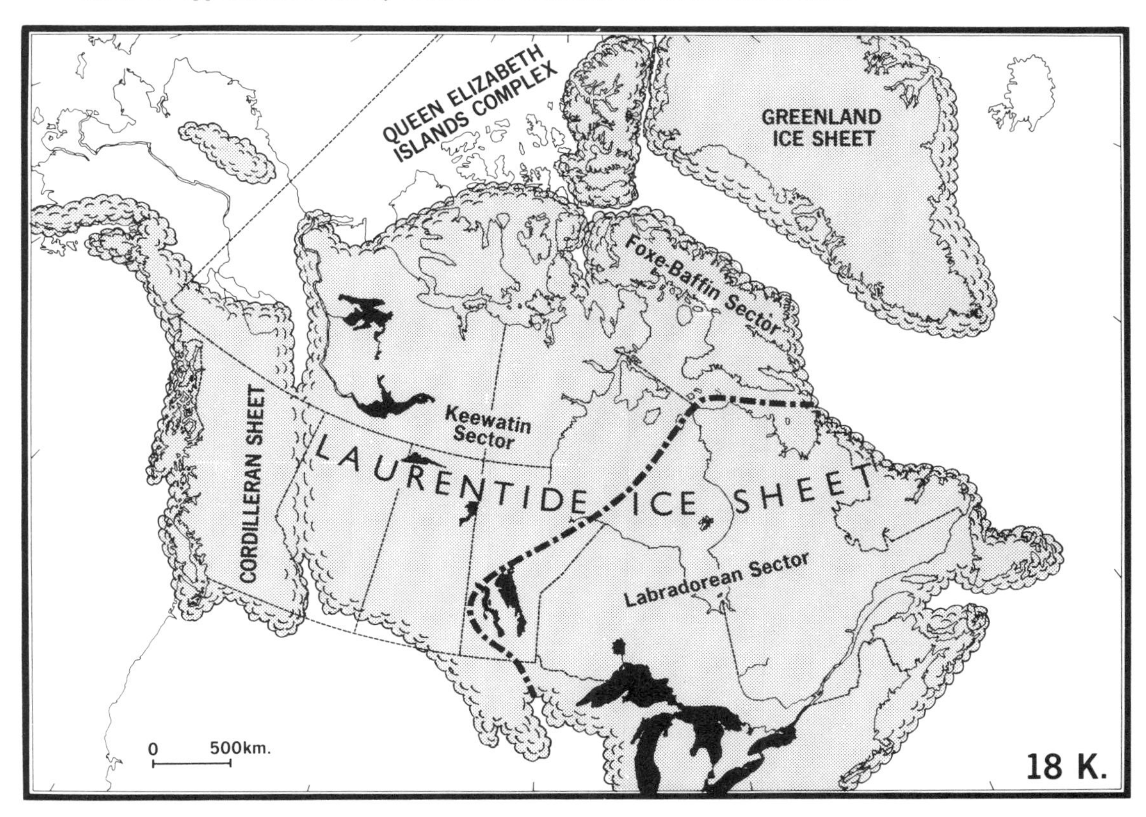

For several millennia centred on 18,000 yr BP, Canada, except for a few areas in the far northwest and a few high mountain peaks, was covered by ice, as shown in Figure 1.1. This massive aggregation of continental and montane glaciers extended into the United States for variable distances between 100 and 600 km beyond the southern border of Canada. In fact, there was not a single glacier mass but several complexes that were extensively confluent at about 18,000 yr BP (Prest 1984). The Laurentide Ice Sheet was made up of three main segments – the Foxe-Baffin Sector, centred on west Baffin Island; the Labradorean sector, centred on the Quebec–Labrador uplands and extending west to fuse with the Keewatin Sector and south over the Great Lakes–St. Lawrence basin; and the Keewatin Sector, centred in Keewatin and extending west to the mountains, north to the arctic, and southwest to the plains. The Cordilleran Ice Sheet consisted of a complex of many intermontane, piedmont, and valley glaciers that formed a "2000-m thick mass in the central and southern interior of British Columbia and its surface probably stood higher than the confining Mountain ranges" (Prest 1984, p. 20).

It is important to note that there is considerable disagreement in the geological literature about various limits and extents of Wisconsin ice sheets, such that the recent map by Prest (1984) depicts both a "maximum portrayal limit" and a "minimum portrayal limit." The differences are rarely important from our viewpoint, and, in any case, the scale of the maps used here is such that precise locations of ice front positions cannot be depicted.

In response to climatic warming, the ice began to recede, not uniformly, but with local asynchronous advances and recessions by the ice front position. This complex process, accompanied by the formation of large proglacial lakes and a global scale rise in sea level, began about 16,000 yr BP and ended when the two main residual continental ice caps disappeared from Keewatin and northern Quebec–Labrador about 6,000 years ago. Between the time of maximum ice cover and the present day, when a broadly zonal pattern of vegetation covers Canada (Fig. 1.2), a complex, still incompletely known process of revegetation occurred. This book aims to gather together into a coherent synthesis what is known of this roughly 15,000 years of history.

The total array of Canada's vegetated landscapes comprises five ecological provinces – Tundra,

Figure 1.2. The approximate limits of the five main ecological provinces (Tundra, Boreal, Temperate, Grassland–Parkland, Montane Pacific–Cordilleran) with subdivisions and transitional zones shown where appropriate.

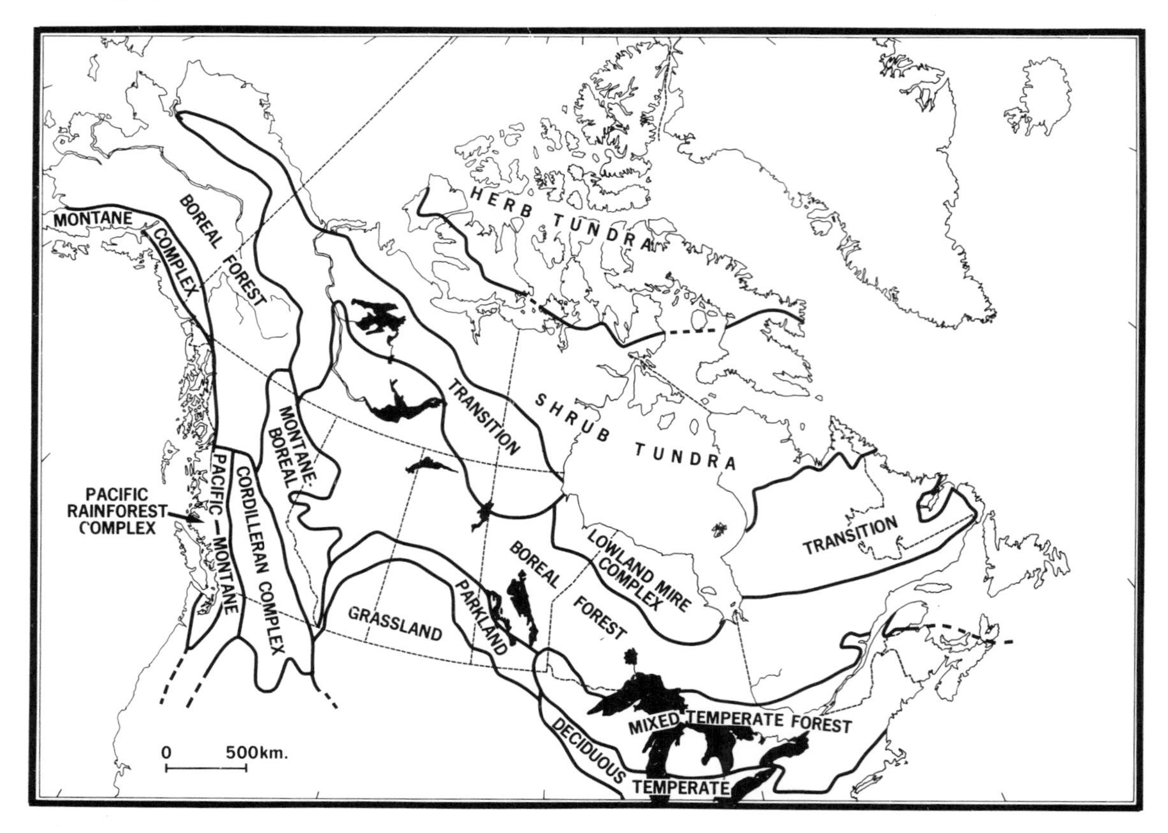

Boreal, Temperate, Montane Pacific–Cordilleran, and Grassland–Parkland. I use the term *ecological province* to conform both in concept and in application with a recent, very effective description of the climates, soils, and vegetation of Canada (Ecoregions Working Group in press). An ecological province is the highest unit of classification of terrestrial landscapes, and it describes an area that has a distinctive array of macroclimates and a distinctive type of plant cover both floristically and physiognomically. My usage of these terms will become obvious in Chapter 2, and it will be clear that they correspond closely with similar terms and groupings used elsewhere or in Canada by other treatments (e.g., the ecosystem regions of the United States of Bailey 1980; the land regions of the ecological (biophysical) classification in Canada by Thie and Ironside 1976). The five classes are subdivided hierarchically, where appropriate, in Chapter 2, under "Bioclimates." However, as was noted above in the comments on the question of scale, subdivision will take place only to the level for which fossil information is available. It would, of course, be desirable to subdivide Canada into small units of about 50,000 to 100,000 km^2 (the type regions of the IGCP Project 158 in Berglund 1986), but the irregular distribution of fossil sites and low density even in the most thoroughly investigated region (southern Quebec), preclude such an endeavour at present.

The five primary, essentially biogeographic groups provide both a framework for organizing the data to be examined and a logical basis for moderate subdivision into smaller units, for example, into various major physiognomic types, from tundra through conifer woodlands and forests, to mixed conifer–hardwood forests, to parklands and grasslands, all with enormous regional diversity of structure and composition along gradients of topography, humidity, thermal regime, and surface materials (Fig. 1.2).

It will be discovered that this classification into five ecological provinces, or bioclimates, meets the requirements of a "good classification" (Egler 1977, p. 451) both in its provision of a coherent hierarchy and, more interestingly, in that major classes appear to have "an equally shared past history."

The size limitation of this book, and the magnitude of the task, preclude a detailed treatment of the modern vegetation of Canada along the lines

Figure 1.3. Geographical range maps of ten taxa to illustrate the more important floristic elements in Canada – Arctic (*Potentilla vahliana*), eastern Boreal (*Abies balsamea*), eastern Temperate (*Pinus resinosa*, *Picea rubens*, *Acer saccharum*), Prairie (*Sphaeralcea coccinea*), Cordilleran (*Pinus contorta*, *Abies lasiocarpa*, *Picea engelmannii*), and Pacific (*Acer macrophyllum*). The data for these and the range maps in Chapter 3 have been abstracted from Fowells (1965), Little (1971), and Porsild and Cody (1980).

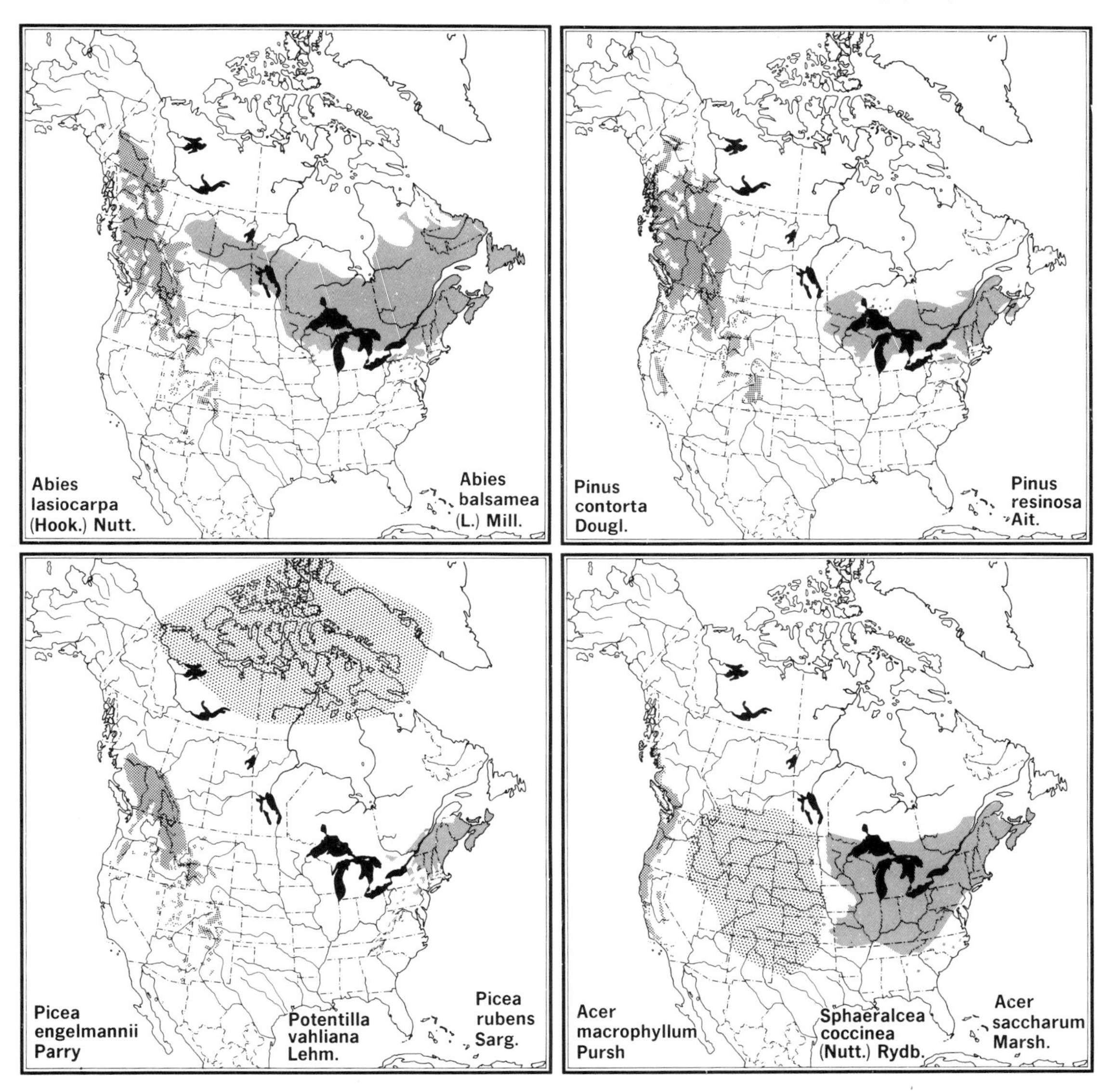

of a previous synthesis of one small segment of the country (Ritchie 1984a). I have chosen instead to integrate a very brief description of the modern vegetation with the account of climate, using a schema of bioclimates borrowed from others (Ecoregions Working Group in press), but I direct the reader to more detailed descriptions of regional vegetation at appropriate points in the text.

The task can be further elaborated by considering briefly the ranges of a few of the species that dominate some of the regional plant communities. Some (*Picea glauca*, *Populus balsamifera*) have achieved continent-wide distribution. Others (*Fagus*, *Tsuga*, *Pinus strobus*) are confined to the southeastern part of the country, extending for variable distances north or south of the Great Lakes–St. Lawrence axis. Some (*Abies lasiocarpa*, *Picea rubens*) have been restricted to western or eastern areas, whereas one interesting group of herbs and shrubs is restricted to the southern plains region and another more or less to the modern arctic–subarctic zone (Fig. 1.3). An attempt will be made to trace the histories of both individual taxa and the aggregations they make up at any particular time past in assembling the record of past vegetation.

Terms and abbreviations

The following terms are used throughout the text.

Brief reference is made in Chapter 2 to the soils that are characteristic of particular bioclimates; only the main categories are used, following the hierarchical classification of the Canada Soil Survey Committee, Subcommittee on Soil Classification (Canada Department of Agriculture 1970). They are the nine soil orders found in Canada, as follows:

Regosols: Soils weakly developed with absence of genetic horizons, usually due to youthfulness or instability

Solonetzic: Soils developed over saline parent materials, usually in semiarid climates, with a characteristic prismatic or columnar B horizon

Chernozemic: Soils of subarid regions with a thick organic A horizon, and lacking the properties of Solonetzic soils

Brunisolic: Soils with weakly developed eluvial and illuvial horizons, with a characteristic diffuse brown B horizon

Luvisolic: Soils developed on base-saturated, often calcareous parent materials, with eluvial and illuvial horizons, the Bt of the latter always within 50 cm of the surface

Podzolic: Soils with well developed eluvial (Ac) and illuvial (B) horizons, the clayey B horizon (Bt) occurring at depths greater than 50 cm

Cryosolic: Soils with permafrost within 1 m of the surface, usually affected by cryoturbation, occupying most of the arctic-subarctic

Gleysolic: Soils with saturated lower horizons that also have reducing conditions, resulting in ferrous mottling effects

Organic: Soils made up of deep peat, muck, or other organic material

The following conventions and abbreviations are used:

PAR, pollen accumulation rate, is expressed as the number of pollen grains per cm^2 per year, and NAP is sometimes used as the short form of nonarboreal pollen.

Système International units are used throughout.

I have tried to avoid using such recent items of ecological vocabulary as catastrophe, crises, disturbances, etc. (Raup 1981) to describe, for example, disease epidemics, because the imagery associated with these words connotes spurious notions, particularly that some factors are intrinsically different from such "natural" factors as climate. The norm is that several major environmental factors, varying greatly in frequency and intensity, cause change in vegetational structure and composition. It is probably more useful to describe that range of factors and to record and quantify their effects in as much temporal and spatial detail as possible, without designating some as catastrophes, disturbances, etc.

Radiocarbon dates are given as years Before Present (BP), using the 1955 Libby half-life value. When a radiocarbon age for an event, trend, or change is given with its standard deviation, a level of precision defined by the indicated probabilities is implied; on the other hand, if no standard deviation is shown, it should be understood to indicate an approximate age, roughly to the nearest millennium. Similarly, when I summarize pollen data in the text, approximate pollen percentages are given, rounded off in most cases to the nearest 5 or 10 percent.

I use the terms *full glacial*, *late glacial*, and *postglacial* in the conventional but informal way of most North American treatments, for example, the recent statement of Porter (1983), and in most cases the radiocarbon age is included to remove any doubts. *Full glacial* is used to describe events associated with the maximum extension of glacier ice and in the present book that means from about 20,000 to 14,000 years ago for many but not all regions. *Late glacial* describes transitional conditions when glaciers were still widespread and, while occasionally expanding, were in general receding. *Postglacial* usually refers to events or times subsequent to either the disappearance of glacial ice or the end of a glacial climate, but clearly the postglacial in northern Labrador, for example, began several millennia later than in New Brunswick. Similarly, the term *Holocene* is used imprecisely to refer to the latest 10,000 or 12,000 years of the Quaternary period, characterized generally as a time of nonglacial climate, or as the beginning of the present interglacial climatic mode. Two recently published companion volumes, one on the late Pleistocene (Porter 1983) and one on the Holocene of the United States (Wright 1983), illustrate the fact that, in the words of one of the editors, "The environmental changes of the latest Pleistocene and the Holocene can, therefore, be viewed as a continuum in part reflecting the ongoing recovery from the glacial age but also reflecting the secondary climatic variations superimposed on the general first-order trend" (Porter 1983, p. xiv). This continuum is reflected in considerable overlap and even duplication of material between the two volumes, in no way detrimental to their great value.

Final revision of the scientific aspects of the manuscript was made in June 1986, which was also the latest date for references cited in the text.

2 The biogeographical setting

Geology, physiography, and surface materials

The physiography of Canada is the result of modification of the bedrock framework by many complex processes. This historical geology of the country is a massive and rapidly evolving body of knowledge, most of which is beyond the scope of this book. I shall assemble here only those data that are directly pertinent to an understanding of the vegetational history. Interested readers can find the larger subject presented effectively in Bird (1980), Douglas (1971), and Fulton (1984).

The following account reviews the basic geological structures and their subsequent modification by Pleistocene events to form the modern physiographic patterns. Finally, the distribution of surface materials is reviewed, described broadly in terms of soil characteristics.

The structural framework

Four structural elements make up the continent of North America, and it will be discovered that they have been of central importance in influencing both the glacial history and the ecological processes that culminated in the present-day pattern of vegetation cover. The basic geological features are the central Canadian Shield, a broken ring of orogenic features, four interior platforms, and certain minor areas of Continental Shelves (Fig. 2.1).

The *Canadian Shield*, covering 4×10^6 km^2, is a complex, dissected peneplain made up of a large variety of rocks with complex and ancient origins. It is the dominant physiographic feature of the northeastern sector of the continent and comprises the major part of the surface area of Quebec, Ontario, and Keewatin. The shield has a core of Precambrian rocks, with younger overlying sedimentary layers forming the periphery. It also underlies parts of Hudson Bay and the Hudson Bay Lowlands. In the northwest, along the western edge, and throughout northern Quebec, Labrador, and Baffin Island, large areas of Shield bedrock are exposed, with little or no mantle of unconsolidated materials (Fig. 2.1). As a result, the bedrock geology has primary influence in these areas on the relief, surface roughness and resulting microclimates, drainage pattern, occurrence and morphometry of lakes, vegetation pattern, frequency and size of lightning fires, and soil development. Relief is generally about 100 m in the western and southern segments, but is as high as 1,000 m in parts of Labrador, Quebec, and Baffin Island. By contrast, in the interior arc of the Shield that makes up roughly the eastern half of Keewatin, much of

Figure 2.1. A generalized map of the structural geology of Canada, based on *The National Atlas of Canada* (Fremlin 1974). 1 Cordilleran Orogen, 2 Appalachian Orogen, 3 Innuitian Orogen, 4 Canadian Shield, 5 Canadian Shield Lowlands, 6 St. Lawrence Lowlands, 7 Hudson Bay Lowlands, 8 Arctic Lowlands, 9 Interior Plains.

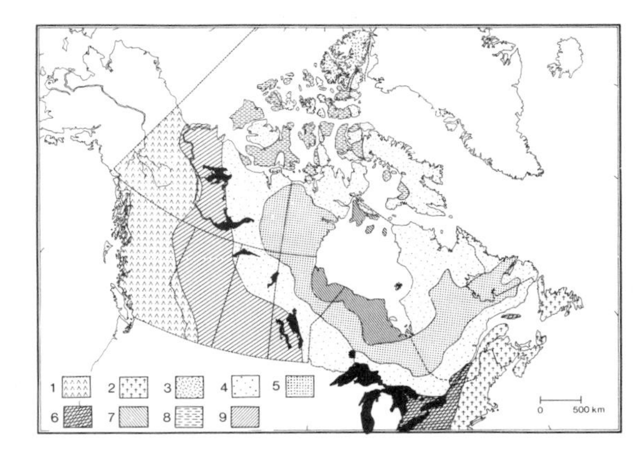

north central Manitoba, parts of northern Ontario and central Quebec and Labrador (Fig. 2.1), the topography has been modified by the addition of thick, variable deposits of unconsolidated materials derived from the Laurentide Ice Sheet and proglacial lakes. In addition, coastal tracts round Hudson Bay and the central arctic were inundated by seas during isostatic and eustatic Pleistocene events, and the residual marine deposits have modified the basic Shield topography. The resulting landforms in these areas display reduced surface roughness and large areas of moderate to poor drainage usually supporting extensive mires. However, the major drainage patterns, expressed as large rivers and lakes with extensive catchment areas, remain controlled by the bedrock structure.

The *Orogenic Ring* consists of three separate areas of mountains – the marginal geosynclines – and from our restricted viewpoint, the salient point is that Canada is flanked to the west, to the southeast, and in the far north by mountain chains that not only influence their own regional ecosystems, but have an important role in modulating the macroclimate of the interior of Canada. The longest and highest mountain chain is the Cordilleran, which borders the entire Pacific side of the country. It consists of linear complexes of mountains – Coast, St. Elias, Columbia, Selwyn, Mackenzie, and Rocky, to mention only the more extensive and familiar ones – interior plateaus (Yukon, Fraser, Kamloops, for example), and intervening trenches, basins, and valleys. The interaction of this massive axis of montane landforms with the adjacent Pacific Ocean and the predominant westerly atmospheric circulation results both in a complex array of Cordilleran vegetation types, from coastal rain forests and interior grasslands to montane tundras and woodlands, and in a significant climatic effect eastward and southward into the interior of the continent, as will be seen later in this chapter.

The Appalachian Mountains chain is the second part of the broken ring of orogenic belts. It runs from northern Florida to Newfoundland and also consists of linear spines of mountains and uplands (in Canada, the Notre Dame Mountains of southern Quebec, the New Brunswick Highlands, and the Atlantic Uplands of Nova Scotia and Newfoundland) and intervening trenches and basins. It will be discovered later that this major physiographic complex had an important influence on both the floristic diversity of the temperate forest biome and its full-glacial refugial distribution.

Finally, the Innuitian Orogen in the high arctic is the third and smallest mountain complex in Canada. It consists of several small ranges and intervening basins that comprise the northwestern two-thirds of Ellesmere Island and all of Axel Heiberg Island. The mountains exceed 2,000 m and several support large ice caps.

The third structural category of *Interior Platforms* consists of thick, generally horizontally bedded Phanerozoic (roughly 570 to 60 million years ago) rocks that overlie large parts of older shield structures in four regions of Canada, forming the Interior, St. Lawrence, Hudson, and Arctic Platforms (Fig. 2.1). The physiographic expression of these structural units is four distinct Lowlands. The largest is the *Interior Plains*, extending in a broad belt from the Mackenzie Delta region southward to the Canada–United States border and beyond to Mexico. The underlying rocks are predominantly Cretaceous sedimentary shales, siltstones, and sandstones. The *St. Lawrence Lowlands* extend from Manitoulin Island in Lake Huron eastwards to the immediate south of the Shield, through southern Ontario and southern Quebec, including Anticosti Island in the Gulf of St. Lawrence. The Hudson Platform is the structural basis of a tract of lowlands that occupies northeast Manitoba and adjacent northern Ontario and a very small area in Quebec as well as the southern half of Southampton Island, collectively referred to as the *Hudson Bay Lowlands*. Finally, the *Arctic Lowlands* occupy much of the central Canadian arctic (Fig. 2.1).

The fourth structural category, the *Continental Shelves*, is of limited relevance to our immediate interests for obvious reasons. The Shelves consist of broad submarine accumulations that extend beyond the present coastline for varying distances in the Pacific, the Beaufort Sea, all along the Baffin and Labrador coasts, and off Newfoundland and Nova Scotia. Their primary relevance here is their possible role during the full glacial as refugia or migration routes at times of lowered sea level.

Pleistocene geology

The following summary of the aspects of the Pleistocene geology of Canada that are pertinent to our current enquiry depends heavily on the publications of Prest (1970, 1984). Only two aspects concern us among the complicated and still uncertain sequence of events involved in the advance, recession, and final disappearance of the latest North American Pleistocene ice sheets. The first is the point made in Chapter 1, that as over 95 percent of the surface of Canada was covered by late-Pleistocene ice, the previous plant cover was almost completely removed, and, therefore, the present vegetation is entirely a product of migration from adjacent refugia and of various processes of subsequent adjustment to external and internal environmental factors. The second is that the present and postglacial landscapes, including features of macro- and microtopography, local climate, drainage pat-

Figure 2.2. Four sketch maps indicating the approximate positions of glaciers and proglacial lakes at 18,000, 12,700, 10,000, and 8,000 yr BP, after Prest (1970) and *The National Atlas of Canada* (Fremlin 1974).

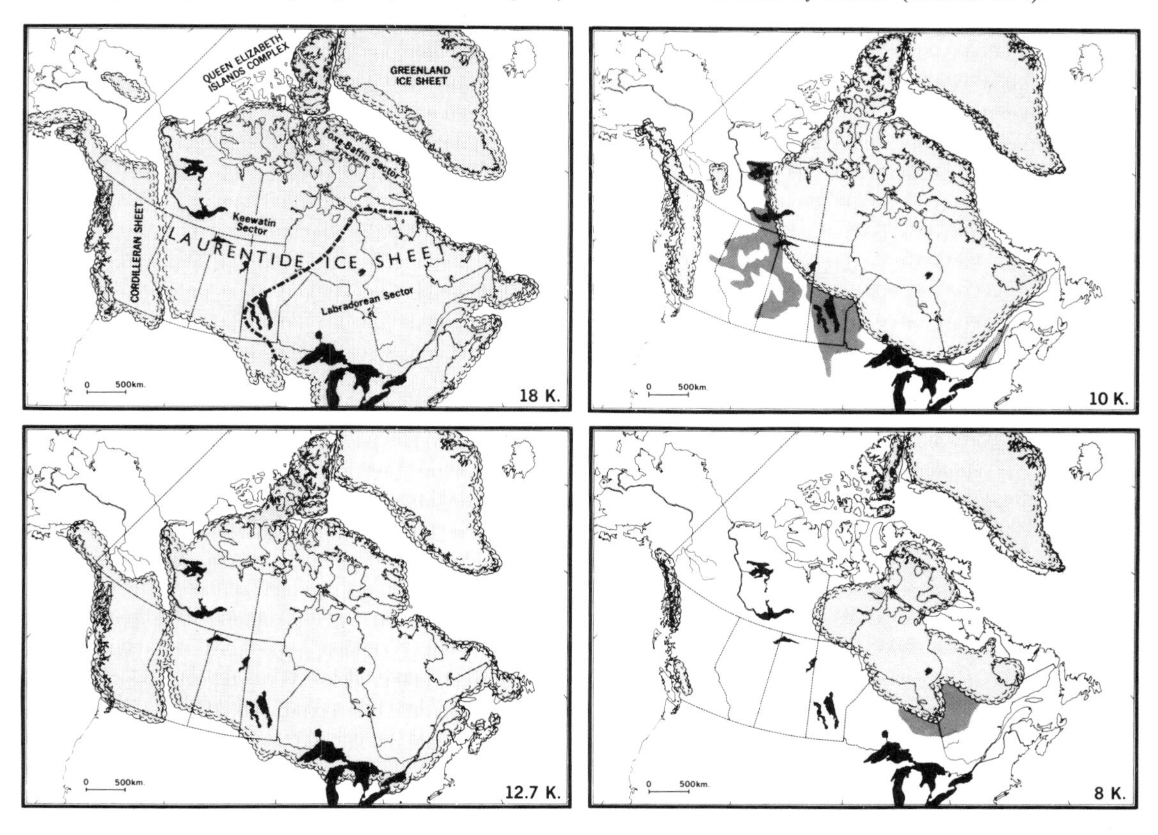

tern, soil development, and impediments or expedients to plant dispersal, are products of the glaciation and deglaciation events.

The entire complex of glaciers consisted of a large Laurentide Sheet that covered the interior mainland of Canada and extended into the adjacent United States, an Appalachian glacier complex that included several coalescent ice caps in the Maritime Provinces, a Cordilleran glacier complex that covered the mountains from the Coast Range to the Rocky Mountains, and a Queen Elizabeth Islands glacier complex that covered parts of the arctic (Fig. 2.2).

The extent of the latest glaciation, the Late Wisconsinan, has been mapped recently by Prest (1984) and Figure 2.2 shows the "maximum portrayal limit" of the two possible configurations shown by him. Disagreement about limits persists in the literature, but the issue is of little relevance to our discussion.

Recent findings collated by Prest (1984) indicate that the maximum extent of the Laurentide Ice, particularly in the Great Lakes basin, was not a synchronous event but rather "several ice lobes . . . appear to have reached their termini at different times between 19 and 21.5 ka." The Cordilleran Ice extended only about 150 km to the south into what is now the western United States, and it spread to the north only into the southern part of the Yukon, leaving the central and northern regions completely free of glacier ice. Similarly, different interpretations result in larger or smaller "ice-free corridor" reconstructions for the area between the Laurentide and Cordilleran Ice Sheets (Prest 1984, p. 28).

The sequence and chronology of glacier recession are of interest in the present context because they set out the spatial and temporal framework for the revegetation process. I have followed Prest (1970) and the map in the *National Atlas* (Fremlin 1974, pp. 31–32) in reconstructing the ice front position during the recession for 12,700, 10,000 and 8,000 yr BP.

The Cypress Hills and adjacent areas of southern Saskatchewan and southern Alberta and large areas of the northern Yukon and adjacent Northwest Territories were ice-free at the maximum extent of the Late Wisconsinan glaciation, which I take to have occurred at 18,000 yr BP. In addition, all of lowland Alaska and a large part of the central United States remained free of glaciers.

Parts of southern Newfoundland, southern Ontario, southern Alberta, and the far northwest of the Northwest Territories and the Yukon were ice-free by between 12,000 and 13,000 yr BP (Fig. 2.2). By about 10,000 yr BP, the ice margin was roughly coincident with the southern limit of the Canadian Shield (Fig. 2.2) and several large proglacial lakes were forming in the Interior Lowlands all along the margin of the Laurentide Ice, near Great Bear Lake, Great Slave Lake, the Peace and Athabasca rivers in central Manitoba and Saskatchewan (Teller and Clayton 1983), and in Ontario and Quebec (mapped in Prest 1970, pp.714–25). The expansion and recession of these large glacial lakes was a complex, still poorly understood process that lasted for a few millennia before and after 10,000 yr BP, and the approximation of their extent depicted in Figure 2.2 should be viewed with caution. Few of the major glacial lakes were synchronous in the phases of their growth and recession. Thick beds of sediment were deposited in these proglacial lakes to form flat or gently undulating surfaces. Also, during the recession of the Laurentide Ice, sea level varied, primarily in response to isostasy, and large areas of marine incursion occurred in the St. Lawrence Lowlands and the Hudson Bay region. By 8,000 yr BP (Fig. 2.2), the residual ice was centred on Hudson Bay, the Tyrrell Sea occupied a large area of northern Manitoba and northern Ontario, and only small local ice caps persisted in the Cordilleran and Appalachian segments. In Chapter 8, an attempt will be made to relate these broad stages of ice recession to the vegetational history as deciphered from the fossil record.

The glaciers, both as they expanded and in the stages of their disintegration, effected important changes on the underlying structural framework of the landscape itself already modified in largely unknown ways by multiple glaciations earlier in the Quaternary. The Shield rocks were scoured and eroded by glaciers and large areas of bedrock remained exposed, particularly near the outer edge of the Shield area, while thicker deposits of glacial drift were deposited in a broad arc from central Keewatin, northern Manitoba, a narrow belt adjacent to the Hudson Bay Lowlands in Ontario, and in an irregular central area of Quebec and Labrador, to as far east as the Hamilton Inlet. Clayton, Teller, and Attig (1985) show that the tills deposited over the Shield area in northern Manitoba and Saskatchewan, and adjacent Northwest Territories are predominantly sandy, while those on the prairies to the southwest are silty and clayey, and they suggest that these differences were responsible for rapid ice movement along the southwestern margin of the Laurentide Ice Sheet. The sandy till area on the Shield coincides with a sector of the boreal forest that is dominated by jack pine. The Interior Plains were modified by the addition of till sheets on top of tills from earlier glaciations, forming great thicknesses (>100 m) of unconsolidated materials, morainal features (both recessional and terminal), and extensive glacio-lacustrine sediments (Teller 1985). As a result of the last process, large areas of mire deposits developed in northeast Alberta and the adjacent Northwest Territories, as well as in central western Manitoba and adjacent Saskatchewan. The St. Lawrence Lowland region in southern Ontario has thick deposits of glacial drift, morainal and fluvio-glacial deltaic sediments, whereas a large tract of southern Quebec flanking the St. Lawrence River, from Ottawa to Quebec City, consists of thick clays and silts deposited during the postglacial incursion of the Champlain Sea (Gadd 1971).

Bioclimates

The following account of the climates of Canada is written from an ecological rather than a climatological viewpoint, for two reasons. First, my primary interest is to examine the climatic controls of plant communities and taxa as possible analogues of past climate, so the emphasis is on the measures and descriptions of climatic factors that appear to be important in regulating vegetation dynamics and plant growth and reproduction, and, second, the climatological and more especially the dynamical aspect of climate, while of crucial significance in explaining the resultant patterns of regional macroclimate, are beyond the scope of the present volume partly because "attempts to wrap up the dynamic climate in plain language descriptions always oversimplify the complicated reality" (Hare and Thomas 1979, p. 64), and in part because synoptic palaeoclimatology is properly the domain of the professional climatologist.

Of course there are a few generalizations about the dynamic climate of Canada that can be marshalled here as a preface to a review of regional bioclimates, and the interested reader can pursue that topic in detail in such treatments as those of Hare and Thomas (1979, Chapters 3 and 4) and Hare and Hay (1974). These authorities emphasize the central facts: (a) the climate of Canada is controlled by the global atmospheric circulation, a complex balance of energy and moisture exchanges, and (b) the particular juxtaposition of the Cordilleran axis and the vast Interior Plains and Shield regions plays a major role in determining the climate of the continent. The montane axis along the Pacific coast "is a most significant obstacle to both zonal westerlies and trade winds, while the plains of the north and east provide an unobstructed path of low thermal admittance for great meridional excursions of both arctic and tropical air masses" (Bryson and Hare

1974, p. 1). One result of this major impediment to the penetration of unmodified maritime (Pacific) air is that the climate of almost all of Canada is unusually severe by comparison with areas at similar latitudes elsewhere in the Northern Hemisphere except eastern Siberia. Winter conditions in particular are extreme, in terms of both cold and deep snows. The major eastern cities (Toronto, Montreal) are on the same latitudes as southern Europe where the recent (1985) week or two of below freezing was a rare event in the normally equable winter climate, whereas the normal winter of southeastern Canada has several months with almost continuous freezing temperatures.

The climate at the surface depends on the characteristics and behaviour of airstreams and the intervening fronts. Four major airstreams dominate the circulation patterns of Canada: the arctic airstream, originating in the arctic region and characterized by cold, dry air in winter and cool, moist air in summer; the Maritime arctic and polar streams from the northern and southern Pacific, characterized by cool and moist, and mild and moist air, respectively; the tropical airstream, originating in the Gulf of Mexico; and the Atlantic and Pacific trade winds, with warm and humid air. However, it should be noted that "the airstreams do not move about the maps as solid masses . . . " and " . . . it is quite misleading to think of them as being homogeneous masses drawn from source regions" (Hare and Hay 1974, p. 55). They are schemes of descriptive convenience, subsuming extremely complex three-dimensional interactions that occur throughout the atmosphere.

It has been suggested that some major boundaries, for example those between forest and tundra and between forest and grassland, are determined by, or at least coincide with, the mean positions of these air masses and their frontal zones (Bryson 1966). The clearest example is the occurrence of the forest–tundra ecotone at the conjunction, in summer, of arctic and southern moist air masses where thermal gradients are markedly steeper than to both the south and north of the frontal zone (Hare and Ritchie 1972). However, the role of climate in determining the occurrence and abundance of plants is measurable at the level of bioclimatic rather than synoptic factors. In the absence, with a few rare exceptions, of experimental evidence to show what climatic factors control the range and abundance of one particular taxon, we are dependent on correlations between bioclimatic values and range limits and abundances. Thus, while it is likely that such local nonclimatic factors as soil conditions, fire frequency, interspecific competition, and historical patterns are and have been important in determining local regional community structure and composition, it is probable that the general distribution and abundance of the major species of the vegetation of Canada are controlled by bioclimate. Such an assumption includes the proposition that the taxa are in equilibrium with modern climate – an assumption the validity of which depends largely on the time and spatial scales of the considerations, as discussed recently in detail by Birks (1986), Davis (1986), Prentice (1986), Ritchie (1986), and Webb (1986).

The following grouping of bioclimatic factors is an attempt to identify those influences that appear to be important in regulating the life cycles of the main categories of plants that dominate the vegetation of Canada. Such factors operate throughout the life cycles of plants, and three critical phases are usually recognized for the successful establishment and growth of a population: (1) the seed or other propagule responsible for initial dispersal and spread must germinate successfully; (2) the seedling stage, crucial in many trees, must be survived so that the juvenile or sapling stage (of a tree) can persist; and finally, (3) the plant must continue to grow until it reaches the age of reproduction. The bioclimatic factors that appear to be limiting for many plant community dominants in Canada fall into three major categories: (1) those concerned with the thermal conditions for growth, chiefly but not exclusively in summer, and including such aspects as the length of the growing season, the mean summer temperatures during the period of establishment and growth, and the incidence of extremely low ambient temperatures; (2) moisture, including the annual total, soil moisture deficits and the amount, type and persistence of snow; and (3) interactions between temperature and moisture as they influence potential evapotranspiration and the depth and extent of permafrost. I recognize fully the limitations of this approach. In particular, it overlooks the important biological fact that "sessile organisms respond to sequences of weather events at a point in space rather than to average 'weather' " (Neilson 1986, p. 28). However, the promising new approach to make causal links between high-frequency weather variations and vegetation response is only possible when both long-term data on plant demography and vegetation pattern and long-term weather data are available (Neilson 1986). Neither is available for many northern stations in Canada, and only the latter elsewhere.

Six bioclimatic, or ecoclimatic, regions are recognized in this account of the climates and associated vegetation and soils of Canada, following the groupings used by Hare and Thomas (1979) and Ecoregions Working Group (in press). Some of the regions are further subdivided; the boundaries are shown in Figure 2.3a, along with a set of representative climate diagrams for selected sites. Extensive use is made in this book of climate diagrams, designed originally by Walter and Lieth (1967). The

Figure 2.3a. The bioclimatic regions of Canada, showing their approximate boundaries and representative climate diagrams. The bioclimatic regions are High Arctic (HA), Mid-Arctic (MA), Low Arctic (LA), Subarctic (S), Boreal (B), Temperate (T), Northern and Southern Pacific (NP, SP, respectively), Northern, Interior, and Southern Cordilleran (NC, IC, SC, respectively), and Grassland (G).

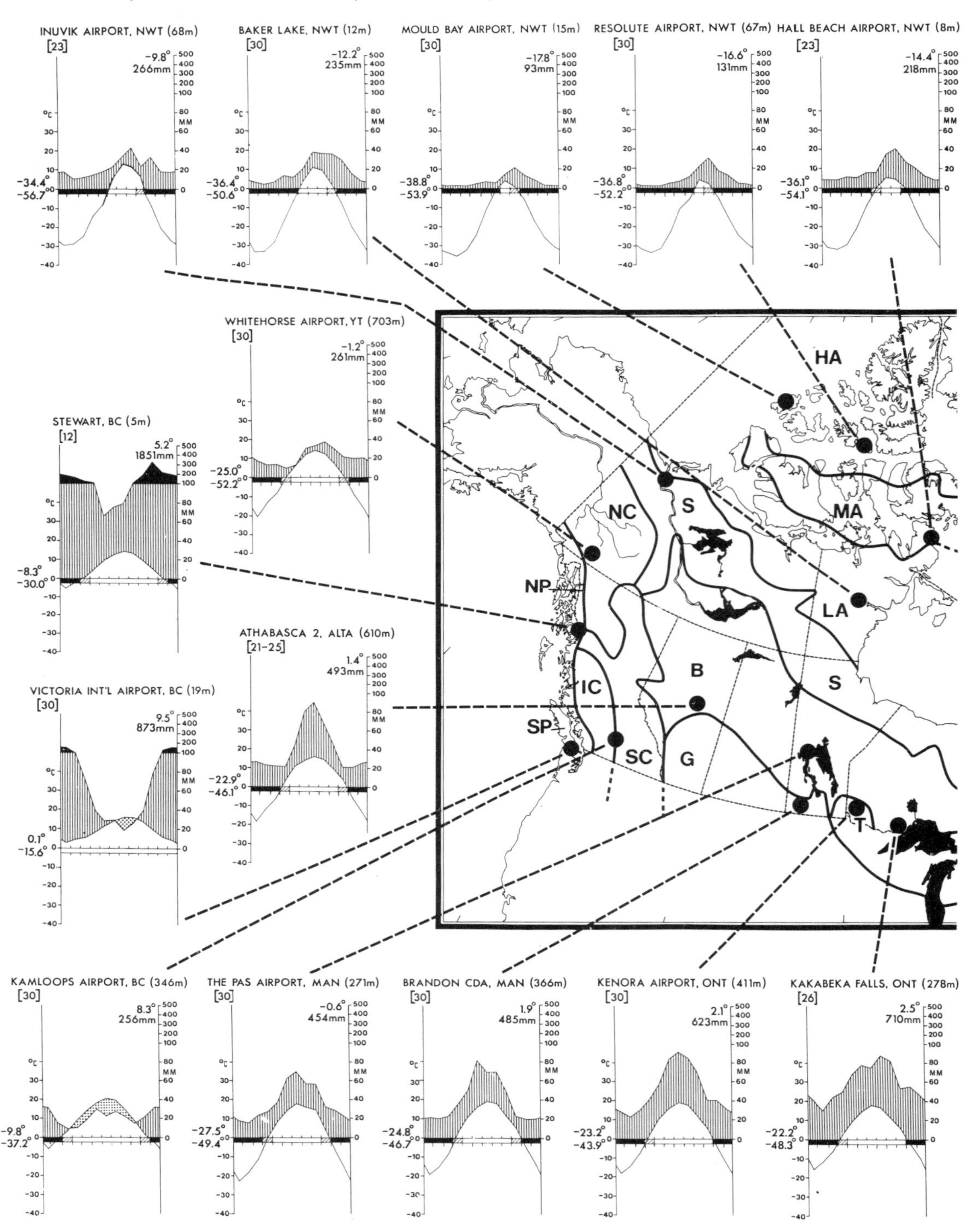

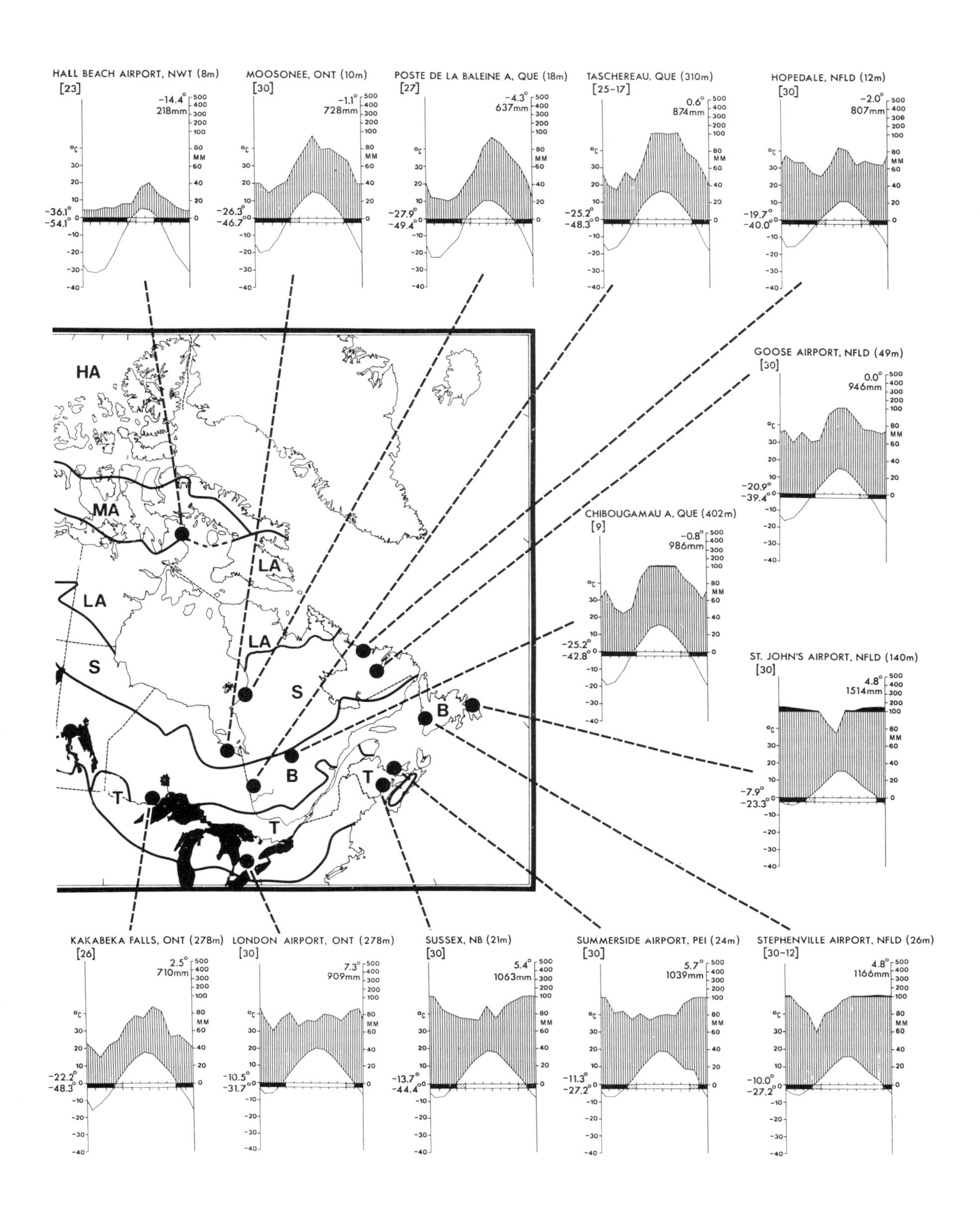
HALL BEACH AIRPORT, NWT (8m)
[23]
-14.4°
218mm
-36.1°
-54.1°
MOOSONEE, ONT (10m)
[30]
-1.1°
728mm
-26.3°
-46.7°
POSTE DE LA BALEINE A, QUE (18m)
[27]
-4.3°
637mm
-27.9°
-49.4°
TASCHEREAU, QUE (310m)
[25-17]
0.6°
874mm
-25.2°
-48.3°
HOPEDALE, NFLD (12m)
[30]
-2.0°
807mm
-19.7°
-40.0°
GOOSE AIRPORT, NFLD (49m)
[30]
0.0°
946mm
-20.9°
-39.4°
CHIBOUGAMAU A, QUE (402m)
[9]
-0.8°
986mm
-25.2°
-42.8°
ST. JOHN'S AIRPORT, NFLD (140m)
[30]
4.8°
1514mm
-7.9°
-23.3°
KAKABEKA FALLS, ONT (278m)
[26]
2.5°
710mm
-22.2°
-48.3°
LONDON AIRPORT, ONT (278m)
[30]
7.3°
909mm
-10.5°
-31.7°
SUSSEX, NB (21m)
[30]
5.4°
1063mm
-13.7°
-44.4°
SUMMERSIDE AIRPORT, PEI (24m)
[30]
5.7°
1039mm
-11.3°
-27.2°
STEPHENVILLE AIRPORT, NFLD (26m)
[30-12]
4.8°
1166mm
-10.0°
-27.2°
HA
MA
LA
LA
LA
S
S
B
B
T
T
T

Figure 2.3b. A key and legend to the climate diagrams. Notable features are the west-to-east trends at both high and low latitudes in increased precipitation and decreased seasonal variation in mean monthly temperature; length of growing season gradients from south to north, modulated in the east by proximity to the Atlantic; contrasts in both the mean temperature of the coldest month and the absolute minimum temperature, between temperate, boreal, and arctic stations in the eastern half of Canada, and between Northern Cordilleran and Southern Pacific stations. The Kamloops Station (Southern Cordilleran) has the warmest and driest climate in Canada, whereas Mould Bay (High Arctic) has one of the coldest and driest.

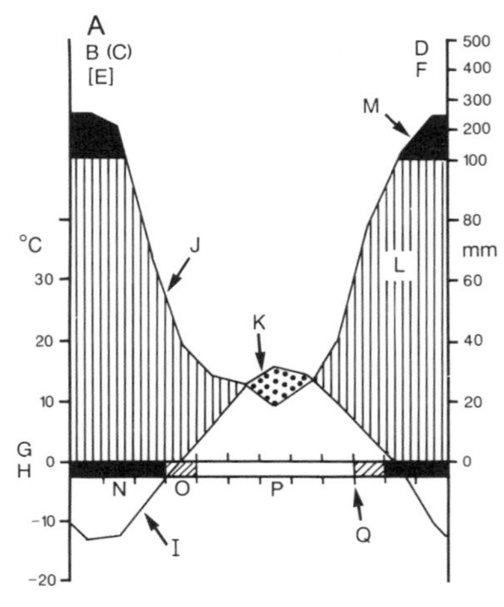

Key to Climate Diagrams

A – Ecoregion
B – Station Name
C – Elevation Above Sea Level (m)
D – Mean Annual Temperature (°C)
E – Number of Observation Years (where two figures are given, the first indicates temperature and the second precipitation)
F – Mean Annual Precipitation (mm)
G – Mean Minimum Monthly Temperature of Coldest Month (°C)
H – Absolute Minimum Temperature (°C)
I – Curve of Mean Monthly Temperature (°C)
J – Curve of Mean Monthly Precipitation (mm)
K – Period of Relative Drought (dotted pattern)
L – Relative Humid Season (vertical pattern)
M – Mean Monthly Precipitation >100 mm (scale reduced to 1/10, shown in black pattern)
N – Months with a Mean Daily Mean Minimum <0°C (black pattern)
O – Months with Absolute Minimum <0°C (diagonal pattern)
P – Mean Duration of Daily Temperatures (>0°C)
Q – Division of the Year (January through December, left to right, respectively)

diagrams provide convenient summaries of several climatic elements that are probably important in determining broad vegetational and species range limits. A brief scrutiny of the key to understanding the diagrams, provided in Figure 2.3b, enables the reader to examine and compare the various climatic diagrams presented throughout the book and abstract the relevant data from among: the monthly means for temperature and precipitation, frost-free periods, the absolute temperature minimum recorded, and other values. Data for Canadian sites were abstracted from the most recent *Canada Climate Normals* (Environment Canada 1982) and for sites in the United States from *Weather of U.S. Cities* (Anonymous 1981).

Arctic and its three subzones

The arctic climate is dominated throughout the year by air masses originating over the Arctic Ocean or Greenland. As a result, the climate is severely cold at all seasons, supporting substantial glaciers and ice caps on the higher elevations of Baffin, Ellesmere, Devon, and Axel Heiberg Islands. Sea ice remains throughout the year between most of the islands of the high arctic subzone, and the mean date of final snow melt from land surfaces is July 1 (Maxwell 1980). The soils have continuous permafrost with shallow active layers, and the prevailing soil type is the cryosol.

The particular climate of the highest latitudes – the *high arctic subzone* – is illustrated graphically by the Mould Bay and Resolute climate diagrams (Fig. 2.3a). The greatly shortened growing season, very cold summers, and extremely cold winters result from the combined influence of low, highly seasonal radiation inputs and the high albedo values in spring over relatively unbroken, snow- and ice-covered parts of the landscape. The effect of latitude is expressed primarily in the duration of daylight hours, and the solar elevation, and the contrast in these respects between high and low latitudes is shown clearly in Table 2.1. The high albedo (up to 80%), results in delayed snow melt until July, with the result that the growing season is no longer than six to eight weeks. In addition, the extremely low precipitation results in many upland surfaces remaining unprotected by snow in winter from both frost and wind and therefore unsuitable for any perennial plant growth.

The vegetation of the high arctic is discontinuous or absent on upland sites, dominated by a small flora of herbaceous plants, often tufted perennials (*Dryas*, *Oxyria*, *Saxifraga*, *Potentilla*, *Oxytropis*), mosses and lichens, comprising collectively no more than 10 percent coverage of the land surface. Shallow upland depressions where snow accumulates are the only habitats that support woody growth, and *Salix arctica* is almost the only common shrub in the high arctic. Extensive, poorly drained lowlands, by contrast, support a more or less continuous cover of herbaceous plants, dominated by *Eriophorum* and *Carex*.

Interested readers can find more details of high arctic bioclimates and vegetation in the excellent accounts by Edlund (1983a and b). She subdivides the high arctic region into four bioclimatic

Table 2.1. *A comparison of the seasonal patterns of daylight hours and solar elevation at different latitudes in Canada*

	Duration of daylight hours			Solar elevation (degrees)		
Latitude	Winter solstice	Spring and fall equinox	Summer solstice	Winter solstice	Spring and fall equinox	Summer solstice
80°N	0	12	24	0	0 to 10	13 to 33
65°N	3	12	22	0 to 1.5	0 to 25	0 to 49
45°N	8	12	15	0 to 21	0 to 45	0 to 69

Source: Abstracted from Maxwell (1980) and Hare and Thomas (1979).

Table 2.2. *A summary of the changes in floristic richness, plant cover, and vegetation type in the bioclimatic zones recognized in the Queen Elizabeth Islands (High Arctic)*

		Moisture regime								
			% Cover			% Cover			% Cover	
Edlund's zones roughly from N to S	Total number of vascular species	Wetlands	Vascular	Total	Mesic	Vascular	Total	Xeric	Vascular	Total
1	<35	Bryophytic mat	<1	<75	Patina of soil lichens and bryophytes	<1	<75	Scattered lichens	<1	<5
		Graminoid-moss meadow	1–5	50–100	Herb-patina tundra	1–5		Herb barrens	1–5	<5
2	35–60	Graminoid-sedge-moss meadow	1–10	75–100	Herb-patina tundra with scattered dwarf shrubs	5–10	<75	Herb barrens	1–10	<10
3	60–100	Sedge-moss meadow	5–15	75–100	Dwarf shrub-herb-patina tundra	5–25	<90	Dwarf shrub-herb barrens	5–15	<20
4	>100	Sedge-moss wet meadow (local willow)	10–25	75–100	Dwarf shrub-patina tundra	15–50	<90	Dwarf shrub barrens	5–25	<30

Source: Edlund (1983a), modified from Table 53.1.

zones that are arbitrary segments of "a continuum in which floristic diversity and percent cover decrease and the importance of the dwarf shrub life-form diminishes, until it is replaced by an entirely herbaceous life-form" (Edlund 1983a, p. 382). She proposes that these trends in "increasing impoverishment" of the vegetation are correlated broadly with mean daily July temperatures, persistence of snow, thawing degree days, and topography. The species numbers and estimated plant cover of each zone are shown in Table 2.2. Her bioclimatic zone 1 represents the northwest extremity of the high arctic as well as the highest elevations of the southern islands (Fig. 2.3a). It is characterized by over 90 percent unvegetated terrain with less than 35 vascular plant species, all herbaceous, grading into zone 4, which is localized at low elevations in the southern parts of the Queen Elizabeth Islands and has the

Figure 2.4. Discontinuous tundra vegetation in the mid-arctic zone of Banks Island. The dark sward is *Arctostaphylos alpina* and elsewhere the dominants are *Dryas* and *Salix arctica*.

richest vegetation with approximately 100 species of vascular plants including several woody taxa (*Salix arctica*, *Cassiope*, *Vaccinium*, *Dryas*). The chief factors regulating the sparse, specialized flora appear to be the short, cold growing season, extensive snow-free surfaces exposed to abrasive winds, very low freezing temperatures for ten months of the year, soil instability, and low precipitation.

The *mid-arctic subzone*, illustrated by the Hall Beach climate diagram (Fig. 2.3a) is an intermediate stage in the climatic progression southward from the polar region. The zone consists of most of the arctic islands that lie south of Melville Sound and north of the mainland. The summers are cold and the mean daily temperature exceeds 0°C during approximately two months. The winters are extremely long and cold and the precipitation is only slightly greater than in the high arctic. However, the somewhat less extreme climate is reflected in a less discontinuous plant cover on upland mesic sites, with more woody species (*Salix* and Ericaceae) and a more varied assemblage of herbaceous perennials, grasses and upland sedges (Fig. 2.4). Xeric sites are sparsely vegetated or bare, but lowlands have a continuous cover of sedges, *Eriophorum*, mosses, and herbs (Edlund 1983b).

The *low arctic subzone* is a wide belt extending from the North Yukon coast to Ungava Bay, and it is characterized by a cool growing season of more than 3 months, a mean July temperature of 8 to 10°C (Baker Lake climate diagram of Fig. 2.3a), and a continuous shrub tundra vegetation on upland surfaces. *Betula glandulosa* with *Salix* and several ericoids (*Ledum*, *Vaccinium*, *Arctostaphylos*) dominate the upland tundras; alder (*Alnus crispa*) is common on lower slopes, whereas various tall (>1 m) willow species are abundant on poorly drained sites. Peat accumulations, often with permafrost features, such as ice wedges and hummocks, are widespread in poorly drained regions. The northern limit of this subzone coincides roughly with the northern limit of *Betula glandulosa* and the southern limit is at the northern treeline (Payette 1983). Krummholz forms of *Picea* occur in the southern fringes of the low-arctic subzone.

Exceptional local microclimates often produce quite spectacular, azonal communities, and Edlund and Egginton (1984) report such a phenomenon from the Minto Inlet of southwest Victoria Island in the low arctic region. Tree-size *Salix alaxensis*, up to 8 m in height and 16 cm trunk diameter, occur in dense thickets on river floodplains in a deep valley, growing on sandy gravels with a deep active layer (>60 cm). The authors suggest that the exceptional, sheltered topographic situation favours higher summer temperatures and deep snow cover in winter, and they speculate that these factors can also maintain local pockets of ice-free soil in the river valley. Age/height analysis of the willows indicates that they are not old relict populations of a former milder climate. Cryosolic soils prevail throughout the arctic zone.

Subarctic zone

Between the southern limit of shrub tundra and the northern limit of the zone of closed-crown, continuous conifer forests – the boreal zone – lies a transitional region intermediate in both climate (Fig. 2.3a) and vegetation. The climate diagrams from Inuvik, Moosonee, Poste de la Baleine, and Goose Airport illustrate the range of climatic conditions of the subarctic zone (Fig. 2.3a). The summer thermal regime is roughly constant with four to five months of growing season and mean July temperatures at just above 10°C. However, there is a gradient of increasing precipitation from Inuvik (266 mm) to Hopedale (807 mm) and mean January temperatures decrease from Hopedale (–19.7°C) to Inuvik (–34.4°C). The highest values in Canada for mean winter snow depth are found in the Quebec–Labrador segment of this zone and the adjacent boreal zone. In the northern parts, the vegetation is a forest–tundra type with scattered individual trees or groves (always of tree form, as opposed to the shrubby or krummholz form of the low-arctic tundra) alternating with shrub tundra communities. Farther south, the groves of trees are more extensive or continuous, but generally of an open-crowned type – the taiga zone, as it is designated by Quebec (cf. Richard 1981a) and European botanists. Payette (1983) defines the forest–tundra as the zone of vegetation with a mixture of tundra communities and discontinuous forest stands in which the trees have arboreal growth form (>5 m). He subdivides the forest–tundra zone in Quebec–Labrador into a northern shrub subzone, where tree species in tree form are confined to local, lowland populations, and a southern forest subzone where tree-form groves occur on uplands alternating with shrub tundra. *Picea mariana* and *P. glauca* are the dominant trees and there is considerable regional variation in their abundances, as shown in Figure 2.5. *Populus tremuloides*, *P. balsamifera*, *Larix laricina* and *Betula papyrifera* are common.

Boreal zone

The boreal zone is distinguished climatically from the arctic by the replacement in summer of arctic airstreams by air masses originating primarily from the Pacific. However, this spring changeover from predominantly cold, dry arctic to warm, moist Pacific airstreams occurs earlier in the west than in the east, resulting in west-to-east delay of the growing season and an extension farther south of the boreal and arctic zones in eastern Canada (Fig. 2.3a). This displacement southward of the arctic–boreal zones is augmented by the influence of Hudson Bay (Fig. 1.2), which remains completely frozen from January to June, and still has ice patches in mid-August. The late beginning of the growing season in the Maritime region (Nova Scotia, Newfoundland) is also due in part to the persistence of pack ice in the Gulf of St. Lawrence.

The climate diagrams from boreal stations illustrate the main features of this, Canada's largest, terrestrial ecosystem (Fig. 2.3a). Two spatial trends are apparent. First, sites near the southern limit (Athabasca, Alberta; Kakabeka Falls, Ontario) have longer growing seasons (> 6 months) and higher mean July temperatures (17 to 18°C) than sites near the northern edge (Cree Lake, Saskatchewan, as shown in Fig. 2.6; Moosonee, Ontario and Goose Airport, Newfoundland, as shown in Fig. 2.3a), where the growing season is five months and mean July temperature about 15°C. Second, there is a major gradient of precipitation from west to east, illustrated by a transect of climate diagrams from the western extremity of the boreal zone at Yellowknife (Northwest Territories) through Cree Lake (Saskatchewan), The Pas (Manitoba), Cameron Falls (Ontario), Taschereau (Quebec), to Harrington Harbour (Quebec) (Fig. 2.6). In the boreal zone of Ontario, Quebec, and Newfoundland, very deep snow accumulates early in the winter and persists, with occasional renewals, until late April, and until late May in central Quebec. Accumulations of over 300 cm are normal in these Maritime and central Quebec regions. Summer precipitation east of cen-

Figure 2.5. The southern forest subzone of the forest–tundra zone on the uplands immediately east of the Mackenzie Delta, Northwest Territories, with a mixed tree layer of *Picea glauca* and *P. mariana*, and a dense shrub layer of *Alnus crispa*, *Betula glandulosa*, and ericads.

tral Ontario is also significantly higher than in the western boreal region because of midlatitude cyclonic storms from the Atlantic and the Gulf. The Maritime influence also maintains higher winter temperatures (mean January about –10°C) than in the interior (mean January about –20°C). Nonetheless, the climate of the Atlantic region of Canada is basically continental, modulated by oceanic influences.

The physiognomy of the boreal forest zone is remarkably homogeneous across the entire 6,000 km of its span. It consists of closed-crown conifer forests, with a conspicuous deciduous element made up of white birch (*Betula papyrifera*) and poplar species (chiefly *Populus tremuloides* and *P. balsamifera*). The dominant conifers are spruces (*Picea glauca* and *P. mariana*), pine (*Pinus banksiana*), larch (*Larix laricina*), and fir (*Abies balsamea*). The proportions of the dominant conifers vary greatly, however, apparently in response to one of, or combinations of, climate, topography, soil, fire history, and pest epidemics. For example, *Abies balsamea* is more abundant in the eastern half of the boreal zone, apparently in response to its lower tolerance of winter and spring droughts, than the spruces and jackpine. It is the dominant conifer in the Atlantic boreal region, where precipitation ranges from 50 to 100 mm per month. In the central boreal zone of Quebec, where the highest snowfall values are found, *Abies* and *Picea mariana* dominate the forests whereas *P. glauca* is more localized in occurrence, as shown in Figure 2.7. The properties of the substratum can be the primary determinant of the composition of the forest. The extensive tracts of coarse-grained, often sandy soils, derived from glacial outwash that occurs in parts of northern Saskatchewan, Manitoba, and adjacent Northwest Territories appear to favour the abundance there of *Pinus banksiana*. *Abies* disappears entirely from the boreal zone west of Lake Athabasca, apparently because of its susceptibility to drought and spring frosts (Bakuzis and Hansen 1965). Natural fires are an important factor in the

Figure 2.6. The range of boreal bioclimates is illustrated by the sketch map showing the approximate limits of the boreal forest region, excluding areas within the Cordilleran complex to the west, together with climate diagrams in addition (except for The Pas Airport) to those shown in Figure 2.2. West-to-east gradients in seasonality, summer precipitation, and winter precipitation (snowfall) are clearly seen, and differences in length of frost-free and growing season between southern limits (Cameron Falls) and central sites (Cree Lake) are obvious.

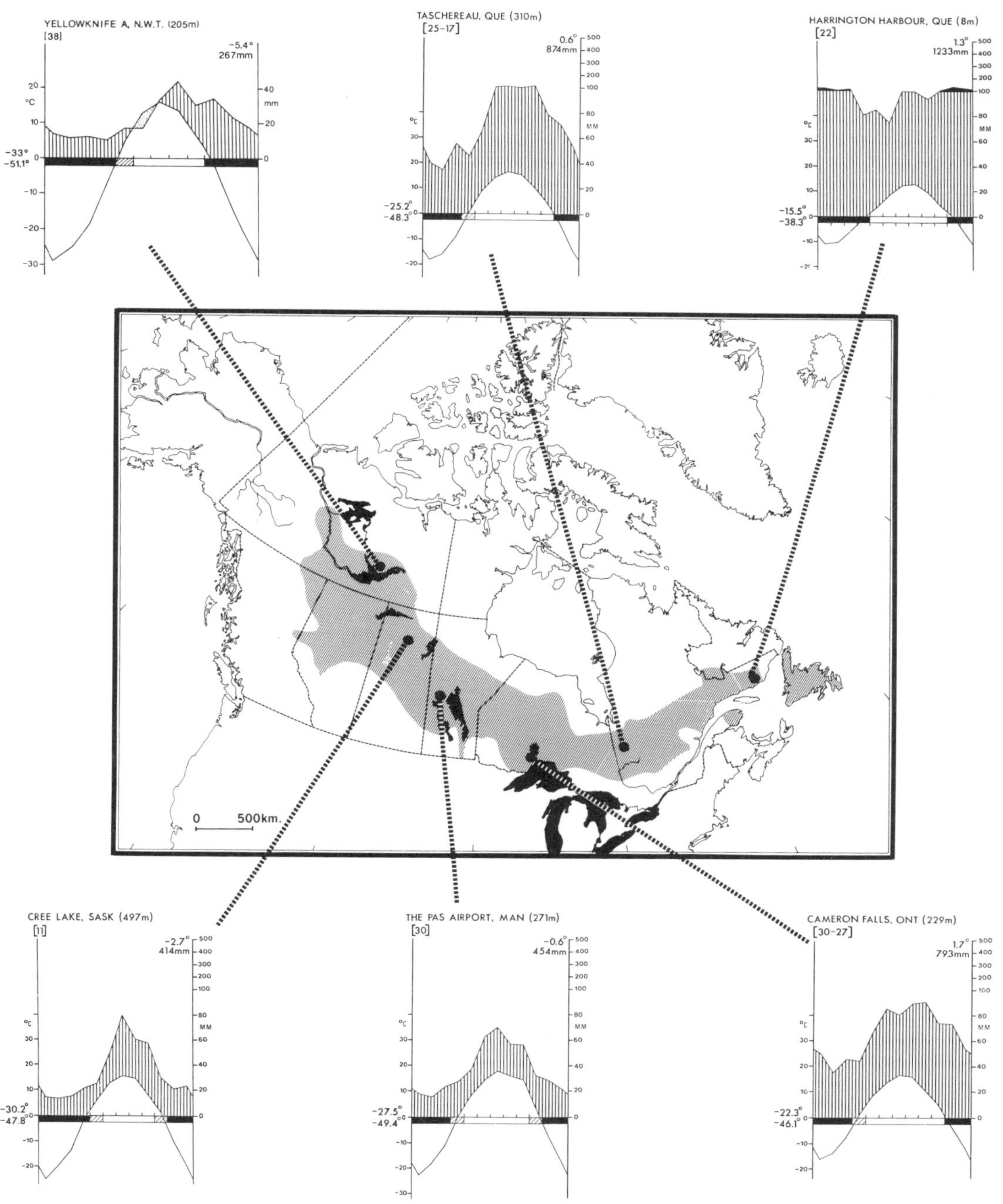

Figure 2.7. Continuous boreal forest (chiefly *Picea*) near the Cascapedia River, Quebec.

boreal zone, causing short-term successional patterns, so that many parts of the forest consist of a mosaic of different successional stages following fire. The faster-growing but shorter-lived trees (aspen and birch), and/or the fire-adapted jackpine, form the initial stages, replaced later by the shade-tolerant spruces and fir that also have greater longevity. Interested readers can gain entry to this recent literature on boreal forest wildfire ecology by reference to the collations by Wright and Heinselman (1973), Heinselman (1981), Wein and Maclean (1982).

In the southeast of Quebec–Labrador, a pronounced climatic gradient from the coast inland is responsible for a distinctive zonation of mire vegetation types, as recently mapped and described by Foster and Glaser (1986). They demonstrate that ombrogenous mires are concentrated in a coastal zone characterized by a less continental, wetter climate (illustrated by the climate diagrams for Harrington Harbour and Goose Airport as shown in Figs. 2.6 and 2.3a, respectively). Within this zone, the concentric, raised bog type of ombrogenous mire prevails along the coastal zone, where oceanic influences are greatest, whereas plateau and eccentric bogs are common inland. They note a distinct boundary along the 1,100-mm precipitation isopleth between the coastal ombrogenous bog zone and the interior, where minerotrophic mires of the patterned fen type prevail. These inland bioclimates are illustrated by the Chibougamau climate diagram (Fig. 2.3a). Foster and Glaser (1986) suggest that the major controlling influence is climate, and, in particular, that minerotrophic mires (= fens) develop in the interior of Quebec–Labrador as a result of excessive spring runoff from deep snow (> 500 cm per annum). The effect of the minerogenic water flow "is to promote decomposition and inhibit the accumulation of peat above the anaerobic water table." By contrast, the less continental coastal climates promote peat growth and the development of raised bogs.

The soils of the boreal zone are predominantly podzolic on mesic sites with organic fibrisols in the extensive poorly drained areas that occur throughout. The latter are particularly extensive in the residual lowlands of proglacial lake basins in central and northwest Alberta, and parts of Manitoba and northern Ontario. Discontinuous permafrost is confined to the northern half of the boreal zone.

In summary, the arctic, subarctic, and boreal bioclimates are distinctive in that they extend zonally across the entire country, whereas the four following bioclimates are regionally restricted. The southern limit of this vast boreo–arctic region coincides roughly with the mean winter position of the arctic air stream, and the transition from boreal to arctic, essentially a vegetational shift from treed to treeless communities, correlates closely with the summer position of the arctic frontal zone (Bryson 1966; Hare and Thomas 1979). The dominants of these zones are plants adapted to short growing seasons, long winters with low air and soil temperatures, high winds, frequent storms, frequent fires in the forest zone, and shallow cryic or podzolic soils.

Eastern temperate zone

The zone is separated from the boreal zone to the north by a broad transition, where intermediate conditions of both climate and vegetation exist, as will be seen in more detail when the bioclimatic transition zones are examined as aids to the interpretation of fossil data. The temperate (eastern) zone is distinguished from the boreal by its longer period with mean daily temperatures greater than 0°C (April to November) and higher mean monthly summer temperatures (July, >18°C). Also, precipitation (700–1,200 mm p.a.) is spread fairly uniformly throughout the year. An examination of the climate diagrams along a transect from the northern boreal (Moosonee, Ontario) through the central boreal (Taschereau, Quebec), the northern temperate (Kakabeka Falls, Ontario) to the southern temperate (London, Ontario), provides a clear depiction of the chief thermal and moisture trends across these two important zones (Fig. 2.3a). The climate of the area, referred to generally as the Great Lakes–St. Lawrence region, is modified locally because of the presence of several large bodies of fresh water and the Atlantic Ocean. The effect of the Great Lakes is to modify winter temperatures by delaying the late winter warming and to produce local increases in

Figure 2.8. A sketch map of eastern North America showing the northern and western limits of *Acer saccharum* (dotted line) and *Tsuga canadensis* (dashed line), the southern limit of *Picea glauca*, and a set of climate diagrams that are representative of the region.

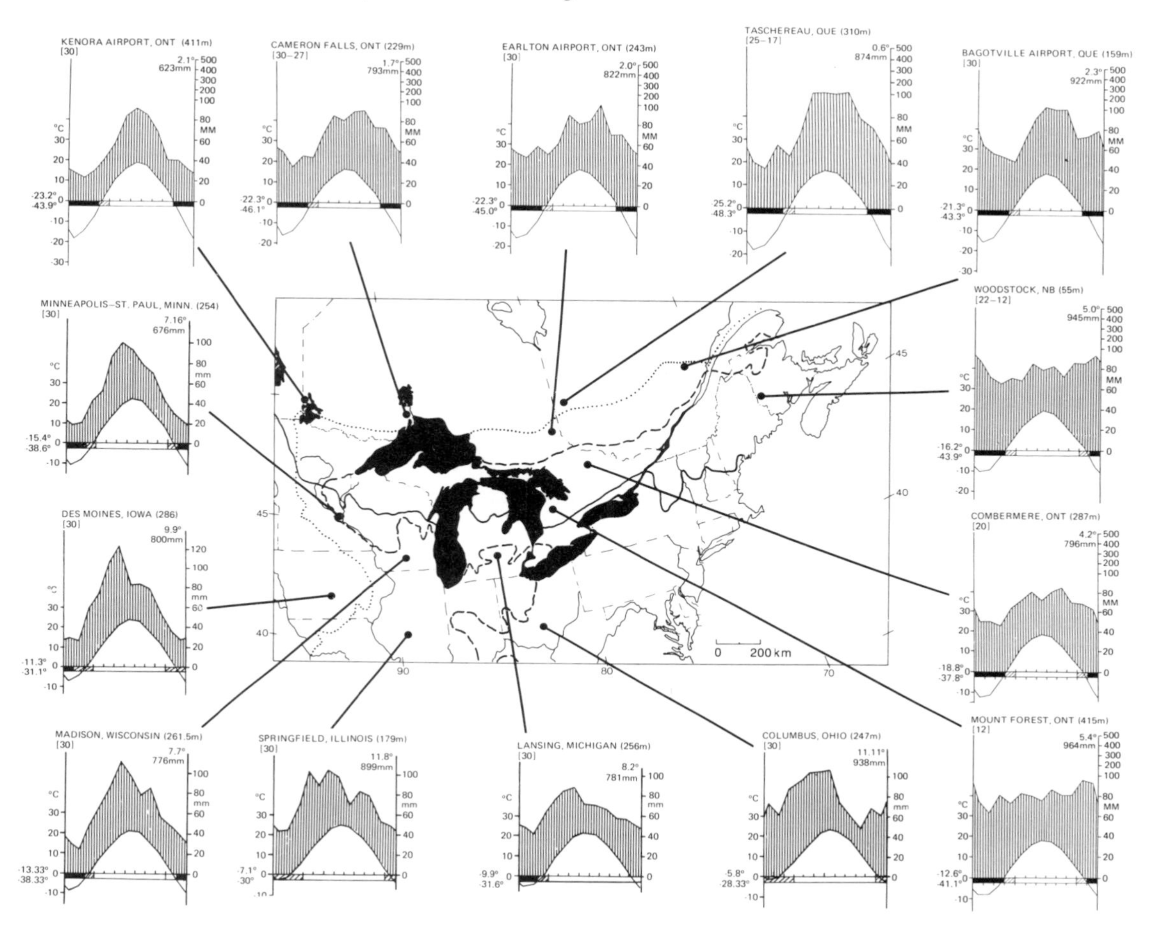

snowfall as cold arctic air passes over the lakes. Winter storms with freezing rain are frequent and have an important effect on several dominant tree species. The main trends in seasonal temperatures and precipitation, length of growing season and winter minimum temperatures can be understood most readily by referring to Figure 2.8, where representative climate diagrams are shown for the entire mixed forest (northern) temperate forest region and the ecotonal regions northward to the boreal forest and westward to the grasslands.

The vegetation of this region is, on one hand, clearly separable floristically and structurally from neighbouring bioclimatic zones but, on the other hand, it is so internally variable that it defies a straightforward descriptive summary in the present, necessarily brief treatment. A useful general account can be found in Rowe (1972), and regional descriptions in Munroe (1956), Hare (1959), and Grandtner (1966, 1972) for Quebec, Loucks (1962) for Nova Scotia and New Brunswick, and Hills (1960), Maycock and Curtis (1960) and Maycock (1963) for Ontario. Only the main features are noted here. The forests are made up of variable mixtures of taxa that are either confined to this bioclimate, or whose main centres of distribution are to the north in the boreal zone or to the southeast in the deciduous forest zone. I will outline only those directions of variation in species composition that are similar in broad quantitative terms to the variations that appear in the fossil record.

The common dominant trees with the greatest shade tolerance and longevity are *Tsuga canadensis*, *Betula alleghaniensis*, *Fagus grandifolia*, *Acer saccharum*, and *Pinus strobus*, and they are variably associated with *Pinus resinosa*, *Acer rubrum*, *Quercus rubra*, *Tilia americana*, *Ulmus americana*, *Thuja occidentalis*, *Fraxinus nigra*, *F. americana*, *Populus grandidentata*, *Carya* species, *Ostrya virginiana*, *Carpinus caroliniana*, *Juglans nigra*, and

Figure 2.9. Eastern temperate forest in Gatineau Park, Quebec, showing a mixed stand of *Thuja occidentalis* (along the lakeshore), *Pinus strobus*, and *Acer saccharum*.

others, as well as taxa from the boreal zone – *Picea*, *Larix*, *Betula papyrifera*, and *Abies* (see Fig. 2.9). The general trends of composition show some relationship to soil moisture and aspect (Hills 1960; Maycock and Curtis 1960; Maycock 1963) and Table 2.3 is a useful summary of some of these relationships as worked out by Hills (1960) for south-central Ontario.

I have already examined the general geographical ranges of some of the dominant tree species of this region, and in Chapter 3 I will attempt a systematic review of the autecology of the main species. For the present, some generalizations about the plant communities are appropriate, although, as two of the authorities on the subject have noted: "Almost any variety of stand composition resulting from combinations of any of the tree species, either in mixtures or in pure stands, is encountered in the Lakes region" (Maycock and Curtis 1960, p.5). As it happens, a detailed quantitative examination of the full range of temperate forests in eastern Canada, while it would be of great interest and value, would be difficult to use in conjunction with the pollen record as both the scale and taxonomic detail of the latter impose severe constraints.

A broad trend from north to south is recognized in both the Hills grouping (Table 2.3) and the typology of Grandtner (1966) for the temperate forests of Quebec as adapted by Richard (1977a and many subsequent papers) for his investigation of the vegetation history of that province. Grandtner's (1966) four groupings are: (1) *Le domaine de la Sapinière à bouleau jaune*, an association of *Betula alleghaniensis* and *Abies balsamea* in the northern part of the St. Lawrence Lowlands region, with *Pinus strobus* and *P. resinosa* on rocky and sandy habitats. (2) *Le domaine de l'Erablière à bouleau jaune*, an association on mesic sites of *Acer saccharum*, *Betula alleghaniensis*, *Fagus grandifolia*, *Tsuga canadensis*, and *Ostrya virginiana*, lying more or less immediately to the south of the previous "domaine," with warm, moist summers and moderate, snowy winters, typified climatically by the Sussex, New Brunswick, climate diagram (Fig. 2.3a). Poorly drained sites have abundant *Fraxinus nigra*, *Thuja*, *Ulmus* and *Abies*. (3) *Le domaine de l'Erablière laurentienne* is floristically richer than (2) by the addition of such species as *Juglans cinerea* and *Fraxinus americana*, but the main dominants on less-disturbed mesic sites are *Acer saccharum*, *Fagus*, and *Ostrya*, with *Quercus rubra* and *Betula alleghaniensis* on drier sites. (4) *Le domaine de l'Erablière à caryers* occupies the warmest climate, typified by the London, Ontario, diagram (Fig. 2.3a). Summers are humid and hot with the period of mean daily temperature greater than 0°C lasting from April to late November, and winters are mild and snowy. The basic composition and structure of (2) and (3) is altered only by the addition of some species with predominantly more southern ranges – *Carya cordiformis* (hickory), *C. ovata*, *Quercus macrocarpa*, and *Acer nigrum*. Interested readers might also consult the descriptive accounts of the forests of the Quebec portion of the Ottawa River watershed by Lafond and Ladouceur (1968) and of the St. Lawrence valley of Dansereau (1959), both of which illustrate that, although classification schemes are useful descriptions of this complex phytogeographical region, the forests are the expression of incompletely overlapping ranges of the main taxa that vary in abundance in response to various environmental factors.

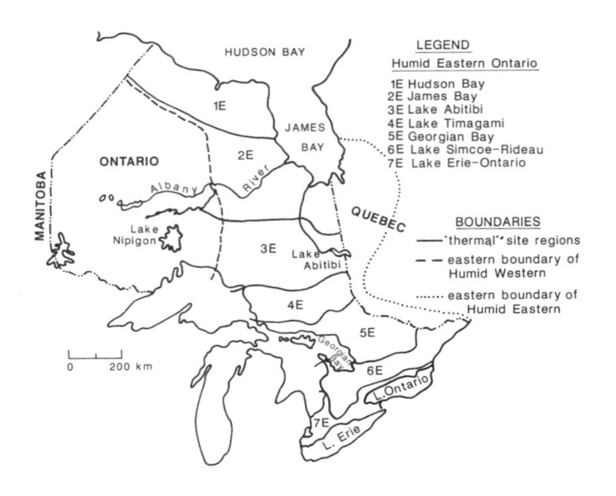

Table 2.3. *Balsam fir and other species in physiographic site classes under least-disturbed conditions in humid eastern regions mapped on page 22*

Ecoclimate soil	Hotter			Normal			Colder		
	Drier	Fresh	Wetter	Drier	Fresh	Wetter	Drier	Fresh	Wetter
Hudson Bay	*Betula p.* *Populus t.* *Picea m.* *Pinus b.*	*Picea g.* *Picea m.* *Betula p.* *Populus t.* *Pinus b.*	Open *Picea m.* *Larix l.* *Alnus s.*	*Picea m.* *Picea g.* *Pinus b.*	Open *Picea m.* *Larix l.*	Open *Picea m.* *Larix l.*		Mosses and lichens	
James Bay	Locally open *Pinus b.* *Betula p.*	*Pinus b.* *Picea m.* *Picea g.*	Open *Picea m.* *Picea g.*	*Pinus b.* *Picea m.* *Picea g.*	*Abies b.* *Picea g.* *Picea m.* *Populus t.*	Open *Picea m.* *Larix l.*	Open *Picea m.* *Larix l.*	Mosses and lichens	
Lake Abitibi	*Pinus b.* *Picea m.*	*Pinus b.* *Picea g.* *Pinus s.* *Pinus r.*	*Populus b.* *Thuja o.* *Ulmus a.*	*Pinus b.* *Betula p.* *Populus t.*	*Abies b.* *Picea g.* *Populus t.* *Populus b.*	*Picea m.* *Abies b.*	*Picea m.* *Pinus b.* *Larix l.*	*Picea m.* *Larix l.*	Locally open mosses and lichens
Lake Timagami	*Pinus s.* *Pinus r.* *Pinus b.* *Betula p.* *Populus t.* *Populus g.*	*Acer s.* *Acer r.* *Betula a.* *Pinus s.* *Abies b.* *Picea g.*	*Betula a.* *Thuja o.* *Ulmus a.*	*Pinus s.* *Pinus r.* *Pinus b.* *Acer s.* *Acer r.*	*Abies b.* *Picea g.* *Pinus s.* *Pinus r.* *Betula p.* *Populus t.*	*Abies b.* *Picea g.* *Picea m.*	*Abies b.* *Thuja o.*	*Picea m.* *Larix l.*	Locally open *Picea m.* *Larix l.*
Georgian Bay	*Quercus r.* *Pinus s.* *Pinus r.* *Quercus m.*	*Acer s.* *Quercus r.* *Fagus g.* *Tsuga c.*	*Ulmus a.* *Fraxinus n.* *Acer r.* *Picea g.* *Picea r.* *Tsuga c.*	*Pinus s.* *Pinus r.* *Tsuga c.* *Fagus g.* *Fraxinus a.*	*Acer s.* *Betula a.* *Tsuga c.* *Pinus s.* *Picea g.* *Abies b.*	*Fraxinus n.* *Acer r.* *Pinus s.* *Picea g.* *Abies b.*	*Picea g.* *Abies b.* *Pinus s.*	*Abies b.* *Picea g.* *Pinus s.* *Thuja o.* *Betula p.*	*Abies b.* *Picea m.* *Picea g.* *Larix l.*
Lakes Simcoe–Rideau	*Fagus g.* *Tsuga c.* *Quercus r.* *Ulmus r.* *Ulmus t.* *Populus g.* *Populus t.* *Juniperus v.*	*Quercus a.* *Quercus r.* *Carya o.* *Ulmus t.* *Juglans c.*	*Acer s.* *Acer s.* *Acer r.* *Betula a.* *Juglans n.* *Prunus s.*	*Acer s.* *Acer r.* *Quercus a.* *Quercus r.* *Fraxinus a.* *Ulmus a.*	*Fagus g.* *Acer s.* *Quercus r.* *Tsuga c.*	*Tsuga c.* *Betula a.* *Ulmus a.* *Ulmus r.* *Thuja o.* *Abies b.*	*Pinus s.* *Pinus r.* *Fraxinus a.* *Ulmus a.* *Ulmus t.* *Thuja o.*	*Picea g.* *Abies b.*	*Abies b.* *Picea m.* *Picea g.* *Larix l.*
Lakes Erie–Ontario	*Quercus m.* *Quercus r.* *Quercus c.* *Quercus b.* *Castanea d.* *Carya g.* *Carya o.* *Juglans c.* *Pinus s.* *Juniperus v.*	*Liriodendron t.* *Juglans n.* *Quercus a.* *Quercus r.* *Carya t.* *Carya g.* *Fraxinus q.* *Fraxinus a.* *Juglans c.*	*Platanus o.* *Liriodendron t.* *Ulmus a.* *Ulmus r.* *Fraxinus s.* *Betula a.*	*Quercus r.* *Quercus a.* *Carya o.* *Carya g.* *Ulmus a.* *Ulmus t.*	*Acer s.* *Fagus g.* *Quercus a.* *Quercus r.* *Carya o.*	*Quercus b.* *Quercus p.* *Fraxinus p.* *Fraxinus n.* *Ulmus r.* *Ulmus a.* *Carya c.*	*Tsuga c.* *Fagus g.* *Betula a.* *Pinus s.*	*Ulmus a.* *Ulmus r.* *Fraxinus n.* *Quercus b.* *Acer r.*	*Picea m.* *Picea g.* *Abies b.* *Acer s.* *Acer r.* *Betula a.*

Source: Hills (1960), modified from Key to Map No. 1.

The chief variables at the scale of regional pollen analysis, as defined in the introduction, seem to be major physiographic features and related soil types, summer thermal conditions, precipitation and drought, extreme low winter temperatures, responses to disease factors, and the life cycle attributes of the dominant taxa as they determine behaviour in response to disturbance by natural mortality, wildfire, windstorms, erosion, or other environmental effects.

Throughout the eastern temperate region, soils are more or less constant as follows: gray-brown luvisols and melanic brunisols on mesic sites, humo-ferric podzols on xeric sites, and gleysols and peats in poorly drained habitats.

It should be noted here that I have included under the heading "eastern temperate forests" two forest regions that are often described separately – the Great Lakes–St. Lawrence and the Acadian forest region (see Rowe 1972, for example). However, I have considered them together because, although there are climatic differences, as noted below, the vegetation is basically similar in composition and structure throughout. Although *Picea rubens* (Fig. 1.3) might be cited as an exception, as it is chiefly a tree of Nova Scotian and New Brunswick forests, the basic theme of *Acer saccharum*, *Tsuga canadensis*, *Betula alleghaniensis*, and *Fagus grandifolia* prevails with the same set of successionally or topographically controlled variations. The climatic distinctions are illustrated by a comparison of the Summerside, Prince Edward Island pattern with those of London, or Kenora, Ontario (Fig. 2.3a), in that the Maritime bioclimates have higher, uniformly distributed precipitation (1,000 mm p.a.), and a cooler summer (mean July temperature of 18°C compared to 20°C at London) with a shorter season of mean daily temperatures greater than 0°C.

It is worth noting the apparent outlier of the temperate region that straddles the southern extremity of the Ontario–Manitoba boundary (Figs. 2.3a and 2.8). The outlier effect is, of course, an artefact of the restriction of the map to Canadian territory. The temperate bioclimate zone, in its North American entirety, forms a triangular wedge running southeast from this outlier in western Ontario as far as Kentucky, then northeast along the eastern Appalachians to the Atlantic coast at the latitude of north Delaware–New Jersey (Fig. 1.2). However, the bioclimate and vegetation of this outlier is at the western limit of the temperate zone and could equally be thought of as transitional to the central boreal zone to the immediate north. The Kenora climate diagram (Fig. 2.3a) illustrates its marginal status as temperate (623 mm precipitation, mean January temperature of –18°C, low winter snowfall). The vegetation lacks certain temperate region dominants whose northwestern limits are to the east and/or south – *Fagus*, *Tsuga*, *Carpinus*, *Fraxinus americana*, and *Betula alleghaniensis*. However, the forests contain *Pinus strobus* and *P. resinosa*, here near their northwestern limits (Fig. 1.3), *Thuja*, *Fraxinus nigra*, *Ostrya*, *Ulmus*, *Quercus macrocarpa*, and *Tilia*, as well as all the dominant trees of the boreal zone. This subsection is of interest because of its transitional position to both the boreal forest of central Canada and the grassland and aspen parkland zones to the west and, as I discuss fully in Chapter 8, because the range limits of the dominant trees have varied during the Holocene.

Grassland zone

A glance at the climate diagram for Brandon, Manitoba, and a comparison with the Kenora diagram (both in Fig. 2.3a) illustrates the two salient characteristics of grassland bioclimates. First, annual precipitation is low, with autumn particularly dry, and, second, there is extreme seasonal variation in mean monthly temperatures for this latitude, from roughly 20°C above to 20°C below zero. This extreme continental climate, with very cold winters, relatively short warm summers, and low precipitation, results in soil moisture deficits, particularly in late summer. Under undeveloped conditions, now virtually extinct, natural grassland, or parkland to the north, prevailed – the prairies of the northern plains region. However, the subhumid climate and

Figure 2.10. Aspen parkland developed on hummocky disintegration till in southern Manitoba. Shallow ponds, referred to locally as potholes, are surrounded by clonal *Populus tremuloides* stands. Elsewhere, prairie or cultivated fields prevail.

chernozem soils are favourable for cereal cultivation, so that this entire subarid central core of the country is given over to agriculture. The northern fringe is variously described and mapped as the aspen parkland or transitional zone. It forms an arc from south-central Manitoba along the southern edge of the boreal forest to the foothills region of Alberta (see Fig. 2.10). Summers are cooler than to the south, winters are long and severe, and soil moisture deficits are less frequent. In the east, in Manitoba and eastern Saskatchewan, continuous woodlands of *Populus tremuloides* and *Quercus macrocarpa* prevail with *Ulmus* and *Tilia* in lowlands (see Fig. 2.11). Farther west the woodlands become discontinuous as groves of predominantly *Populus tremuloides* are surrounded by grassland or, more often, agricultural fields (see Fig. 2.12). The core of the subarid region is near the junction of the borders of Saskatchewan, Alberta, and the United States, where annual precipitation is as low as 300 mm and daily temperatures can reach almost 40°C on individual summer days.

Figure 2.11. Gallery woodland of *Quercus macrocarpa* and *Populus tremuloides* along the Qu'Appelle valley, Saskatchewan, in the grassland zone.

Figure 2.12. Grassland communities near Medicine Hat, Alberta, in the heart of the grassland bioclimatic zone.

Pacific and Cordilleran zones

The western region of Canada, dominated by montane axes running from the 49th parallel to the North Yukon, is the most difficult to classify into both climatic and vegetational groupings, primarily because of the enormous variation in topography over short horizontal distances. I am following Krajina (1969), Hare and Thomas (1979), and Ecoregions Working Group (in press) who recognize two major parts – the Pacific zone, which is the area to the west of the crest line of the Coast range and the Cascade Mountains, and the Cordilleran zone, which is from the same crest line east to the Rocky Mountain and Mackenzie Mountain foothills (Fig. 2.3a).

A wide tract of plateaus and deeply dissected subarid valleys (in the south, the Kamloops plateau; to the north, the Fraser plateau) separates the Coast mountains of the Pacific zone to the west from the Columbia Mountains to the east. They are made up of several complex mountain systems (Caribou, Selkirk, Purcell, and Monashee) that create an orographic climatic effect and vegetation complex similar to, but less humid than, the Coast ranges and with much colder winters.

The Pacific zone is divided here into southern and northern subzones, each of which could be further subdivided elevationally into coastal, subalpine, and alpine sections of quite distinct vegetation and climate (Krajina 1969). However, it seems inappropriate here to extend consideration beyond the north–south subzones because so little is known about the late-Pleistocene and Holocene vegetational history of large parts of British Columbia and adjacent Yukon that more detailed information would be of very little use in later chapters.

The salient climatic feature of the southern Pacific subzone is that, at low elevations, both summer and winter mean monthly temperatures are warm (>2.5°C) and mean daily temperatures less than 5°C are confined to December, January, and February. The maximum precipitation occurs between October and April, the result of an almost

Figure 2.13. The interior of the coastal Pacific (southern) forest on Vancouver Island, dominated by *Pseudotsuga menziesii*.

Figure 2.14. Kootenay National Park, Rocky Mountains, showing the upper elevational zonation of southern Cordilleran vegetation types. Treeline is formed by *Picea engelmannii* and *Abies lasiocarpa*; *Picea glauca* and *Populus tremuloides* are dominant on lower slopes and alluvial surfaces.

unbroken procession of westerly cyclonic storms; it is usually heaviest in late fall. By contrast, the summers are usually warm and dry (Fig. 2.3a). These coastal bioclimates support the most productive forests in Canada, the Douglas-fir (*Pseudotsuga menziesii*) (see Fig. 2.13), western hemlock (*Tsuga heterophylla*), and western red cedar (*Thuja plicata*) forests. *Pinus monticola*, *Abies amabilis*, and *A. grandis* are abundant locally. In Krajina's (1969) classification, this is the Pacific coastal mesothermal forest region, which he subdivides into the coastal Douglas-fir and coastal western hemlock zones. An interesting variant occurs here at the northernmost tip of a more extensive range in Washington: it is the association of *Quercus garryana* and *Arbutus menziesii* with Douglas-fir on coastal sites along southeast Vancouver Island and on islands in the Straits of Georgia. *A. menziesii* also occurs on the mainland coast south of Powell River, always in association with "shore pine" (*Pinus contorta*), the coastal representative of lodgepole pine.

At middle elevations in the south Pacific subzone, from roughly 800 to 1,500 m, the summers are cooler and moister than along the coastal belts, and the winters have abundant snowfall and less than 0°C mean monthly temperatures from October to March. *Tsuga mertensiana*, *Chamaecyparis nootkatensis* (yellow cypress), and *Abies amabilis* are dominants, with *Thuja plicata* becoming abundant on wetter sites, and *Pseudotsuga menziesii* and *Pinus contorta* on drier substrata, the mountain hemlock zone of Krajina (1969). At higher elevations (1,500 to 1,800 m), short summers and cold winters with abundant snow and strong winds restrict or prevent tree growth and an alpine tundra vegetation prevails. *Abies lasiocarpa*, the alpine fir, is the treeline species (see Fig. 2.14). Dystric brunisols and humo-ferric podzols prevail in the coastal and low-elevation sites on mesic sites.

The northern subzone of the Pacific zone is politically divided between Alaska and British Columbia. The Stewart (British Columbia) climatic diagram (Fig. 2.3a) illustrates aspects of the subzone as a whole, although it represents a low-elevation site. The summers are cool (12°C in July) and humid, and the winter months have very high precipitation. The low-elevation forests are dominated by *Tsuga mertensiana* and *Thuja plicata*, with local abundance of *Abies amabilis*, *Picea sitchensis*, *P. glauca*, and *Alnus rubra*. At higher elevations, a subalpine forest of *Picea glauca* with *Populus* occurs. The soil type of the low-elevation mesic sites is podzolic.

The *Cordilleran zone* also has two latitudinal subzones, a northern and southern, and an interior subzone. The Cordilleran climate in general is influenced by the orographic interactions of the Pacific westerlies with the mountain ridges, producing humid western flanks and dry interior valleys, and with continental airstreams from both the arctic via the plains to the east, and from the southeastern Great Plains region of the United States. The result is extreme variation in conditions. For example, a

Figure 2.15. A more detailed analysis of the complex bioclimates in the southern Cordilleran and Pacific regions can be seen in this sketch map of southern British Columbia and adjacent Alberta and Washington. The main topographic regions are shown along with representative climate diagrams from two Pacific stations (Victoria and Alert Bay) and five Cordilleran stations. All except Victoria and Kamloops are in addition to those included in Figure 2.3a.

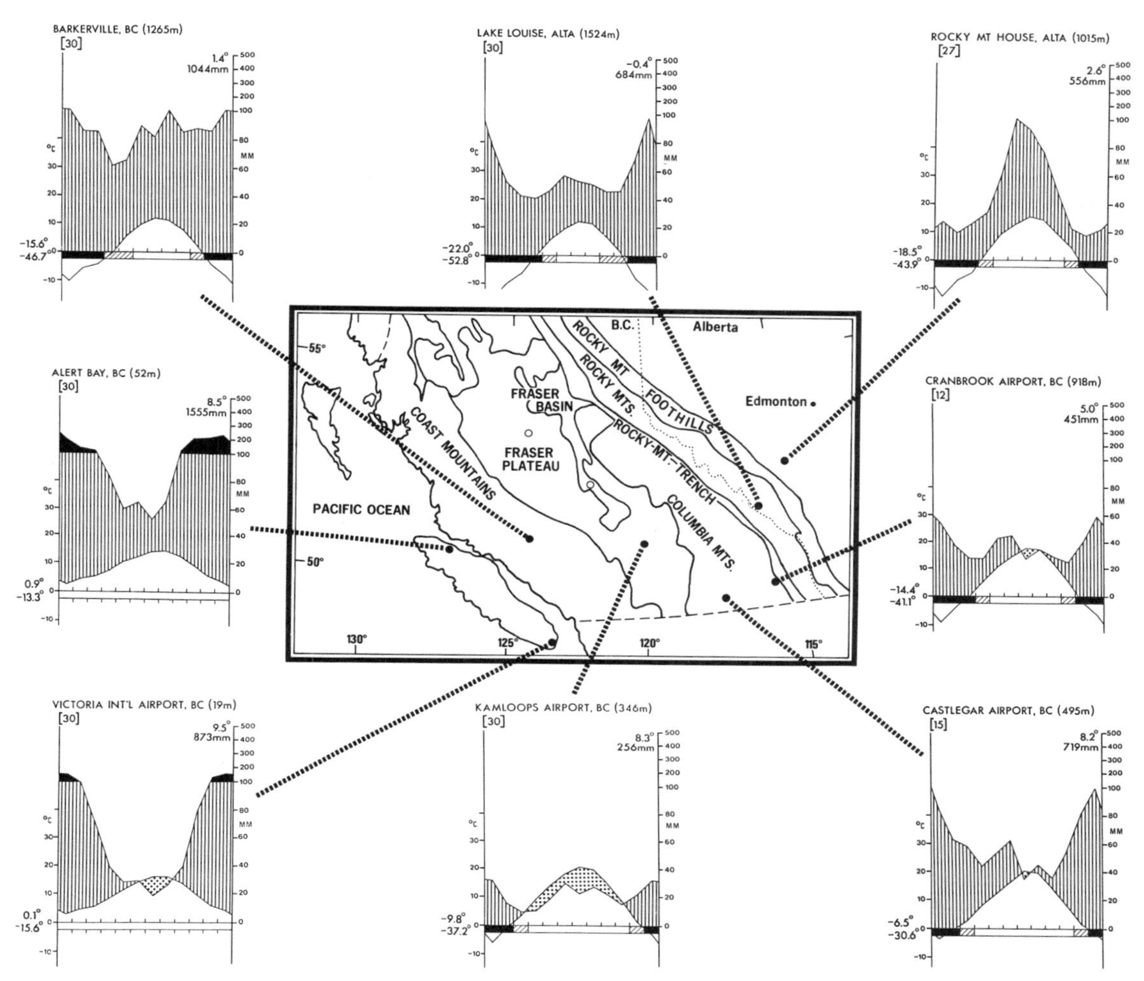

transect of climate diagrams at roughly latitude 50°N from the Pacific coast to western Alberta illustrates the extreme variation occurring over short horizontal distances in the southern Cordilleran zone. Kamloops (see Fig. 2.15) is in the interior subzone and lies in a broad valley about 360 m above sea level, with a subarid climate characterized by some of the driest and warmest summers in Canada as a whole. The vegetation is a *Pinus ponderosa* savanna on upland slopes, with *Populus*, *Thuja plicata* and *Betula* along valley bottoms, the ponderosa pine–bunchgrass zone of Krajina (1969). Farther south in the Okanagan Valley (see Figure 2.16) where the dominant air is still of Pacific origin, hot spells occur when dry air from the Columbia Plateau and other parts of the Basin and Range region drifts northward. The summers are dry and warm with mean July temperatures of 20°C, and winters are mild. The vegetation is predominantly grassland – the intermontane palouse – with open stands of *Pinus ponderosa* on mesic soils, as shown in Fig. 2.17. The regional soils in these subarid interior valleys are brown chernozems. Farther east on the Columbia Mountains, in the southern Cordilleran subzone, orographic precipitation from the Pacific westerlies produces a relatively high winter rainfall region (Castlegar, Fig. 2.15), and the forests are very similar to those of the Pacific zone, with *Tsuga heterophylla*, *Thuja plicata*, and *Abies*, and an interior race of *Pseudotsuga*, *Pinus monticola*, and *Larix occidentalis* (western larch) on drier and successional sites. The grouping of Krajina (1969) distinguishes the interior western hemlock and interior Douglas-fir zones. The soils are humo-ferric podzols.

Figure 2.16. Ponderosa pine (*Pinus ponderosa*) and sagebrush (*Artemisia*) savanna near Penticton, British Columbia, in the Okanagan River valley.

Figure 2.17. Interior montane grasslands with *Pinus ponderosa* on the valley flanks in the southern Cordilleran interior subzone, at Cache Creek near Clinton, British Columbia.

Figure 2.18. Treeline in the southern Cordilleran subzone of the Kananaskis valley, Alberta, showing *Abies lasiocarpa* in both tree and krummholz form.

Farther east, on the eastern slopes of the Rocky Mountains inside the Alberta border, a climate with a cool, humid summer and cold, snowy winter influenced chiefly by arctic continental air (Lake Louise, Fig. 2.15) supports a distinctive forest zone at elevations from 1,500 to 2,200 m, the Engelmann spruce–subalpine fir zone of Krajina (1969). *Picea engelmannii* (Engelmann spruce) and *Abies lasiocarpa* dominate mature stands (see Fig. 2.18), and *Pinus contorta* (lodgepole pine) dominates successional, post-fire forests. The eastern end of the transect is in the lower foothills of the Rocky Mountains and is essentially a transitional bioclimate to the boreal zone (Rocky Mountain House, Fig. 2.15). Summers are humid and cool with a relatively short growing season, whereas the winters are cold. The elevations of this lower foothills region are between 1,200 and 1,700 m, and the vegetation is similar to the boreal forest with the notable addition of *Pinus contorta*, a successional species that becomes increasingly important at lower elevations where it dominates most upland stands following fire (Horton 1956).

The northern Cordilleran subzone includes the plateaus and mountains of central northern British Columbia and adjacent Yukon Territory, and the cool climate with very cold winters (Whitehorse, Fig. 2.3a, represents the northern part of this subzone), and low precipitation supports a conifer forest on lower and middle slopes dominated by *Picea glauca*, *P. mariana*, and *Pinus contorta*, with secondary stands dominated by aspen and white birch.

3 Autecology and pollen representation

Introduction

The purpose of this chapter is to assemble concisely information on the autecology and pollen representation of the main taxa encountered in the late-Quaternary fossil record. It is hoped that this conflation will equip the reader to interpret the pollen and macrofossil diagrams. The effectiveness of these interpretive tools depends upon the adequacy of the investigations from which they are derived. Autecological data on the major taxa are very variable in quality and for the most part are based on simple correlative studies of species distributions and environmental measurements (e.g., Adams and Loucks 1971; Damman 1976). However, such an approach is entirely congruent with both the level of detail and the spatial and temporal scale of the fossil record of Canada. The next phase will undoubtedly involve a more rigorous approach to the interpretation of the fossil record as quantitative palaeoecology advances from the descriptive and narrative phases into the analytical phase (Birks 1985).

In the following catalogue, the taxa are arranged in four major phytogeographic groups. A brief narrative describes for each taxon the geographical range; community types; climatic characteristics, particularly at the range limits; landforms and soils; life cycle characteristics; ecology, including shade tolerance; susceptibilities to disturbance factors; and, finally, aspects of the pollen identification and representation. Many of these ecological characteristics are summarized in Table 3.1., pp. 52–3. Except where indicated otherwise, the distributional, ecological and other data for tree taxa were abstracted from Fowells (1965), Krajina (1969), Franklin and Dyrness (1973) and Minore (1979).

Transcontinental, primarily boreal taxa

***Abies balsamea* (L.) Mill.**

Balsam fir is one of the most important and abundant conifers in the boreal and temperate forests of central and eastern Canada (Fig. 1.3). It reaches its maximum development in eastern regions, particularly in the Maritime provinces, southern Quebec, and east central Ontario (Halliday and Brown 1943). Its climatic range is illustrated by the transect of climate diagrams in Figure 2.6, showing that it extends from humid boreal climates with eight months of growing season, 1,000 mm of precipitation evenly distributed throughout the year, and mean July and January temperatures of 10° and –10°C, respectively, to the highly continental boreal interior with 400 mm of precipitation, five to six months of growing season, and July and January means of 15° and –24°C, respectively. It is probably moisture limited at its southern and northwestern range limit and thermally limited (summer growing season) at its northern limit (Bakuzis and Hansen 1965). In western Canada, it is edaphically restricted to alluvial soils and occurs as discontinuous stands in valley bottoms, often associated with *Picea glauca* and *Populus balsamifera*. In the eastern half of its distribution, it occupies a wider range of podzolic and brunisolic soil types. It is often a codominant with *Picea glauca* and *Populus tremuloides* in the boreal forest, and as a codominant in the "Sapinière à bouleau jaune" (*Betula alleghaniensis*) (Grandtner 1972) of the St. Lawrence Lowlands and adjacent regions. Balsam fir forms an altitudinal zone in the northern Appalachian Mountains between 750 m and 1,500 m. It is also a treeline species in these mountains, with black spruce (Reiners and Lang 1979). In the southern parts of its range, it is a minor, but frequent, associate of *Tsuga canadensis*, *Acer saccharum*, and *Thuja occidentalis*, and locally in better sites with *Ulmus americana*, *Fraxinus nigra*, and *Ostrya*.

Hills (1960) has provided a detailed analysis of the distribution of balsam fir in Ontario in relation to soil conditions (Table 2.3), and Bakuzis and Hansen (1965, p. 54) provide a useful summary of these and other data as "a progressive shift in its adaptability from warm dry sites in the north to cool wet sites in the south," and "a wide range of adaptability over many sites in central Ontario in contrast to its narrow range of adaptability on few sites at the northern and southern ends of the Province."

Balsam fir has a shallow root system and is very susceptible to windthrow, particularly in the Appalachian uplands (Sprugel 1976). It is also prone to damage from hail storms and icing (Bakuzis and Hansen 1965). It is particularly susceptible to budworm attack, but its high shade tolerance enables it to restore almost pure stands quickly by self-seeding (Blais 1983).

In eastern and central Canada, *Abies* pollen can be assumed to represent balsam fir as it is the only fir species. It is significantly underrepresented in pollen spectra, by a factor of five according to Webb and McAndrews (1976), based on their analysis of vegetation–pollen ratios for east-central North America. It is represented in the early Holocene record from eastern Canada by higher percentages than in any modern spectra, even from areas where it is abundant, but this fact has been widely overlooked in the literature. Modern spectra from lake sites in western Canada within its range, but where it is not a dominant element in the landscape, often lack any records of fir pollen. Its range overlaps slightly with that of the closely related *Abies lasiocarpa* in west-central Alberta (Fig. 1.3), with the result that confusion in the interpretation of the pollen record is probable in the absence of macrofossils.

Figure 3.1. Range maps of *Pinus banksiana* (jackpine), a boreal species with range restrictions at both its western and eastern extremities, and of *Picea glauca* (white spruce), the most wide-ranging boreal conifer in Canada.

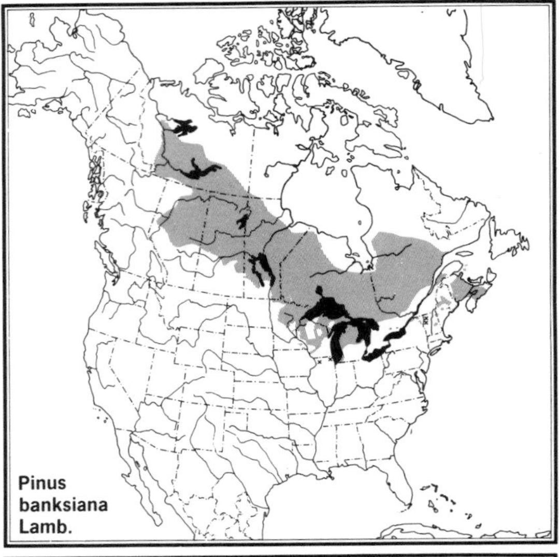

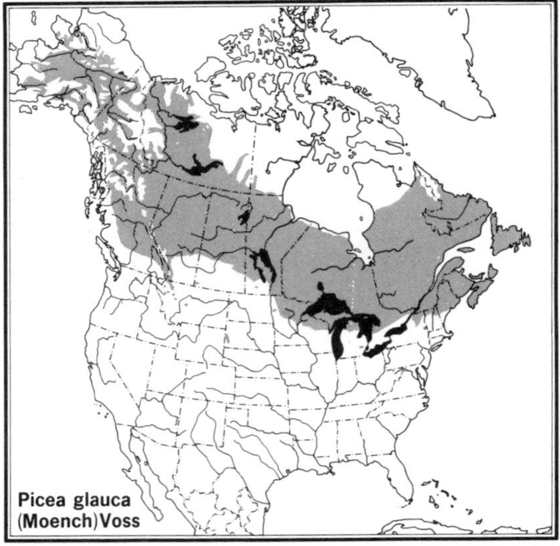

Alnus crispa (Ait.) Pursh

Green alder is a transcontinental boreal shrub, represented in southwestern Yukon, British Columbia, and Alaska by the subspecies *Alnus crispa* ssp. *sinuata*. It grows slightly beyond the northern limit of *Picea* in the subarctic (Gilbert and Payette 1982) and 5 to 20 m above the elevational limit of trees in the northern Cordillera. Its southern boundary in central and eastern Canada coincides with that of *Picea glauca*, as shown in Figure 3.1. It is an upland shrub, intolerant of shade, and a common pioneer on recently disturbed mesic soils where its nitrogen-fixing root nodules may provide an advantage. It reproduces both by abundant winged seeds and vegetative sprouting from basal stems. It has been recorded as both pollen and macrofossils in Holocene sediments throughout boreal–subarctic North America. The distinctive, predominantly four-pored pollen has been recognized separately by many investigators. It is generally regarded as overrepresented in pollen spectra (Ritchie 1974; Richard 1977a; Birks 1980), but some investigators have found it to be more or less equally represented. It is a conspicuous member of regional and extra-regional pollen sources, as illustrated by the 27 percent value recorded by Birks (1980, p. 108) in a sample from the Klutlan Glacier ice surface, although the plant is rare in the area. Birks (1977, p. 2375) notes that modern pollen assemblages from *Picea glauca* forests in the St. Elias Mountains, Yukon, include high *Alnus* values (56%) although the shrub is rare in the area.

Alnus rugosa (Du Roi) Spreng.
Speckled alder has a very similar range to that of *Alnus crispa*, except that it is absent from Pacific coastal zones. It is a lowland shrub, characteristic of alluvial soils that are flooded seasonally. It is intolerant of shade, and, in the northern boreal zone, it is usually replaced by *Populus balsamifera* in successions on alluvial deposits. It has been separated palynologically by its predominantly five-pored grains and has also been recorded as macrofossils.

Betula glandulosa Michx.
Dwarf birch is a boreal–subarctic species that ranges from the southwest tip of Greenland through the boreal–subarctic zone to Alaska. It has been recognized in sediments by its macrofossil remains (leaves, stems, wing-margined nutlets, and bracts), but the identification of its pollen depends on size statistical separations and not all investigators have found such analyses completely convincing (e.g., Ives 1977). Others use the method routinely and with reasonable effectiveness, particularly as macrofossil evidence has so far always corroborated the pollen data, (Ritchie 1977; Richard 1981a; and others). The northern limit of *Betula glandulosa* is often used to define the boundary between low-arctic and mid-arctic bioclimates (Richard 1981a; Edlund 1983a and b; Jacobs, Mode, and Dowdeswell 1985, and others). It extends beyond the northern limit of the forest–tundra (defined by Payette, 1983: see Chapter 2, p. 17) in western Canada, to southern Banks Island and Victoria Island, across Keewatin to Southampton Island. Its range extends to the northern tip of the Ungava Peninsula, and to South Baffin Island (Porsild and Cody 1980, map 444, p. 258). The pollen and macrofossils of dwarf birch occur in many late-glacial deposits, often dominating the assemblages in particular units of sediment, and characteristically it succeeds assemblages dominated by herb taxa, often of arctic species. The late-glacial role of *Betula glandulosa* is examined in more detail in Chapter 8. Some investigations show that dwarf birch pollen is overrepresented in modern samples by a factor of two to four (Birks 1977; Ritchie 1977; Cwynar 1982) but Rampton (1971) found that it was underrepresented at sites with abundant *Alnus* and *Picea* nearby.

Betula papyrifera Marsh.
White or paper birch is a transcontinental boreal species ranging from western Alaska to Newfoundland and south into both the Pacific–Cordilleran and the eastern temperate forest regions. It has been subdivided into many taxonomic units, chiefly varieties, several of which are geographically distinct. However, none has been separated in macrofossil material. The separation of white birch pollen from that of dwarf birch and yellow birch is not exact, being based on grain diameter measurements among taxa whose pollen have overlapping size-frequency curves. It grows in a range of climates that is very similar to that of white spruce, from the dry, cold continental northwest extremity (Inuvik, Fig. 2.3a) to the temperate Cordilleran (e.g., Castlegar, British Columbia, Fig. 2.15), the perihumid, oceanic boreal climates of Newfoundland (St. John's, Newfoundland, Fig. 2.3a) and throughout the temperate mixed forest region. White birch is a short-lived (70 years), shade-intolerant tree that plays a secondary role in forest succession, rarely persisting for more than one generation. In the boreal region, it occurs commonly on lighter textured soils, very often on slopes, and in the southern parts of its range, it is restricted to cooler, north-facing slopes (Fig. 3.2). It produces abundant winged seeds that are usually released in winter and spread widely. It regenerates vigorously by sprouting from stumps, often producing multiple stems from a single stock.

Figure 3.2. *Betula papyrifera* is common on upland slopes in the northern parts of the boreal forest. Here it is codominant with *Picea glauca* on a site that was burned roughly 80 years ago in the Lower Mackenzie River basin.

Larix laricina (Du Roi) K. Koch
Tamarack is a nearly transcontinental species (see Fig. 3.3) centred on the boreal area but extending north to the subarctic, to the treeline in the northern

Figure 3.3. Range maps of four dominant, transcontinental boreal taxa.

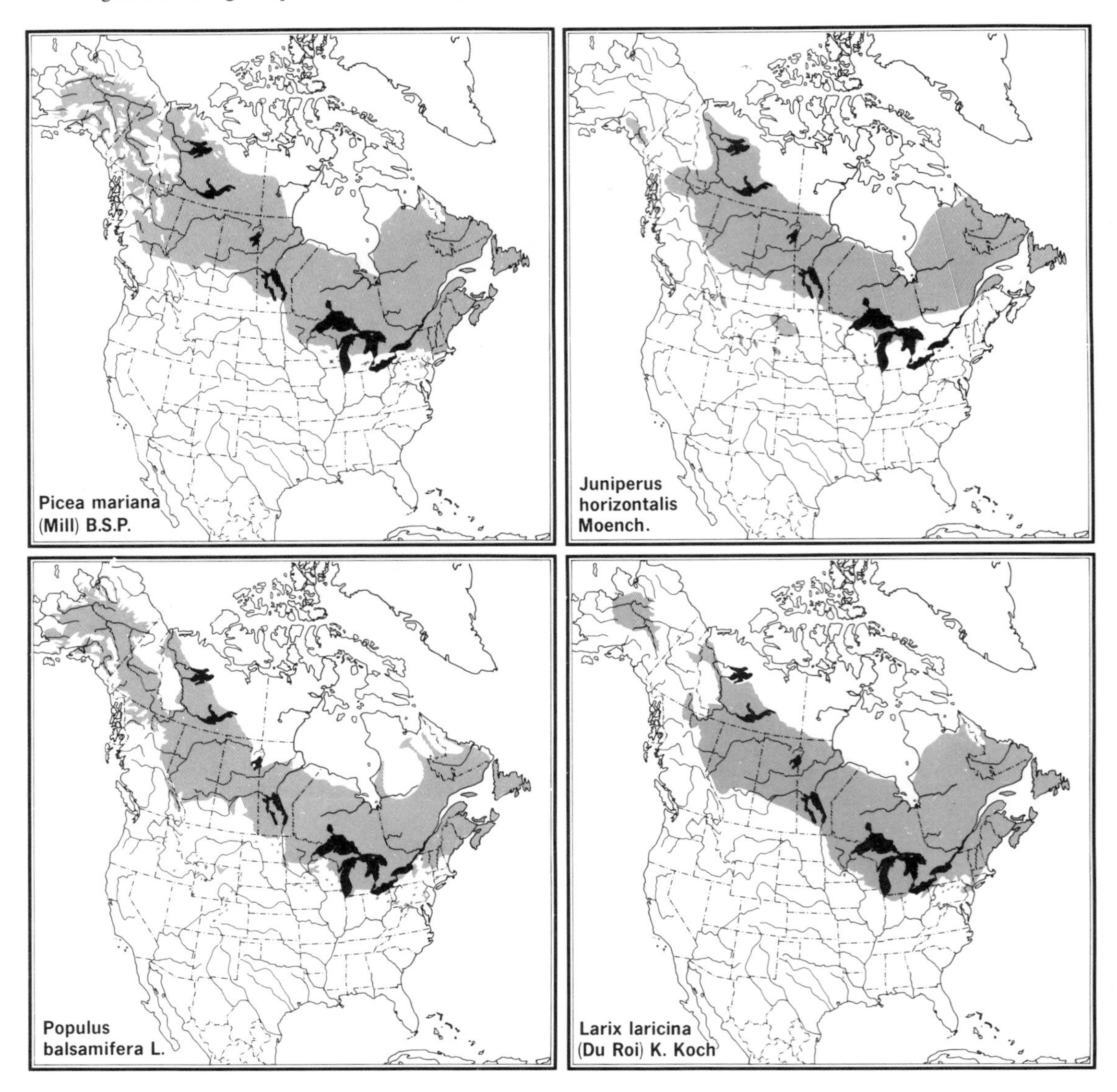

Cordillera (see Fig. 3.4), and south into the Great Lakes–St. Lawrence region. It has an unusual, irregular pattern in northern Yukon and Alaska (mapped in detail by Zoltai 1973). Its northern climatic limit coincides roughly with the 900-degree day isotherm, the 10°C July isotherm, and a 3.5-month growing season. It occupies a wide range of precipitation regimes from mean annual values of 250 to greater than 1,000 mm. It occurs over a wide range of soil conditions, in poorly drained fens and nutrient-rich bogs in the central and southern parts of its range, whereas in the subarctic and subalpine environments, it occupies upland, moderately drained sites in addition to wet habitats. Its precise occurrence at the treeline in northern Quebec has been reviewed recently by Payette (1983) and I will return to that topic in Chapter 8. It associates commonly with *Picea mariana*, *P. glauca*, and *Abies balsamea* in the boreal zone, and with *Fraxinus nigra*, *Ulmus americana*, and *Acer rubrum* in the temperate zone forests. Tamarack has a shallow, extensive root system. It reproduces vegetatively by both stem and root shoots. Maximum height and age in the south-central parts of the range are 25 to 30 m and 150 to 250 years, respectively. It is shade-intolerant and is usually replaced by *Picea mariana* in fen-bog succession. *Larix* pollen appears to have been overlooked in some analyses. It is a moderate pollen producer,

Figure 3.4. *Larix laricina* form the treeline locally in the northern Cordillera, particularly on calcareous substrata. Here it is on a north-facing slope in the Richardson Mountains, Yukon Territory.

but appears to be underrepresented in lake sediments. On the other hand, macrofossil studies at the treeline in northern Quebec have been particularly effective (Gagnon and Payette 1981).

***Picea glauca* (Moench) Voss**

White spruce has a transcontinental boreal range (Fig. 3.1), and tolerates absolute January temperature minima of –60°C. It is less tolerant than black spruce of shallow active layers in the soil and the site differentiation of black and white spruce in the subarctic is often determined by the depth of the active layer (Viereck et al. 1983). White spruce is intolerant of highly acidic, base-deficient soils, where it is replaced by black spruce. In the boreal region, it is common on alluvial soils, brunisols, luvisols, and deep gleysols. Along the east coast of Hudson Bay, it is the main species in the humid climates of a narrow (2- to 10-km) coastal strip, replaced inland by black spruce and tamarack. It is relatively uncommon in much of interior Labrador where base-deficient humo-podzolic soils prevail, supporting extensive black spruce–lichen woodlands. Along the east coast, it forms the treeline in association with tamarack and black spruce (Payette 1983). In Keewatin and the Mackenzie District, white spruce extends into the forest–tundra zone along rivers on alluvial soils and on the coarse soils of esker ridges. It is the only conifer in the extensive woodlands of the Mackenzie Delta, and extends to the arctic along north-flowing rivers such as the Anderson, Horton, and Coppermine. In the Cordilleran region, white spruce is locally common in submontane and lowland habitats associated with *Populus tremuloides*, but it is rare in, or absent from, the subalpine Engelmann spruce–subalpine fir communities. It occurs on alluvial lowlands in the interior Douglas-fir and western hemlock zones. In the northern sectors of the Cordillera, it becomes abundant, particularly on alluvial and well-drained upland soils, where it forms the montane treeline. It extends southwards into the grassland zone along river valleys, particularly on north-facing slopes (see Fig. 3.5). Buxton et al. (1984) report that seedlings of white spruce have superior drought resistance to those of either black spruce or jackpine under experimental conditions of low soil moisture. White spruce competes more effectively with other species in the southeastern sector of its range than does black spruce, partly because of its greater shade tolerance, height growth, and longevity (see Fig. 3.6). For example, mature trees can reach heights greater than 50 m on alluvial soils in northern central Alberta, greater than 25 m in interior Alaska and greater than 30 m at 150 years in central Ontario. Trees older than 300 years have often been recorded by Sutton (1969), and others have reported specimens older than 350 years from Churchill, Manitoba. It is ineffective in competition with hardwood species (maple, beech, yellow birch) but can outlive *Abies balsamea* and *Picea mariana*. It is susceptible to attacks from the insect pest, spruce budworm (Blais 1983). White spruce produces an annually shed crop of winged seeds that are reasonably dispersed by wind. It reproduces by layering at high latitudes, but less commonly than black spruce. I will examine its reproductive and dispersal capacities more fully in Chapter 8 when its postglacial history is examined. The species is well differentiated ecotypically (Sutton 1969) and there is evidence for geographically differentiated genotypes that might have relevance to the historical biogeography of the species (Wilkinson et al. 1971; Tsay and Taylor 1978; Von Rudolph, Oswald, and Nyland 1981; Khalil 1985).

***Picea mariana* (Mill.) B.S.P.**

Black spruce is a transcontinental, boreal species (Fig. 3.3) that forms the arctic treeline across much of northern Canada (Payette 1983) and is the chief tree limit species in krummholz form of the shrub tundra (see Fig. 3.7). The vast range of climates in which it is found can be seen in Figure 2.3a: it ranges from the cold, extremely dry continental regimes of the far northwest (e.g., Inuvik, Fig. 2.3a) to the perhumid, oceanic conditions on Newfoundland (St. John's, Fig. 2.3a). The southern limit in the east coincides approximately with the 21°C July isotherm. It tolerates sites with permafrost soils in which the active layer is less than 25 cm. Black spruce tends to be restricted to organic soils in the southern boreal and mixed temperate forests, but in the central and northern sectors of the boreal forest, it forms pure stands on uplands, usually occurring on heavier-textured, cool sites (clays and clay loams derived from glacial till) as well as on organic soils in lowlands (Viereck et al. 1983). It is the dominant tree in the northern Quebec and Labrador forests often forming extensive open lichen woodlands. West of Hudson Bay, on sites that have been undisturbed for several decades, it forms extensive stands of closed "feathermoss" forest in the central boreal region (Corns 1983) and open lichen woodland farther north in the subarctic (Rowe 1984). In the subarctic, it often forms clumped, multiple-stemmed growth forms (Fig. 3.7) apparently in response to winter snow and wind effects (Payette et al. 1985). Farther north, in the shrub tundra, it can persist indefinitely

Figure 3.5. *Picea glauca* at its southern limit in the western interior, extending along river valleys on north-facing slopes near Drumheller, Alberta.

Figure 3.6. Regenerating *Picea glauca* in a 50-year-old *Populus tremuloides–Betula papyrifera* stand in the southern boreal forest zone, Riding Mountain National Park, Manitoba.

as vegetative, krummholz shrubs. It produces small, winged seeds in semiserotinous cones that are retained on the trees for up to five years (Fig. 3.8), thus providing a significant seed supply that usually survives fire (Black and Bliss 1978, 1980). Cogbill (1985) suggests from his composition and age analysis of stands in the Laurentian Highlands of Quebec that the black spruce–feather moss type dominates because it is adapted to reestablishment after short-cycle wildfires, (see Fig. 3.8).

Figure 3.7. *Picea mariana* in characteristic krummholz habit near the northern treeline in the forest–tundra zone on the uplands east of the Mackenzie Delta.

Figure 3.8. *Picea mariana* after fire retains seed-bearing cones in the dense, compact terminal shoots.

Black spruce is the dominant of many forests and woodlands in the northern boreal region, but in the southern portion of its eastern range, it is usually replaced by *Abies balsamea*, *Thuja occidentalis*, *Ulmus*, *Fraxinus nigra*, and other trees.

The recent demonstration that pollen of *P. mariana* can be reliably distinguished from that of *P. glauca* (Hansen and Engstrom 1985) has accelerated a trend among palynologists to make the separation that was begun by the statistical size-difference technique introduced by Birks and Peglar (1980). Recent investigations in Labrador have made effective use of this tool, as can be seen in more detail in Chapter 5 (Engstrom and Hansen 1985). Brubaker, Graumlich, and Anderson (1987) have provided a thorough assessment of the morphological and statistical methods of distinguishing between modern and fossil pollen of white and black spruce from Alaska, and they propose that the use of the maximum likelihood method provides the most reliable estimates of the proportions of each taxon in mixed populations.

***Pinus banksiana* Lamb.**

Jackpine occupies a curtailed continental, boreal range, being confined almost entirely to Canada (Fig. 3.1). Its eastern limit is south-central Quebec and Nova Scotia, and its western limit is the Mackenzie River valley to the immediate west of Great Slave Lake. Its southern limit in the west coincides roughly with those of the spruces and tamarack and is presumably related to soil moisture and summer warmth (Zoltai 1975). However, its southern limit in eastern Canada lies to the north of those of the other boreal coniferous trees (Figs. 3.1 and 3.3). Its northwest limit is not obviously related to any climatic boundary, but the length of growing season varies across the boundary from four months at Cree Lake (Fig. 2.6) to three months at Inuvik, Northwest Territories (Fig. 2.3a). Also, there is a steep gradient of mean annual degree-day totals (above 5°C) across its northern limit in the upper Mackenzie Valley from 2,000 at Fort Simpson, within its range, to 1,600 at Norman Wells, only 400 km to the north, beyond its limit. Its northern limit appears to be thermally limited in Saskatchewan and Manitoba but in the Maritimes and eastern Quebec, it is assumed that it is related to an intolerance of deep snowfall. It is abundant on glacial outwash sand plains and raised beaches. It grows with all other boreal species except tamarack and rarely balsam fir, and in Alberta and adjacent Northwest Territories, its range overlaps that of *Pinus contorta* (lodgepole pine). Rudolph and Yeatman (1982) provide evidence of introgressive hybridization between jackpine and lodgepole pine in this region near Great Slave Lake and speculate that their area of range overlap was greater in postglacial times than it is today.

In the northern part of its range, its seeds are produced in serotinous cones that often remain closed until after a fire. Germination and establishment is optimum in full light on mineral soil, so the species is well adapted to respond to forest removal by fire. It is very intolerant of shade, with the result that it is quickly replaced by other conifers (white spruce, white pine, hemlock) and hardwoods (birch, maple) on all except the most nutrient-deficient soils. Buxton et al. (1984) conclude from experiments with seedlings that jackpine "was able to exploit soil-water reserves more effectively than either black or white spruce, presumably because of a larger and more rapidly growing root system." Jackpine has the highest degree of pollen overrepresentation of any boreal tree.

***Populus balsamifera* L.**

Balsam poplar is a boreal transcontinental species whose range extends slightly farther north than that of aspen, but it is absent from the Pacific–Cordilleran region and does not occur as far south as aspen in the eastern temperate sector (Fig. 3.3). It is taxonomically differentiated into various subspecific categories, mainly varieties, that are variously recognized or ignored in different manuals (Viereck and Foote 1970). Balsam poplar has a more restricted landform–edaphic range than aspen. It occurs chiefly on lowland mesic–hydric habitats, in swales, or on alluvium, and in the northern parts of its range, it is frequent on gravelly lake shorelines where a high water table appears to be a critical requirement (Murray 1980). It is of palaeoecological interest because it has been recorded frequently in late-glacial sediments as both pollen and macrofossils.

Balsam poplar has the distinction of occurring "farther north in North America than any other tree species, occurring as distinct stands in river drainages on the Alaskan North Slope" (Edwards and Dunwiddie 1985, p. 271), and also of forming the treeline at the eastern extremity of the continent, in the Saglek Fjord, on the Labrador Coast, where four isolated groves occur (Payette 1983, pp. 15 and 18). In addition, a clonal palisade of balsam poplar in the Old Man Lake area (see Fig. 3.9) adjacent to the Eskimo Lakes, Northwest Territories, represents the northernmost tree-form arboreal species in the Mackenzie Delta region.

Balsam poplar propagates vegetatively by suckering vigorously and commonly forms clonal groves. Comtois and Simon (1985) have analyzed the sex ratios and allozyme frequencies of northern populations in Quebec and shown that there is considerable genotypic variation that appears to be related to both latitude and postglacial routes of migration.

Figure 3.9. Shoreline clonal palisade of *Populus balsamifera* on Old Man Lake, Northwest Territories, beyond the arctic treeline.

Edwards and Dunwiddie (1985) demonstrate the dendrochronological potential of the species and show that, while abundant pollen is produced, it is poorly preserved in sediments even within the clonal groves sampled at the treeline in northern Alaska. On the other hand, Birks (1977) recorded 67 percent *Populus* pollen in one moss polster sample within a balsam poplar forest in the southwestern Yukon.

Balsam poplar is a short-lived (120 years), medium-sized (20 to 25 m), shade-intolerant species that plays a successional role in boreal forest and northern mixed forest landscapes. For example, it forms a characteristic intermediate stage in the succession on extensive deltaic deposits in the northern boreal zone, eventually being replaced by *Picea glauca* (Jeffrey 1961; Viereck 1970; Gill 1973; Van Cleve and Viereck 1981). It produces abundant plumed seeds that travel far but are short-lived.

***Populus tremuloides* Michx.**

Trembling aspen, in common with the other five Canadian species of poplar, is an amentiferous, wind-pollinated tree that produces abundant pollen (at sites in the Canadian forests where it has been measured), but has an irregular pattern of preservation in sediments. An investigation of the fate of poplar pollen immediately after its discharge from the tree, in the southern boreal forest of Manitoba where the tree is abundant, showed that the main degradation and disappearance of the grains occurred in the water column, at a rapid rate (Hadden 1978). However, as Mott (1978) and others have shown, significant amounts of well-preserved poplar pollen have been recorded from late-glacial sedi-

Figure 3.10. *Populus tremuloides* forming clonal stands in the aspen parkland zone south of the Riding Mountain National Park, Manitoba.

ments throughout Canada and adjacent United States, often associated with macrofossils. It is of interest that Cwynar (1978) found that the poplar grains were "excellently preserved" in the late-Holocene sediments from a very deep (65 m) meromictic lake in central Ontario. Richard (1970) and Brubaker, Garfinkel, and Edwards (1983) separate the fossil pollen of *P. tremuloides* type and *P. balsamifera*. However, the Holocene record of poplar pollen is so erratic that it has contributed little to our knowledge of the history of the vegetation.

Aspen occurs in every forested part of Canada except the Pacific coastal region. It is a short-lived, shade-intolerant species that reproduces effectively by wind dispersed, plumed seeds. It also forms large clonal stands by its aggressive vegetative propagation (Maini and Horton 1966), as shown in Figure 3.10. It occurs on upland, well- to moderately drained sites, typically with brunisolic or luvisolic soils. It competes ineffectively with all other tree species, so it plays a role in early succession after fire or other disturbance.

***Shepherdia canadensis* (L.) Nutt.**

This shrub is widely distributed throughout the Boreal–Cordilleran regions of Canada (Porsild and Cody 1980, map 824, p. 466). Although entomophilous, its distinctive pollen has been recorded in late-glacial sediments throughout Canada and the adjacent United States, often in appreciable percentages (5 to 10%), and records of macrofossils serve to confirm its presence. However, it is never found in comparable percentages in modern spectra, except when it is locally very abundant in the vegetation adjacent to a sample site (Birks 1980). Its possible palaeoecological significance has been examined by several investigators (Lichti-Federovich 1970; Ritchie and Hare 1971; Richard 1974; Mack, Bryant, and Pell 1978; and Birks 1980) and the consensus is that it is shade-intolerant but capable of colonizing immature, mineral soils, in part possibly facilitated by its nitrogen-fixing root nodule system.

In summary, the transcontinental–boreal taxa recorded commonly in the fossil record are trees or shrubs in all cases, except Cupressaceae, distinguishable as pollen to the generic level, and in some to species. Their northern limits fall in a band characterized by mean annual total degree days (above 5°C) of 1,400 to 450, mean July temperatures from 13 to 15°C, and absolute minimum January temperatures as low as –58°C. Their southern limits in the Interior Plains all fall close to the average annual potential evapotranspiration and water deficit values of 52 and 15 cm, respectively. In the Great Lakes area, their southern limit is roughly the 3,500-degree day isotherm and the 20°C mean July isotherm, with the exception of *Pinus banksiana* (Fig. 3.1). The climate data for each taxon are summarized in Table 3.1, pages 52–3.

The four important conifer genera (*Picea*, *Pinus*, *Larix*, and *Abies*) are described by Sakai (1982, 1983) as the tree groups that show the maximum resistance to low temperatures. He has shown experimentally that their tolerance range extends to –70°C and that the adaptation is a greatly reduced primordial shoot permitting rapid ice segregation provided the cooling occurs gradually. These genera appear able to withstand the resulting freeze-dehydration. On the other hand, the deciduous broad-leaved genera (*Betula*, *Salix*, and *Populus*) have been shown to survive freezing to –70°C by extracellular freezing in the xylem cells. Sakai (1983, pp. 2328–9) suggests that "winter desiccation is also an important limiting factor" and that, while the hardy boreal–subarctic conifers are very resistant to severe winter drought, *Pinus strobus*, *Thuja occidentalis*, and *Tsuga canadensis* are very susceptible. He ascribes this important difference to the fact that "In boreal or subboreal conifers shoot and flower primordia are surrounded by 40 or more scales and the bud surface is covered with resin . . . " whereas in the temperate conifers, there are "only a few bud scales or green scalelike tissues."

Eastern, primarily temperate taxa

***Acer saccharum* Marsh.**

It is almost paradoxical that the sugar maple, one of Canada's most important trees both symbolically and economically, and one of the dominant taxa in the entire temperate forest region, should be one of the least known with respect to its postglacial history. The primary reason is that it is insect pollinated and is therefore significantly underrepresented in regional pollen rain. A minor additional difficulty in abstracting its postglacial history from the pollen literature is that not all investigators have made the relatively slight effort needed to distinguish the main species in the genus *Acer*. It ranges in Canada from the west shore of Lake Superior to Gaspé and Nova Scotia, southwards in the Appalachians to Tennessee at higher elevations (>1,000 m), and west to roughly the prairie–forest boundary zone in Missouri, eastern Iowa, and Minnesota (Fig. 1.3).

It appears to be thermally limited along its northern boundary, which coincides roughly with the 2,000-degree day isotherm and the northern limit of the temperate bioclimate as described in Chapter 2, characterized by the Kakabeka Falls, Ontario (Fig. 2.3a), Earlton Airport, Ontario, and Bagotville, Quebec (Fig. 2.8), climate diagrams. The critical limiting factors might be the length of the growing season (six months), the July mean temperatures (>18°C), and absolute minimum January temperatures (<−45°C). The western limit is, apparently, determined by a summer moisture deficit, whereas in the southern Appalachians, its restriction to cooler north-facing slopes and along streams and ravine bottoms indicates that it is limited by high summer temperatures.

Sugar maple is an important codominant of several forest types in the temperate region (Table 2.3), and its large size (30 m at 300 years of age), vigorous vegetative reproduction, prolific seed production, and shade tolerance enable it to compete effectively and persist in many mature stands on a wide range of upland, mesic soils in the mixed conifer–hardwood forest belt of the Great Lakes–St. Lawrence and Acadian regions (Jarvis 1956; Woods 1984 for a recent review). It has similar competitive attributes and life-cycle characteristics to beech and hemlock, and these are the codominants of mature stands in the region. Sugar maple is relatively resistant to windthrow, but is susceptible to a wide range of animal depradations (von Althen 1983).

***Acer rubrum* L.**

Red maple has a slightly wider geographical range than sugar maple, occurring in Newfoundland, southeastern Manitoba, and throughout the southeastern United States from eastern Texas to Florida. Its northern and northwestern climatic limits are very similar to those of sugar maple.

***Acer saccharinum* L.**

The silver maple has a more constricted range than sugar maple, and in Canada, it is of very local occurrence in western Ontario; elsewhere, it extends from about Sault Sainte Marie to central New Brunswick. It grows in lowland, often alluvial, soils associated with elm and basswood.

Both silver and red maple are low pollen producers and their records in the postglacial history from eastern North America are inadequate for any coherent reconstruction of their past community roles or routes of spread. Both are relatively intolerant, short-lived trees.

***Betula alleghaniensis* Britton**

Yellow birch differs markedly from white birch in its range and its ecological behavior, and it is unfortunate that many investigators have not attempted, as Richard (1977a) has, to separate the pollen of the birches by size measurements following the initial investigations by Leopold (1956). In eastern Canada, the ranges overlap and the two species often occur together. The area of yellow birch is centred in the southeastern Great Lakes and northern Appalachia, extending in the north from the southeast corner of Manitoba to Newfoundland. It appears, in common with other important eastern temperate trees, to be thermally limited in the north (length of growing season, winter extreme low temperatures, total degree days) and limited by excessive moisture deficits and summer warmth along its western and southern boundaries. It occurs on podzolic and brunisolic soils developed on well-drained uplands of both calcareous and noncalcareous parent material. It is a constant associate of the mature and intermediate successional stages of mixed forest communities variably dominated by beech, hemlock, and sugar maple. Yellow birch is particularly common as a codominant of mature upland forests in the St. Lawrence Valley, where, along with *Abies balsamea*, it forms a recognized forest type, "La Sapinière à bouleau jaune" (Grandtner 1972). This type occurs near the northern limits of hemlock and beech, where their role in the forests is diminishing, and it includes a common *Pinus strobus* type on rocky and sandy soils. The northern limit of yellow birch lies between that of *Pinus strobus* to the north and *Tsuga* and *Fagus* to the south. Yellow birch reproduces by both prolific production of winged seeds that are readily dispersed in early winter across snow-encrusted landscapes and by active growth of basal stem sprouts. It has moderate shade tolerance, so schema of forest succession, for example, by Shugart, Crow, and Hett (1973) for the western Great Lakes region, place it at intermediate successional stages following aspen, white birch, and jackpine, and preceding sugar maple, beech, and hemlock. Tubbs (1973) has shown experimentally that *Betula alleghaniensis* seedling growth is repressed in the presence of *Acer saccharum* seedlings, and that the maple roots secrete an inhibitory substance. He suggests that this allelopathic relationship might be important in the apparent competitive advantage of sugar maple over yellow birch. Detailed macrofossil analyses have largely confirmed the pollen separations made on the basis of size, for example, by Richard (1977a) and his co-workers in Quebec. In Gaspé, Labelle and Richard (1984) provide an excellent demonstration of the power of combining detailed pollen and macrofossil analyses from two adjacent sites in different topographical situations. They are able to separate *Betula papyrifera* and *B. alleghaniensis* using both pollen measurements and macrofossils to work out the forest development in detail.

Carya

The genus hickory is represented in Canada by six species, of which two are relatively common. The pollen of the individual species cannot be distinguished, so its use in palaeoecology is dependent on the presence of associated taxa. The most widespread species in Canada is *Carya cordiformis* (Wong.) K. Koch, which extends from near Quebec City along the upper St. Lawrence Valley into southern Ontario. Along with *C. ovata*, which has a similar distribution, and other deciduous trees, it characterizes the deciduous forest region of Rowe (1972), equivalent to le domaine de l'Erablière à caryers of Grandtner (1966). *Carya* pollen is more or less equally representative of the trees in the landscape (Delcourt, Delcourt and Webb 1984).

***Fagus grandifolia* Ehrh.**

The American beech is the only North American species in the genus so it provides one of the few examples of a taxon whose pollen can be associated with one species. Its range (Fig. 3.11) is eastern. Its northern limit coincides roughly with that of other temperate forest dominants, and George et al. (1974) and Burke et al. (1976) suggest that the –40°C minimum winter temperature is a critical boundary for beech and several other species, beyond which they are unable to maintain the intercellular water in a supercooled, liquid state. However, several sites well within the range of beech have January absolute minima as low as –45°C, although it is obvious that meteorological station records and conditions in forests might differ significantly. In any event, it appears that the northern boundary, whose climatic attributes can be seen in Figures 2.3a and 3.12, is determined by thermal conditions, while its western limits are probably controlled by moisture and summer warmth. Near its northern limit, it occupies only dry-mesic sites (Hills 1960, p. 415, and Table 2.3), but in the deciduous forest zone of the southern Great Lakes region, it occupies a wide range of soil moisture regimes. In the northern half of its range, it is a codominant in mature forests with *Acer saccharum*, *Betula alleghaniensis*, and *Tsuga canadensis*, and with *Pinus strobus* near its northern limit on dry sites. It has high values of shade tolerance, longevity (200 to 300 years), and height (25 to 30 m), and reproduces both by seed and vegetatively, so it is an effective competitor in most landscape types. It is very susceptible to fire because of its thin bark, but is moderately resistant to wind and icing. Episodes of drought followed by extreme cold in winter have been shown to cause severe mortality. Stem sprouting and root suckering are well-developed in beech with the result that it recovers successfully from overbrowsing by deer (Whitney 1984). Held (1983) recorded the pattern of regeneration in stands widely dispersed along the western range limit of beech, from Wisconsin south to Tennessee and North Carolina. He found a significantly higher frequency of root sprouts at a high-elevation site in the Appalachians and at the northwestern extremity of the range (Milwaukee, Wisconsin) and concluded that root sprouting "is possibly a mechanism by which the species survives locally within the range," where such climatic stresses as freeze-thaw occur in early spring and late spring frosts.

Beech is wind pollinated and is generally described as roughly proportionally represented by its pollen in lake sediments (reviewed by Bennett 1985).

Fraxinus

Fraxinus is represented in Canada and the adjacent United States by four species, none of which can be identified to species from its pollen. *F. nigra* and *F. quadrangulata* have tricolpate pollen, but, as the latter is a species of the deciduous forest and has only a few stations in southern Ontario (Fox and Soper 1953), most investigators assume that the tricolpate ash pollen they record belongs to *F. nigra*. The occasional instances of associated macrofossils have confirmed such an assumption. However, no such assumption is possible in the case of the tetracolpate pollen characteristic of both *F. pennsylvanica* and *F. americana*.

***Fraxinus americana* L.**

The white ash is the most common and ecologically important true ash in the temperate region, with northern and western limits similar to those of beech and red maple. It is an upland species of relatively nutrient-rich soils. Its intermediate shade tolerance, height (25 to 30 m), and longevity result in its playing a role in the early and middle successional stages. It is differentiated into several geographically distinct

Figure 3.11. Range maps of three important trees of the temperate forest region and one eastern boreal species (*Thuja*).

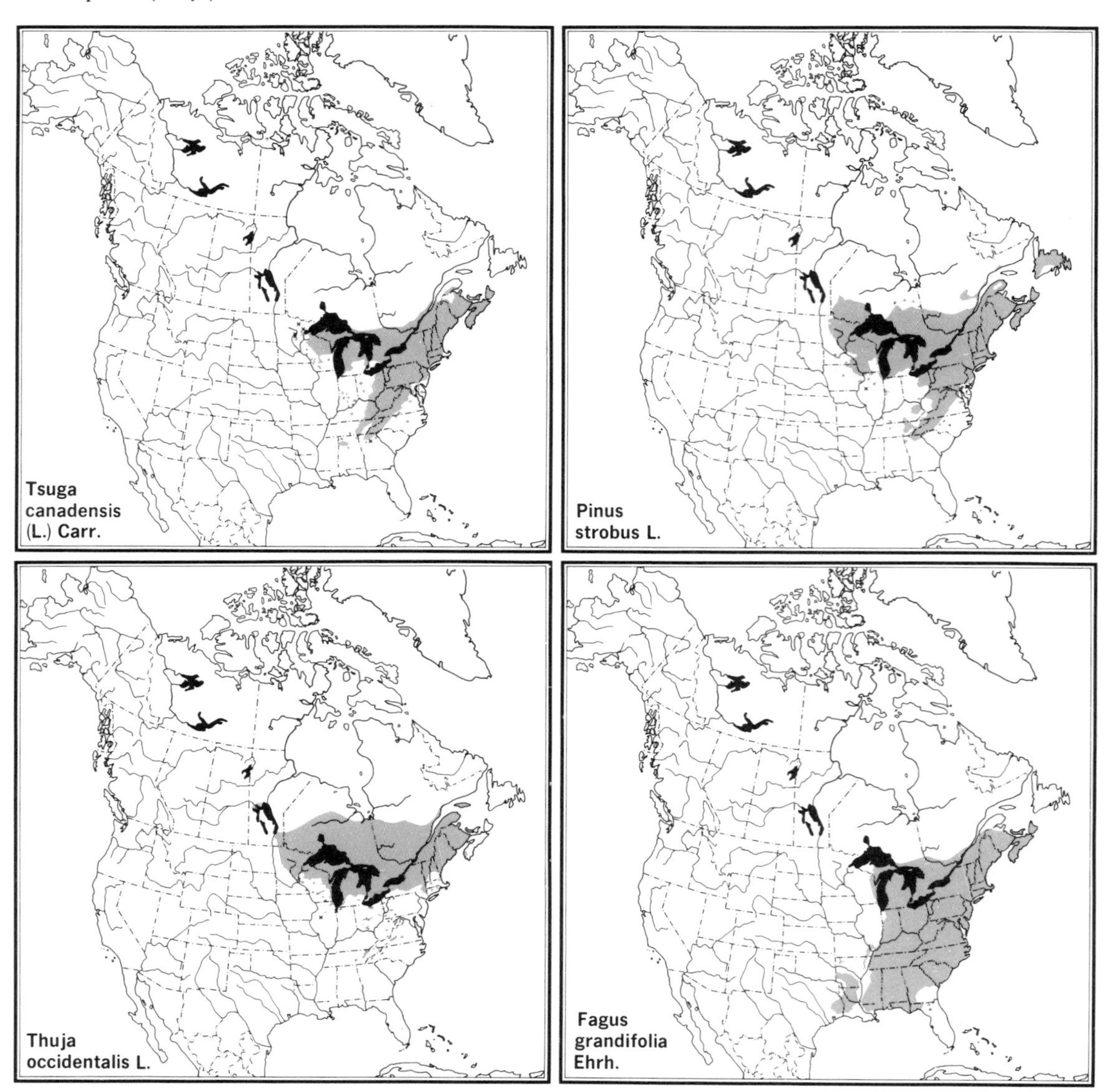

polyploid and ecotypic races, indistinguishable in the fossil record, further reducing the palaeoecological value of the species. Most interpretive conclusions in the literature based on *Fraxinus* cf. *americana* pollen are at best optimistic, at worst vacuous.

Fraxinus nigra **Marsh.**

Black ash is a northern temperate-boreal species (see Fig. 3.13) that appears to be thermally limited at its northern boundary. It coincides roughly with the 1,700-degree day isotherm (Fig. 3.12), the 5.5-month growing season, and a mean July temperature greater than 15°C, whereas the western limit is probably controlled by such aspects of the highly continental climate as soil moisture deficits, high summer temperatures, and extreme winter tem perature minima. Black ash has a narrow range of soil types, confined to hydric sites in swamps or on mineral soils with high water tables. In the northern part of its range, it is commonly associated with black spruce, northern cedar (*Thuja*), *Ulmus*, and *Acer rubrum*.

Figure 3.12. A sketch map of Canada showing the approximate positions of various isohyets (100, 300, 500, 700, 800, 900, 1,200, and 1,600 mm mean annual precipitation) and the approximate limits of the 3,000, 2,000 and 1,000 mean annual degree days (above 5°C) limits. Both data sets are derived from *The National Atlas of Canada* (Fremlin 1974).

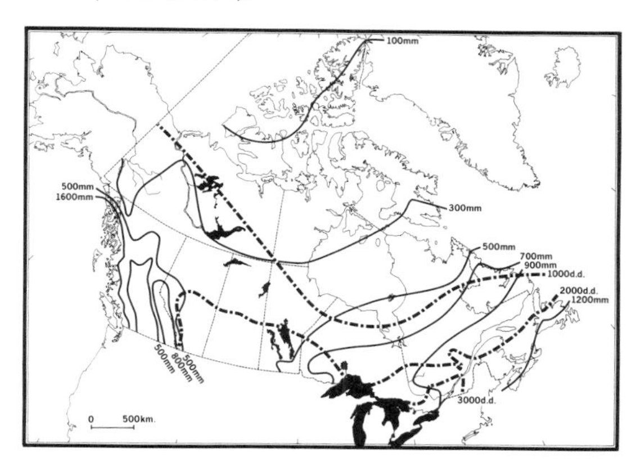

Fraxinus pennsylvanica Marsh.

The red ash has the most extensive range of the true ashes, reaching eastern Alberta and throughout the plains to Texas (Fig. 3.13). As it has been shown to include several important, climatically differentiated ecotypes and as its pollen cannot be distinguished from that of *F. americana*, its use in palaeoecology is very limited. In Canada, it is primarily a tree of alluvial soils, and it occurs on upland, disturbed sites as an early successional species.

Juglans

Two native species of walnut occur in Canada: *Juglans cinerea* L., the butternut, a scattered, minor, shade-intolerant member of the deciduous forest region and the southeastern part of the Great Lakes–St. Lawrence region; and *J. nigra* L., the black walnut, which is confined to the deciduous forest region in southern Ontario, where it is a minor element in the landscape. Although an amentiferous genus, the pollen of *Juglans* is never a significant component of either fossil or modern spectra.

Ostrya–Carpinus

A frequent pollen type in sediments from eastern Canada, *Ostrya–Carpinus*, is so designated because the ironwood, *Ostrya virginiana* (Mill.) Kech, and the hornbeam, *Carpinus caroliniana* Walt., produce pollen that is indistinguishable under a light microscope. The modern ranges (maps in Little 1971) overlap extensively from Minnesota east to Quebec and Maine, south to Florida, west to Texas, and north to Minnesota along latitude 95°W. *Ostrya* extends farther west into the United States prairies and as far as the Red River in Manitoba, and farther east into New Brunswick and Nova Scotia. Both species are scattered in occurrence as minor components in the mixed or deciduous forests and neither exceeds 10 m in height

Picea rubens Sarg.

Red spruce is closely related to black spruce and they are reported to hybridize where they occur together. Unfortunately, although its pollen has been distinguished from other *Picea* species by some investigators (Richard 1970; Watts 1979; Birks and Peglar 1980), only Watts (1979) has found it in fossil material. Davis (1983) reports some difficulty in distinguishing fossil *P. rubens* pollen from *P. glauca* and *P. mariana*, and urges caution in accepting specific identifications. No macrofossils have been recorded so far and its postglacial history is totally unknown. This is regrettable because its restricted range (Fig. 1.3), but locally important role in vegetation, suggests that it would be a useful palaeoecological indicator. Siccama, Bliss, and Vogelmann (1982) and Scott et al. (1984) describe a major decline in red spruce abundance in several parts of its range during the past two decades. The causal factor is unknown, but these investigators speculate that droughts in the mid-1960s, possibly combined with the effects of acid deposition and other industrial effects, might be implicated. Its most important role in forest communities is in Maritime Canada, particularly in the Acadian sector of the mixed conifer–deciduous forests.

Pinus resinosa Ait.

Although criteria have been described for distinguishing freshly prepared red pine pollen from that of *P. banksiana* and *P. rigida*, they are difficult or impossible to apply to fossil material because of the poor preservation of the critical features (Amman 1977). As a result, in the absence of macrofossils, pine pollen of the Diploxylon type (verrucae absent) cannot be identified to species in sediments from eastern Canada. As a result, I will not consider its autecology in detail here. It is an eastern species (Fig. 1.3) with a similar range in Canada to that of white pine. It grows predominantly on sandy, podzolic soils intermediate between the poor, nutrient-deficient podzols tolerated by jackpine and the brunisols characteristic of white pine. Red pine reaches greater age (200 to 300 years) and height (30 to 35 m) than jackpine. It is only slightly less shade-intolerant than jackpine and is usually replaced by white pine except on nutrient-deficient sands.

Figure 3.13. Range maps of four important deciduous trees with roughly similar northern limits in eastern Canada, but widely differing southern and western boundaries.

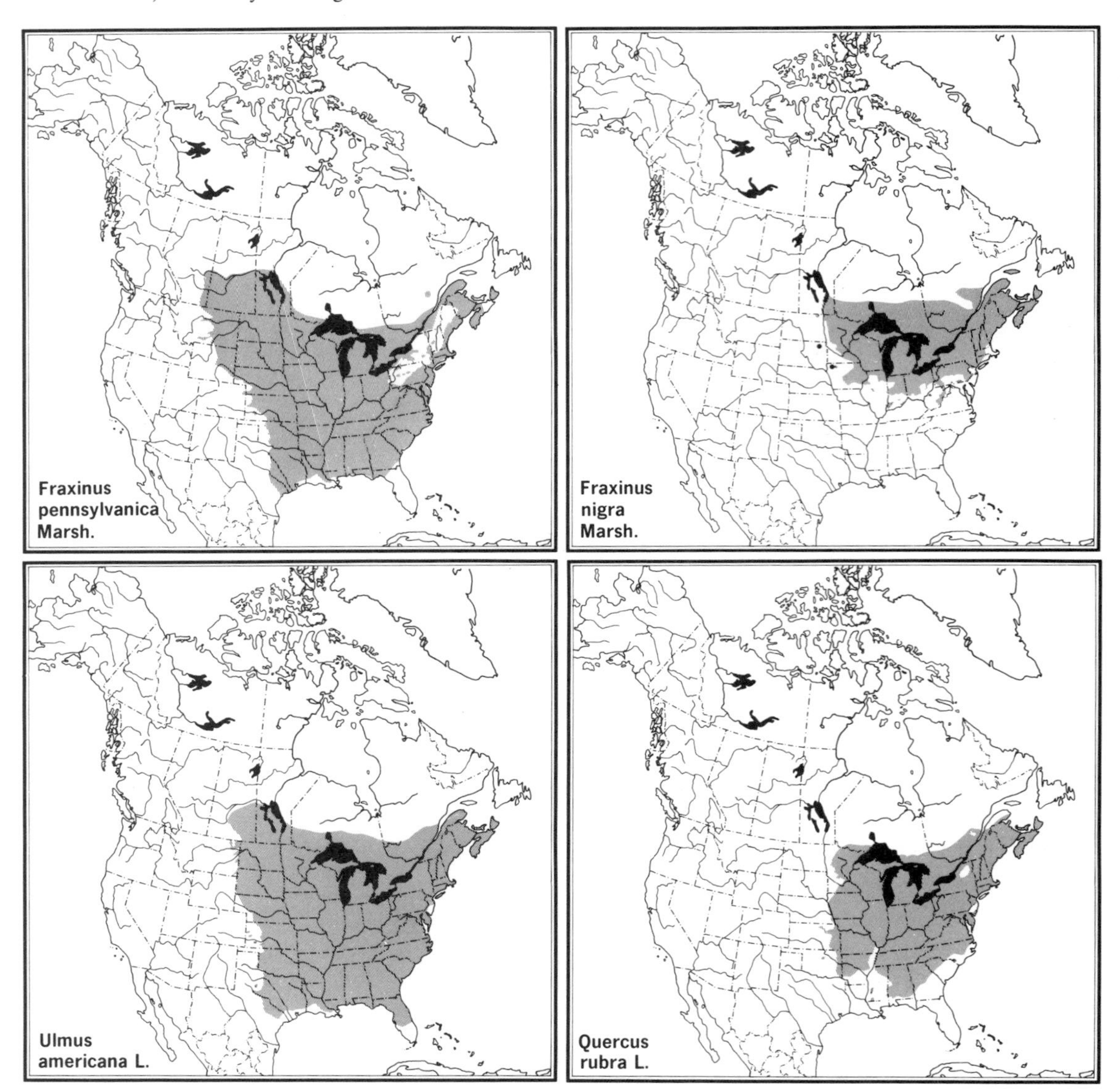

***Pinus strobus* L.**

Eastern white pine has its main area in Canada (Fig. 3.11) centred on the upper St. Lawrence and Lake Ontario and extending into the boreal zone of Quebec and Ontario from southeastern Manitoba to Newfoundland and south through New England, down Appalachia to northern Georgia. The western boundary is widely regarded as limited by precipitation/evaporation (Jacobson 1979) while the northern boundary is probably related to thermal conditions, in particular summer growing season, accumulated degree days, and low winter temperatures.

White pine is an upland species, found mainly on coarse-textured brunisolic and weakly podzolic soils. In the temperate mixed-forest zone, it is most common on rocky and sandy ridges, but, in the absence of hardwood competition in its early stages of growth, it occurs also on heavier textured soils – silts and clay loams. In the southern boreal forest from western Ontario to Quebec, eastern white pine is common on rock and sand ridges (Stiell 1985). There is abundant evidence from the historical records that white pine was more abundant last century, but its large size, straight bole, and desirable timber qualities resulted in severe reductions in its abundance, being replaced by jackpine, white birch, and aspen (Stiell 1985). In the Great Lakes–St. Lawrence Forest region it is a species of intermediate shade tolerance, so that on loamy soils it tends to be replaced by hemlock, beech, and maple, whereas on xeric sites, it often is the dominant of the most mature, stable successional stage (Shugart et al. 1985). In the drier, western parts of its range, white pine has maintained dominance as the result of frequent wildfires (Spurr 1964; Loucks 1983; Peet 1984).

Pinus strobus pollen can be distinguished from that of other eastern pines (*P. banksiana*, *P. resinosa*, and *P. rigida*) by the presence of distal verrucae on the spore body.

Quercus

Ten oak species grow in Canada and approximately another fifty in the United States. Several are important members of the eastern temperate and deciduous forests. However, despite attempts to detect reliable diagnostic features, for example, in the red oaks by Solomon (1983), fossil pollen of *Quercus* cannot be identified below the generic level. As macrofossils are extremely rare, little useful or reliable palaeoecological information can be adduced from *Quercus* records in pollen diagrams unless one makes certain large assumptions about which species might be represented. The three most common species are *Q. rubra* (Fig. 3.13), the red oak; *Q. macrocarpa*, the bur oak; and *Q. alba*, the white oak. Bur oak extends farther west than red oak, reaching eastern Saskatchewan and the western boundary of the Dakotas; it is less common east of the Appalachian axis. Red and white oak have intermediate shade tolerance and red oak in particular competes ineffectively with the dominant trees of the mixed forest – beech, hemlock, and sugar maple. Red oak is most common on xeric sites, particularly bedrock uplands or gravel soils where it often forms stable associations with *Pinus strobus*. *Quercus alba* is less shade-intolerant and is relatively common on mesic uplands, associated with *Tsuga*, *Acer saccharum*, and *Fraxinus* in the mixed forests, and with *Acer*, *Fagus*, *Tilia*, *Juglans*, and *Carya* in the deciduous forest zone. However, several other species of *Quercus* occur in these eastern lower Great Lakes forests, and the problem of interpreting *Quercus* pollen data is aggravated by the fact that it is among the overrepresented pollen taxa (Delcourt, Delcourt, and Webb 1984).

Tilia americana L.

Basswood, or linden, is a temperate forest species in Canada, occurring throughout the Great Lakes–St. Lawrence region and west into south-central Manitoba, south along the Mississippi Valley to northern Arkansas, thence along the Appalachian axis to western New Brunswick. Its northern limit in Quebec and Ontario is near the northern limit of the temperate bioclimate, as described in Chapter 2, and its western limit coincides with a steep gradient westward of increasing continentality and decreasing annual precipitation. *Tilia* occurs on a narrow range of landforms and soil types in Canada, showing its optimum growth on fine-textured silty and clay parent materials. Basswood is an important tree in southern Ontario and southern Quebec, associated with *Fraxinus pennsylvanica*, *Acer saccharum*, and *Betula alleghaniensis*, but its pollen registration is very low and irregular even in these areas, and only a few sites in southern Canada have been recorded that show significant pollen curves for *Tilia* pollen. On the other hand, its postglacial history in regions of the deciduous forest well to the south of our area of interest is reasonably well known (Grimm 1983).

Tsuga canadensis (L.) Carr.

Eastern hemlock is one of the most interesting and important trees in the temperate region of eastern North America, and, although a large and varied literature has addressed its ecology, history, silviculture, migration, and pollen representation, more problems remain than have been solved. Along with beech, sugar maple, and yellow birch, eastern hemlock is a dominant of mature forests throughout the Great Lakes–St. Lawrence and Acadian forest regions (Figs. 3.11 and 2.8) and it extends west to southwestern Ontario and adjacent eastern Minnesota; its southern limits are in the southern Appalachians of Tennessee and Alabama; Olson, Stearns, and Nienstaedt (1959a and b) have published a relatively detailed map of its relative abundance in North America. Its northern climatic limit appears to be roughly coincident with those of its shade-tolerant hardwood associates (beech, sugar maple, and yellow birch) with respect to minimum winter temperatures (George et al. 1974). These authors suggest that January extremes of lower than −42°C are beyond the tolerance of these trees. In general, the thermal characteristics of the northern limit of hemlock is a growing season of six months and mean July temperatures greater than 17°C (Fig. 2.8). Kavanagh and Kellman (1986) suggest, however, that near its northern limit, "competition may limit establishment," although they offer no conclusive evidence to support such an assertion. It is confined in the northwestern sector of its range to bioclimates with greater than 700 mm annual precipitation, but the effective limiting factor appears to be the presence of "rooting environments that are dependably moist" (Rogers 1978, p. 847) as demonstrated by Adams and Loucks (1971) in Wisconsin. There they found it confined to the lower, cooler slopes and valley floors because of its susceptibility to drought. In Ohio, at its southwestern range limit, hemlock is found as isolated stands in canyon valley floors, on steep north-facing slopes, and, rarely, on sand hummocks in a lowland area with a high water table (Black and Mack 1976). Miller (1973a, p. 77) reviews the historical documentation of the response of hemlock to drought in Wisconsin, New York, Connecticut, and Pennsylvania and concludes that drought years can have a severe effect on hemlock populations. Conversely, the more recent slight increase in precipitation and very slight lowering in summer temperatures in the highlands region of southeastern New York State appears to have favoured increased hemlock densities in the ecotonal region between the northern hardwoods region of the mixed forest and the oak–chestnut region of the deciduous forest. Charney (1980) has documented seedling and sapling populations in hemlock–sugar maple–oak–yellow birch stands and shown that "seedling and sapling percentages of *Tsuga* exceed 70% of the total reproductive sample along the hemlock–broadleaf communities transect." Hemlock is moderately susceptible to hurricane-force windstorms (e.g., Hemond, Niering, and Goodwin 1983) and these authors suggest that past hurricanes "appear to be an important factor in perpetuating hemlock."

However, generalizations about the autecology of hemlock should be tempered by the conclusions of Kessell (1979) from a wide-ranging study of the habitats, growth responses, and morphological variation of many populations. He found that populations in southern Canada, Maine, and Massachusetts could be separated into two ecotypes, high-response and low-response types, distinct in both morphology and responses to moisture and temperature. He also analyzed disjunct stands in central Indiana, beyond the western limit of continuous occurrence, where hemlock is confined to steep north-facing slopes and ravine bottoms; these populations were not differentiated ecotypically and could be interpreted as "either the fusion of the two response types, or alternatively, relic populations which never differentiated into the two response types." The successional role of hemlock in the eastern temperate forests has been the subject of several investigations (McIntosh 1972; Nicholson, Scott, and Breisch 1979; Woods 1984, among others), and, while broad generalizations are not possible as different regional environments and ecotypes are likely to produce "varying interpretations in different areas" (Nicholson, Scott, and Breisch 1979, p. 1252), a few patterns seem to be emerging. In general, hemlock, along with beech, is one of the longest-lived species, extremely shade-tolerant, and tends to produce soil conditions inimical to the regeneration and growth of its competitors. In particular, its needles produce a highly acidic litter, producing acidic, base-deficient soil. Woods reports that among beech, maple, and hemlock, "hemlock canopy seems to extend the strongest influence on the seedling and sapling strata . . . " (1984, p. 102).

Eastern hemlock does not reproduce vegetatively, but produces abundant seeds that are small, winged, and released from the hygroscopic cone scales in dry weather, favouring rapid, effective dispersal. Germination and establishment require a moist, shaded site, and there is evidence from areas near the drier, western limits of its range that effective establishment is dependent on years of above-average rainfall. Mature trees are several centuries old and reach heights of 25 to 50 m with trunk diameters of 0.75 to 1.5 m. Aspects of the ecology of the tree are well-illustrated by the results of a recent reanalysis of stands with abundant hemlock in Pennsylvania that had remained undisturbed by fire or felling since an initial analysis in 1929 (Whitney 1984). Whitetail deer populations have increased significantly in the region as a result of various forestry practices and protective legislation. Hemlock seedlings and small saplings are among the favoured browse of deer, and Whitney's (1984) recent census shows that the populations of hemlock seedlings and small saplings (30 cm) have declined significantly between 1929 and 1978, whereas beech, which responds to browsing by increased sprouting, was not seriously affected. However, while "sustained grazing pressure over the past 40 years has prohibited the establishment of any advance regeneration, creating a serious gap in the size–class distribution of many species" (Whitney 1984, p. 406), the author points out that the slow growth of hemlock, and to a lesser extent beech, "will be able to maintain (their) dominance in the canopy for at least the foreseeable future," but the longer-term status is "more problematic" if the browsing pressure continues. It is of interest to note that the prediction of the original investigator of this site (Lutz 1930), that the *Pinus strobus* and *Quercus alba* trees already established in the original stands after local disturbance would be eliminated in five decades, has been confirmed. The role of deer in affecting the successional relationships of hemlock and sugar maple in northern Wisconsin was investigated by Anderson and Loucks (1979). The poor regeneration of hemlock in the area, in dense hemlock and maple stands, had been ascribed earlier by Curtis (1959) to a superior competitive vigour in successional processes by sugar maple. Anderson and Loucks (1979) showed, however, that the effect of greatly increased browsing by whitetail deer, protected by legislation, was to inhibit hemlock regeneration severely, while sugar maple resprouted after browsing and was not suppressed.

In the absence of such disturbance, the normal successional trend is that hemlock, variably associated with beech and yellow birch, assumes dominance in the Great Lakes (Shugart, Crow, and Hett 1973) and northern sectors of its range (Rogers 1978). Rogers (1978, p. 850) proposes that "the very limited distribution of hemlock dominated forests today" can be explained adequately as a response to the various effects of European settlement – clearing and burning, direct exploitation of the bark for tanning, and the effects of greatly increased deer populations.

The importance of shade and moist seedbeds in the autecology of hemlock is well illustrated by a recent investigation of hemlock regeneration after clear-cutting in upper Michigan (Hix and Barnes 1984). The sites analyzed were dominated by hemlock prior to clear-cutting between 1924 and 1947. Hemlock has been almost completely replaced by red and sugar maple, yellow birch, and balsam fir, and even the existing advance regeneration is being "suppressed by faster-growing hardwoods." The authors suggest that the critical factors are the absence of suitable moist microsites, such as rotting logs and the absence of fire, which would create many suitable seedbed habitats for hemlock regeneration. The same investigation showed that the clear-cutting and consequent replacement of hemlock by hardwoods caused a decrease in soil acidity and an increase in nutrient cycling. It is well established that persistent hemlock-dominated forests produce a thick, very acidic soil litter that inhibits nutrient release into the soil. The accumulation of a highly acidic litter horizon has been suggested as a cause of poor hemlock regeneration, as well as a variety of other factors, but Ward and McCormick (1982) conclude from experimental germination tests that hemlock "is allelopathic to its own regeneration." Water extracts of hemlock litter and plant material were used in germination experiments and 100 percent mortality in six-day old seedlings resulted, while seedlings older than two weeks were unaffected.

The representation of hemlock pollen in modern (presettlement) samples has been estimated by a few investigators. Webb and McAndrews (1976, p. 278) concluded from their extensive survey that the pollen limits "coincide in general with the range of hemlock trees." Delcourt, Delcourt and Webb (1984, pp. 95–6) describe the modern pollen–vegetation relationship in *Tsuga* as "problematic" in that the regression line slope of pollen percentage against vegetation percentage suggests overrepresentation of the pollen, while the y-intercept value, a rough measure of "pollen dispersability of the taxa, relative to the area of vegetation sampled," suggests that hemlock is an underrepresented taxon. An investigation of pollen deposition around individual hemlock trees by Ibe (1983) led to the conclusion that dispersal distances in this species are short, but no comparative data are available for other taxa. An investigation by Miller (1973a), more restricted geographically, but more precise analytically than the Delcourt, Delcourt, and Webb (1984) study, shows that hemlock in both pre- and postsettlement measurements from northern New York is slightly overrepresented, a conclusion reached also by Davis and Goodlett (1960) from work in Vermont and by Webb (1974b) and Webb et al. (1981) from Wisconsin and Michigan.

Ulmus americana L.

American elm is an eastern species extending into the subarid plains as far as south-central Saskatchewan, throughout the Dakotas, and south to Texas (Fig. 3.13). At its western limits, it is confined to alluvial and lower slope habitats, whereas in the eastern and central parts of its range, it occurs on a wide variety of soils. Its climatic limits at the northern and western extremities of its range are not known by any critical analysis or experimentation, but its northern limit in the boreal forest of Canada coincides roughly with areas with more than 1,900 degree days, more than 5 months growing season, and mean July temperatures greater than 16°C. Its western limit in Canada is characterized by annual precipitation exceeding 480 mm. It is a tree of intermediate shade tolerance, longevity (200 years), and height (20 m), with the result that it rarely forms pure stands but rather occurs as a major codominant in a few site types and a minor element in many forests of the temperate and southern boreal regions.

It reproduces by vegetative sprouting and its prolific seed crops germinate readily on a wide variety of substrata. The seedling and sapling stages are intolerant of half to full shade, and, as a result, it is a common secondary species in disturbed forests, but absent from closed, mature forests on mesic sites. It is an important codominant with *Fraxinus nigra* and *Acer rubrum* on alluvial sites in the temperate region. However, its abundance throughout its range has decreased rapidly in recent decades due to Dutch elm disease.

Elm is wind pollinated. Its mapped modern pollen range in North America coincides closely with its range in the landscape, and it appears to be roughly equally represented in modern pollen samples.

In summary, the important eastern temperate taxa in the fossil records are all trees that share a basically similar range centred on the lower Great Lakes and northern Appalachia. Their northern limits in Canada lie within a relatively narrow band from Lake-of-the-Woods, Ontario, across the top of Lake Superior to the St. Lawrence estuary near Quebec City. This broad band coincides roughly with the 2,000 degree day, 15°C July, and the –45°C absolute January minimum isotherms. The western limit in the United States varies among taxa from those that range onto the prairies to the more restricted *Tsuga* and *Fagus*. The limiting factor appears to be related to moisture and summer warmth. Table 3.1 and Figures 3.11 and 2.8 provide a summary of these climatic characteristics.

Pacific–Cordilleran taxa

***Abies amabilis* (Dougl.) Forbes**

The Pacific silver fir is restricted to the Pacific sector, extending from roughly 55°N south to 43°N in the Cascade Mountains of Oregon and at scattered localities in northern California. It is a coastal species and occurs in only two of the biogeoclimatic zones of the region – the coastal western hemlock zone, the wettest zone where it is associated with *Pseudotsuga menziesii*, *Tsuga heterophylla*, *Thuja plicata*, and *Picea sitchensis*; and the mountain hemlock zone, a coastal subalpine complex that occurs from 900 to 1,200 m, made up chiefly of *Tsuga mertensiana* and *Chamaecyparis nootkatensis* in addition to *Abies amabilis*.

A. amabilis is dependent on year-round moisture, so its occurrence in the mountain hemlock zone is confined to sites with deep snow cover (roughly 180 cm per month) that maintains the soils unfrozen throughout the winter. A year-round growing season occurs in coastal environments, while in the subalpine mountain hemlock zone, it is about seven months. It is the most shade-tolerant of all Canadian conifers (Krajina 1969). However, in the upper elevations (1,200 to 1,800 m) of the mountain hemlock zone in the northern Cascade Range of Washington, Oliver, Adams, and Zasoski (1985) report that Pacific silver fir was the most common species to recolonize avalanche sites, where it "apparently had a competitive advantage after disturbances such as an avalanche, which remove the overstory but do not destroy the advance regeneration" (p. 231). The same authors note that in other disturbed habitats, for example, following glacial retreat or large-scale soil erosion, it is at a competitive disadvantage with the light-seeded mountain hemlock and sitka alder.

A. amabilis is tolerant of a wide range of soils from acidic humo-ferric podzols to base-rich brunisols, but it requires moist substrata at all seasons. It achieves its maximum growth (45 m) in the coastal western hemlock zone. It reproduces by seeds that are large, winged, and dispersed by wind over short distances. Its pollen is indistinguishable under light microscopy from other *Abies* species, but it has been found as macrofossils, chiefly needles, so it does feature in the fossil record from the Pacific region.

***Abies grandis* (Dougl.) Lindl.**

The grand fir is found at low-elevation sites in both the Pacific and Cordilleran sectors in a precipitation range from 480 to 2,800 mm; it has its optimum development in the coastal Douglas-fir zone and in the interior (Cordilleran) western hemlock zone, but also occurs in the drier sectors of the coastal western hemlock zone and in the wetter sectors of the interior Douglas-fir zones. It is temperature limited, requires a long growing season, and has low frost resistance, so it is absent from the subalpine zones. It is tolerant of dry soils provided they are nutrient-rich, and it is particularly sensitive to calcium and magnesium deficiencies. In the interior Douglas-fir (wetter) subzone, *A. grandis* is associated with *Pinus contorta*, *Pinus monticola*, *Thuja plicata*, and *Pseudotsuga menziesii*, and, in the drier subzone of the coastal western hemlock zone, it is associated with *Pseudotsuga*, *Tsuga heterophylla*, *Thuja plicata*, *Picea sitchensis*, and *Pinus monticola*. By contrast to the south in Oregon and southern Washington, in the southern Cascade Range and Blue Mountains, it forms "the most extensive midslope forest zone" (Franklin and Dyrness 1973), presumably in response to less cold winters. It is shade-intolerant on most sites in the coastal western hemlock zone, but shade-tolerant elsewhere. It achieves maximum growth to 54 m in height in the coastal western hemlock zone. It occurs mainly on brunisols. *Abies grandis* pollen is indistinguishable from other *Abies* species and so far no *A. grandis* macrofossils have been found, so the fossil record of this important tree depends largely on assumptions about its presence by association with other taxa. Earlier distinctions of *Abies* species using pollen size differences have been discounted by modern palynologists (Mathewes 1973, Mack 1971).

***Abies lasiocarpa* (Hook.) Nutt.**

The subalpine fir occurs widely in the Pacific–Cordilleran region at high elevations (Fig. 1.3), except near its northern limit in the central Yukon and Alaska. It forms krummholz in the alpine zones of the Pacific and interior mountain ranges. It is an important tree in the mountain hemlock zone, where it shares dominance with *Tsuga heterophylla*, and in the Engelmann spruce–subalpine fir zone, where it is codominant with *Picea engelmannii*. Alpine fir is frost resistant and occurs in climates with relatively short growing seasons (five months) and cool summers (mean July temperature 11 to 17°C). It requires greater soil moisture than Engelmann spruce and is shade-intolerant on hydric soils.

Abies lasiocarpa is indistinguishable from other *Abies* pollen, but several sites have produced macrofossils, particularly from full-glacial sediments (Barnosky 1981; Hicock, Hebda and Armstrong 1982; Cwynar in press), often in association with *Picea engelmannii* and *Tsuga mertensiana*.

***Alnus rubra* Bong.**
Of the five native alder species in Canada, only this one, the red alder, reaches tree dimensions. It is a Pacific species, restricted to the coastal Douglas-fir and western hemlock zones (see Fig. 3.14). It is shade-intolerant and functions as a pioneer on areas cleared by fire or felling, or on freshly eroded, often alluvial soil. It is most common on deep alluvial soils.

***Picea engelmannii* Parry**
Engelmann spruce is a Cordilleran species (see Fig. 3.15) found chiefly in southeast British Columbia and adjacent Alberta and southwards to northern California, New Mexico, and Arizona (Fig. 1.3). It is the chief tree of subalpine forests in the Rocky Mountains, and it occurs in Washington and Oregon on the east slopes of the Cascades. However, its pollen cannot be separated from that of other spruces, particularly *P. glauca*, with which it often grows at the conjunction of the Cordilleran and western boreal regions. In the absence of macrofossils, pollen is identified as *P. engelmannii* on the tenuous basis of the presence or absence of certain other taxa (Barnosky 1984, pp. 620–1).

***Picea sitchensis* (Bong.) Carr.**
Sitka spruce is a Pacific species with a restricted distribution along the west coast, between Alaska and California, and, although its pollen cannot be distinguished from that of any other spruce, it has been widely identified in fossil records from the Pacific region, confirmed by macrofossils in a few cases, but more frequently by assumption on the basis of such associated pacific taxa as *Tsuga heterophylla*. It is confined to the humid, oceanic, western region, being particularly abundant in the coastal western hemlock zone (Krajina 1969), characterized by the Alert Bay climate diagram (Fig. 2.15). It is tolerant of maritime environments, influenced directly by sea spray, where it appears to flourish in the absence of western hemlock, which is intolerant of these conditions. It grows well on alluvial soils. It rarely occurs above 650 m, and normally is confined to the lower 300-m elevations.

Figure 3.14. *Alnus rubra* in the southern Pacific coastal forest zone.

Figure 3.15. Treeline species in the southern Cordilleran zone – *Pinus contorta*, *Picea engelmannii*, and (prostrate) *Juniperus communis*.

The western pines
Five species of *Pinus* occur in Pacific–Cordilleran Canada, but none can be distinguished palynologically at the species level (Mack 1971). Two have pollen of the Diploxylon type (*P. contorta*, lodgepole pine; and *P. ponderosa*, ponderosa pine). The others (*P. albicaulis*, the whitebark pine; *P. flexilis*, the

limber pine; and *P. monticola*, the western white pine) all have Haploxylon pollen. The general practice of modern investigators is to assume that when Diploxylon pollen occurs in Pacific sites it should be ascribed to *P. contorta* because *P. ponderosa* is a Cordilleran species of dry climates and, although some investigators make a similar distinction between dry (*P. ponderosa*) and moist (*P. contorta*) to separate the species in Cordilleran sites, there is a large element of uncertainty in such an assumption. Hansen and Cushing (1973) developed a key to *Pinus* species of the southwestern United States, which includes a separation of *P. contorta* and *P. pondersosa*, readily seen on most reference material, but unreliable with fossil grains. Others, Barnosky (1985a and b), for example, identify pollen in this category as *P. ponderosa*/*P. contorta*. The Haploxylon group cannot be further separated, but the presence of macrofossils at some sites (e.g., Mathewes 1973; Cwynar in press) provides some basis for making a species separation.

Pinus contorta **Dougl.**

Lodgepole pine is a widespread Cordilleran species (Fig. 1.3) that occurs throughout British Columbia, except in the far northwest, and extends south into the adjacent United States. It displays the greatest amplitude of climatic tolerance of any western tree. It occurs in the coastal western hemlock and Douglas-fir zones, the mountain hemlock, Engelmann spruce–subalpine fir, interior Douglas-fir, interior western hemlock, and boreal zones (Krajina 1969). It even occurs rarely as alpine krummholz (Fig. 3.15). It is absent from the *Pinus ponderosa*–bunch grass zone, but Krajina (1969, p. 91) suggests that the operative factor is the absence of acidic soils because it does occur to the south in Oregon in dry climates (annual precipitation 250 to 300 mm) and on acidic soils derived from tephra deposits. Lodgepole pine is most abundant on well-drained brunisols, but occurs also on poorly drained podzolic soils. It retains seeds in serotinous cones for up to twenty years and the cones open in response to the heat of wildfire. It is a shade-intolerant species of intermediate longevity and stature (maximum 30 m at 100 years) in coastal regimes. It is usually replaced by competitors, chiefly *Pseudotsuga menziesii* in the interior, in less than 100 years. Lodgepole pine is an important pioneer species that establishes rapidly after fire or felling.

The postglacial history of this species in the northern part of its range has been studied recently, as may be seen in Chapter 7.

Pinus ponderosa **Laws.**

Ponderosa pine is confined to the southern Cordilleran region of British Columbia, extending southwards to Mexico. It is restricted to subarid montane and grassland bioclimates, depicted by the Kamloops and Cranbrook climate diagrams (Fig. 2.15). The summers are warm and semiarid and the winters mild and relatively snowy; the growing season is long (six to nine months). Ponderosa pine is the chief tree in the ponderosa pine–bunchgrass zone (Krajina 1969), represented by the Kamloops climate diagram, where it forms mature stable forests on mesic, gray luvisols derived from glacial till. Fine-textured, drier parent materials in this zone support *Agropyron* steppe with chernozemic soils.

Pinus monticola **Dougl.**

While the pollen of the western white pine is indistinguishable from that of other Haploxylon pines, it occurs as macrofossils and is ecologically distinct from the following two species. It is represented by two populations, a coastal one, in both the western hemlock and Douglas-fir zones, and a spatially separate interior population, in western British Columbia in the interior Douglas-fir and western hemlock zones. In the coastal zones, while it is shade-tolerant, long-lived (up to 500 years), and tall (50 to 60 m), it seldom persists as the dominant on stable, undisturbed sites in these humid climates because of an intolerance of acidic, calcium-deficient soils (Krajina 1969). It is most abundant in the interior western hemlock zone, but also occurs in the Engelmann spruce–subalpine fir and mountain hemlock zones.

Pinus flexilis **James,** ***P. albicaulis*** **Engelm.**

The limber pine and whitebark pine are not recorded in fossil material. The former is of very restricted occurrence in the 1,500- to 1,800-mm montane belt of the Cordilleran region and the latter is found mainly in the Engelmann spruce–alpine fir zone at 2,000 m, but also in the mountain hemlock zone near the coast.

Pseudotsuga menziesii **(Mirb.) Franco**

Douglas-fir is the most productive and economically important tree in Canada. It is a Cordilleran species, made up of coastal and interior races, usually distinguished as separate varieties. Most investigators consider that its pollen is indistinguishable from that of *Larix* under the optical microscope, so the degree of confidence in its recognition in the pollen record is usually based on the associated species, and, less commonly, on adequate macrofossil evidence. However, Tsukada (1982, p. 162), maintains that "the triradiate scar in *Pseudotsuga* and the annular thickening in *Larix* become conspicuous after a long acetolysis treatment . . . and both genera can be positively identified and distinguished in most cases." The coastal variety occurs from roughly 55°N on the British Columbia coast to 35°N in California, where mean July and January temperatures range from 15 to 18°C and 0 to 8°C, respectively, and annual precipitation from 600 to 3,000 mm (Fig. 2.3a). The range of the interior variety is centred on the Rocky Mountains and it grows in regions with lower annual precipitation (about 400 mm) and a greater range of summer and winter mean temperatures.

In the Pacific sector, Douglas-fir occurs commonly in the coastal Douglas-fir and the coastal western hemlock zones (Krajina 1969). In the Cordilleran (interior) sector, it is found in a wide variety of forest types – the interior Douglas-fir, interior western hemlock, ponderosa pine, and Engelmann spruce–subalpine fir zones.

Douglas-fir is most common in the drier parts of its coastal range, the coastal Douglas-fir zone of Krajina (1969), where 100-year-old trees are 45 to 48 m tall on mesic to hydric sites. In these and the coastal western hemlock zone, Douglas-fir is shade-intolerant in relation to its immediate competitors, western hemlock (*Tsuga heterophylla*), western red cedar (*Thuja plicata*), and Pacific silver fir (*Abies amabilis*), so, while it achieves the maximum growth of any Canadian tree in these environments (54 to 60 m at 100 years), it is ultimately replaced in stands that remain undisturbed for a very long time. In practice, the effects of fires, lumbering, and insect pests keep the forests in a youthful condition with extensive areas of young, even-aged Douglas-fir. The long-term effects of these "disturbance" factors have been studied recently by a simulation model (Dale, Hemstrom, and Franklin 1986). The model compares the long-term effects of fire, windthrow, and insect damage on an initial 500-year-old typical stand of Douglas-fir, silver fir, and western hemlock. The results confirm the generally held opinion "that the western hemlock–silver fir is truly a self-replicating and stable community in the absence of disturbance." In the Engelmann spruce–subalpine fir zone of the interior, Douglas-fir is shade-intolerant, performing as a pioneer species replaced after about 100 years by spruce or fir. On the other hand, in the ponderosa pine zone of the Columbian valleys, *Pseudotsuga* is less light-demanding than the pine, and is a common dominant on coarse-grained soils.

***Tsuga heterophylla* (Ref.) Sarg.**

The western hemlock is a Pacific species ranging from southern Alaska and northern British Columbia (latitude 55°N) to California and northwestern Montana. It is the climax species in the coastal western hemlock and the interior western hemlock zones (Krajina 1969) and it occurs as a secondary species in several interior zones. The optimum development, as the dominant of mature forests, occurs in the southern and northern Pacific bioclimates, represented by the Alert Bay climate diagram (Fig. 2.15), where annual precipitation is 1,200 to 1,500 mm, January temperatures are lower than –5°C, and the growing season is eight to twelve months. It has low frost resistance and requires snow-covered soils in areas of freezing temperatures. It shows intolerance to drought, particularly in the Cordilleran parts of its range, where these conditions occur relatively often. In common with other hemlock species, it is extremely shade-tolerant and performs at its maximum growth and competitive performance in humid, coastal climates on podzolic soils. It forms the stable, "climax" forests in the southern Cordilleran (interior) regions, on the east side of the central plateau, where it is the dominant on mesic sites in the lower elevations (1,200 m), with a bioclimate exemplified by the Castlegar climate diagram (Fig. 2.15).

The pollen representation of *Tsuga heterophylla* is poorly understood, but it appears to be neither excessively over- nor underrepresented. It is a very important pollen taxon in Pacific records.

***Tsuga mertensiana* (Bong.) Carr.**

The mountain hemlock is a Pacific species of humid, subalpine bioclimates with a short growing season and deep winter snow. It is the dominant tree of the mountain hemlock zone of Krajina (1969), equivalent to the coastal subalpine of Rowe (1972), an ecoregion normally above 900 m in coastal British Columbia and Washington, although stands of mountain hemlock occur in low valleys on Vancouver Island and farther north in British Columbia and Alaska. Typically, the bioclimate has an annual precipitation of 1500 mm of which there is abundant snow (180 cm per month); July and January mean temperatures are 10°C and lower than –10°C, respectively. It occurs less commonly in the Engelmann spruce–subalpine fir communities of the Cordilleran region, for example, in the Selkirk Mountains of southeastern British Columbia, but always in local regions of high snowfall. In the coastal western hemlock zone, it is confined to exposed ridges and mountain tops. Its production of abundant small, light seeds apparently confers a competitive advantage in invasion of exposed mineral soil following the recent recession of valley glaciers or soil mass erosion (Oliver, Adams, and Zasoski 1985). Its occurrence in the fossil record has been used as an indicator of coastal subalpine climates (Barnosky 1984).

Cupressaceae (*Thuja*, *Juniperus*, and *Chamaecyparis*)

Cupressaceae is a problematical taxon in the pollen record both from the Pacific–Cordilleran region and the eastern temperate area. Pollen of three important genera in this family are inseparable as fossil grains using the light microscope – *Thuja*, *Juniperus*, and *Chamaecyparis*. The last is represented in the Pacific region by the yellow cedar, *Chamaecyparis nootkatensis* (D. Don) Spach. It is common at higher elevations in the coastal western hemlock zone, where it grows frequently with *Thuja plicata* L., the red cedar, another Pacific tree, but with a wider range than the yellow cedar. Pollen of Cupressaceae from sites in the Pacific region cannot be identified to genus.

In western and subarctic Canada, only a single genus of the family occurs, *Juniperus*, represented by two transcontinental boreal species *J. horizontalis* Moench (Fig. 3.3) and *J. communis* L., both low shrubs of open habitats. However, *J. scopulorum* Sarg. occurs in the Cordilleran zone, where it overlaps in range with *Thuja plicata*. *Juniperus virginiana* L., the eastern red cedar, occurs in southern Ontario, where it overlaps with both the eastern white cedar, *Thuja occidentalis* L., a widespread eastern tree (Fig. 3.11), and the transcontinental species of *Juniperus*. *Thuja occidentalis* has been found as macrofossils (Liu 1982). In the absence of such evidence, pollen identification of this group must be left at the family level unless one makes assumptions about past history. For example, pollen grains of Cupressaceae in late-glacial and Holocene sediments from the Western Interior of Canada are always assumed to be of *Juniperus* because possible alternative genera (*Thuja*, *Chamaecyparis*) have modern ranges and tolerances quite different from those of any taxa or pollen assemblages associated with the Western Interior.

It is difficult to provide a tidy summary of the phytogeography and ecology of the main taxa represented in the fossil record from the Pacific–Cordilleran region. The reader is referred to Table 3.1, where the more important taxa are listed with some indication of their climatic limits, longevity, height, and tolerance scale. The most important conclusion to be drawn is that, in the absence of macrofossils, the interpretation of the fossil pollen record is seriously hampered by the impossibility of distinguishing the fossil pollen grains of many of the conifer species and genera.

A significant number of trees and shrubs found in the modern vegetation of Canada have been omitted from the above survey, either because they have not yet been found in the fossil record, or, if they have, because the records are fragmentary; their pollen is indistinguishable from other genera; their modern range is restricted and their ecological role is very minor. The omitted taxa are several species of *Acer*; *Betula occidentalis*; *B. populifolia*; *Castanea*; *Celtis*; *Cercis*; *Hamamelis*; *Larix occidentalis* and *L. lyallii*, two Cordilleran species whose pollen is indistinguishable from that of *Pseudotsuga*, which is often a codominant of the first; *Liriodendron*; *Nyssa*; *Platanus*; several uncommon *Populus* species; *Rhamnus*; *Salix*; *Sorbus*; *Taxus*; *Ulmus thomasii*; and *U. rubra*. A few of these taxa are encountered when the record from particular sites is examined in later chapters.

Arctic taxa

Several taxa that are found today exclusively or mainly in arctic or arctic–montane regions have been recovered at many sites as pollen and in some cases as macrofossils, primarily from full- and late-glacial sediments. Only taxa that are both common in the fossil record and of predominantly arctic or arctic–alpine range are considered here. Others that are common both in modern pollen spectra from high latitudes and elevations and in fossil assemblages are not included because they are made up of such a large number of species of widely different ranges and ecological amplitudes that they provide no direct basis for palaeoecological interpretation; Gramineae and Cyperaceae are the main taxa in this category, but Leguminosae, Cruciferae, and Umbelliferae are also included.

Artemisia

The genus *Artemisia* is recorded as pollen from almost every fossil site in Canada considered in this book. Twelve species occur in the Canadian arctic and a further ten elsewhere in Canada. Attempts to identify even groups of species palynologically have failed, so the palaeoecological information accessible from this taxon remains minimal, which probably accounts for the endless, rather vacuous interpretive commentaries that attend the registration of the genus in fossil spectra.

Dryas

The genus *Dryas* includes nine species in Canada and the adjacent United States of widely different geographical ranges but broadly similar ecological amplitudes (distribution maps on pp. 427–8 in Porsild and Cody 1980). Pollen is indistinguishable to species, but macrofossils occur relatively frequently in late- and full-glacial sediments from the plains regions and are identifiable to species. All species are calciphilous to varying degrees.

Dryas integrifolia has a palaeoecological indicator value that is very similar to that of *Oxyria*. It has a wide geographical area, in the entire arctic–subarctic region, in the Cordillera, and in Newfoundland, Labrador, Gaspé, and the northern shore of Lake Superior (map 719, Porsild and Cody 1980); and it is one of the dominants of upland sites in the high arctic region of Canada, particularly on "weakly to moderately calcareous materials" (Edlund 1983a, p. 384). It is only in modern spectra from these high latitudes that it makes up a significant (1 to 3%) proportion of the regional pollen recorded in lake sediments. On the other hand, it comprised up to 10 percent of the spectra in moss polsters collected from the montane *Dryas* tundra in the St. Elias Mountains (Birks 1977). The southern outlier populations of *Dryas*, for example, on the northern shore of Lake Superior and in southeastern New Brunswick, are regarded as evidence that the species is not climatically restricted to arctic regions, but is intolerant of shade and acidic soils, and therefore persists, from a possibly wider late-glacial range, in sites with little competition from forest trees (Birks 1976; Miller and Thompson 1979; Given and Soper 1981).

Dryas is entomophilous so it is underrepresented in spectra from lake sediments. For example, *Dryas integrifolia* is codominant with *Salix arctica* on extensive upland tundras in the Melville Hills region of northwest Northwest Territories near the mouth of the Horton River, but both percentage and concentration values in a series of lake sediment samples were extremely low (Ritchie, Hadden, and Gajewski in press).

Ericales

Fredskild (1967), Birks (1976), and Cwynar (1982), among others, have investigated the pollen of the Ericales and the following groups have been identified as pollen in fossil material.

Cassiope

Two species, indistinguishable palynologically, are found in Canada, *C. tetragona* (L.) D. Don, a circumpolar taxon found throughout the North American arctic (Porsild and Cody 1980, map 870, p. 495), and *C. hypnoides* (L.) D. Don, an amphiatlantic species confined in Canada to the Hudson Bay, Ungava, Labrador–Gaspé, and Baffin Island region. The former is common in lower-slope situations, where snow lies deep and late, from the low to the high arctic, and the latter has similar local habitat characteristics, but is less widespread.

The other groups in the Ericales have been distinguished only to generic level and they all have broadly low-arctic or low-arctic–boreal ranges and have not been recorded often enough so far to justify treatment here. They are *Arctostaphylos*, *Andromeda*, *Ledum*, *Rhododendron*, *Loiseleuria*, and *Pyrola*.

***Oxyria digyna* (L.) Hill**

The mountain sorrel is a circumpolar, perennial, anemophilous herb that is frequently recorded as both pollen and macrofossils, occurring in Canada and the adjacent United States throughout the arctic, along the Cordillera, and in Labrador, Newfoundland, Gaspé, and northern Appalachia (Porsild and Cody 1980, map 454, p. 267). It occurs both above and below the treeline in the Northern Cordillera (Raup 1947; Ritchie 1982). Mooney and Billings (1961) have investigated several aspects of the physiological ecology of this species and shown that there is a high degree of complex differentiation into races and ecotypes that differ in morphology, pigment content, photoperiod, seed production and germination, photosynthesis and respiration, and reproductive phenology. For example, they found that arctic populations required a long photoperiod (>8 h) for successful flowering, whereas populations from low latitudes (40°) would flower in 15-h photoperiods and montane populations flowered at shorter photoperiods. Many of the genotypes recognized experimentally have distinct geographical areas. Therefore, although *Oxyria* pollen has been shown to be characteristic of high-latitude pollen spectra in Canada (Hyvärinen 1985; Ritchie, Hadden, and Gajewski in press), the presence of *Oxyria digyna* in a fossil assemblage is not conclusive evidence of a past environment similar to the modern high arctic unless associated with other taxa of restricted high-arctic range. An additional problem is that the distinction between *Oxyria* and *Rumex* pollen is sometimes doubtful in fossil material.

Saxifraga

Over twenty species of saxifrage are found in Canada. While most are exclusive arctic or arctic–montane species, others range widely south in the subarctic and boreal regions. Four pollen types can be distinguished, of which *S. oppositifolia* L. is the most frequently recorded in both modern and fossil lake sediments. However, this pollen category includes *S. tricuspidata*, *S. aizoides*, and *S. aizoon*, of which *S. tricuspidata* is as locally common and geographically wide-ranging as *S. oppositifolia*. *S. oppositifolia* is usually recorded as a *typus* to draw attention to the wider possible identity of the pollen. It occurs commonly as macrofossils in late-glacial sediments. The species is abundant in the high arctic, often associated with *Dryas integrifolia* on calcareous parent materials. *S. tricuspidata*, on the other hand, is a wide-ranging arctic–boreal species, common on exposed siliceous rocky or sandy substrata. The *Saxifraga cernua* type includes *S. cernua*, *S. rivularis*, *S. hirculus*, *S. caespitosa*, *S. hyperborea*, and *S. flagellaris*. The *S. nivalis* type includes *S. foliolosa* and *S. stellaris*.

Fredskild (1967) has investigated the extent to which *Saxifraga* pollen can be identified confiden- to subgeneric groups and he concludes that, the above four types can be separated re-

Table 3.1. *A summary of the mean annual values for certain climatic parameters at the range limits of the taxa whose pollen comprise the main part of the postglacial pollen record of Canada. Certain ecological and related attributes are also cited in Chapters 2 (climate data) and 3 (autecological data).*

SPECIES	Floristic type[a]	Mean July temperature range (°C)	Mean January temperature range (°C)	Absolute minimum temp. (°C)	Mean annual growing degree-day range (>5.5°C)	Frost-free period range (days)	Annual precipitation (mm)	Shade tolerance[b]	Height range (m)
Cassiope	A	9–4	–20 to –35	–55	800–25	<5–120	<100–800	1	—
Dryas	A	13–4	–20 to –35	–55	1,200–25	<5–120	<100–1,000	1	—
Ericales	A,B,T	14–4	–20 to –35	–55	1,500–25	<10–180	< 100–1,000	1	—
Oxyria digyna (L.) Hill	A	12–3	–22 to –35	–55	1,200–25	<5–120	<100–800	1	—
Saxifraga	A	13–4	–22 to –35	–55	1,500–25	5–120	<100–800	1	—
Abies balsamea (L.) Mill.	B	18–13	–8 to –24	–49	3,500–800	70–180	500–1,500	3	3–23
Alnus crispa (Ait.) Pursh.	B	18–11	–8 to –32	–58	3,000–400	70–150	250–1,000	1	1–2
Alnus incana	B	20–11	–5 to –32	–58	4,500–450	80–160	250–1,000	1	1–4
Betula glandulosa Michx.	B,A	18–5	–8 to –30	–52	3,000–200	15–150	200–1,000	1	~1
Betula papyrifera Marsh.	B	20–12	–5 to –31	–56	4,000–1,000	90–200	250–1,000	1	4–30
Larix laricina (Du Roi) K. Koch	B	21–12	–4 to –30	–56	3,800–1,000	90–180	200–1,200	1	6–30
Picea glauca (Moench) Voss	B	20–11	–8 to –32	–56	3,500–450	90–180	250–1,000	2	2–34
Picea mariana (Mill.) B.S.P.	B	20–11	–8 to –32	–56	3,500–450	90–180	200–1,200	2+	2–28
Pinus banksiana Lamb.	B	18–15	–12 to –28	–52	2,500–1,600	90–160	250–1,000	1	2–30
Populus balsamifera L.	B	20–12	–5 to –31	–58	4,000–500	50–200	250–1,000	1	3–24
Populus tremuloides Michx.	B	22–13	0 to –31	–56	5,600–600	50–200	250–900	1	5–30
Shepherdia canadensis (L.) Nutt.	B	18–12	–15 to –30	–58	3,500–800	50–150	250–900	1	1
Acer rubrum (L.)	NT	26–17	10 to –15	–45	>10,000–2,000	120–300	750–1,200	2	6–36
Acer saccharinum L.	NT	26–19	12 to –12	–40	9,000–2,800	120–280	650–950	2	8–39
Acer saccharum Marsh.	NT	24–16	0 to –15	–45	6,300–2,000	80–260	500–1,800	3	8–34
Betula alleghaniensis Britton	NT	21–16	–8 to –15	–45	5,200–2,000	100–250	750–1,000	2	6–30

Fagus grandifolia Ehrh.	NT	26–18	12 to –15	–45	6,300–2,300	100–300	750–1,200	3	8–37
Fraxinus americana L.	NT,G	26–18	7 to –12	–40	>8,000–2,500	120–280	750–1,200	2	6–25
Fraxinus nigra Marsh.	NT,B	20–15	–8 to –18	–43	5,200–1,700	80–180	500–1,100	2	5–20
Fraxinus pennsylvanica Marsh.	NT,G	26–17	12 to –20	–43	>8,000–2,000	120–280	350–150	2	5–25
Ostrya-Carpinus	NT	20–16	–8 to –18	–45	>8,000–2,000	80–180	500–1,500	1	3–15
Picea rubens Sarg.	NT,B	20–17	0 to –12	–45	3,500–2,100	100–180	800–2,000	3	5–23
Pinus resinosa Ait.	NT	20–15	–8 to –18	–45	4,000–2,000	80–160	500–1,500	1	6–30
Pinus strobus L.	NT	20–16	–8 to –18	–45	6,000–2,000	100–200	500–2,000	2	8–36
Tsuga canadensis (L.) Carr.	NT	24–17	0 to –15	–45	6,500–2,300	80–200	700–1,500	3	8–36
Ulmus americana (L.)	NT,G	26–16	4 to –16	–46	>10,000–2,000	80–320	400–1,500	2	8–36
Tilia americana L.	ST	26–17	0 to –15	–45	>7,000–2,500	80–180	450–1,500	3	8–42
Abies amabilis (Dougl.) Forbes	P	18–11	0 to –9	–30	3,000–1,500	150–250	1,900–6,500	3+	24–45
Alnus rubra Bong.	P	19–13	4 to –4	–25	3,500–2,000	150–250	800–5,000	2	3–20
Picea sitchensis (Bong.) Carr.	P	20–11	4 to –9	–35	3,000–1,000	50–200	1,400–5,000	1	15–54
Tsuga heterophylla (Ref.) Sarg.	P	21–13	5 to –10	–45	3,500–1,500	100–250	450–1,700	3	15–54
Tsuga mertensiana (Bong.) Carr.	P	18–11	–1 to –15	–45	3,000–<1,000	40–120	500–1,800	3	4–33
Abies grandis (Dougl.) Lindl.	C	21–15	5 to –12	–46	3,500–1,500	100–250	560–1,100	2	12–54
Abies lasiocarpa (Hook.) Nutt.	C	17–11	–10 to –18	–55	3,000–1,000	50–150	300–600	2+	6–33
Picea engelmannii Parry	C	21–12	–3 to –18	–56	3,000–1,000	50–200	410–1,700	2	6–39
Pinus albicaulis Engelm.	C	20–11	–6 to –18	–45	3,000–1,000	50–150	50–150	1	3–15
Pinus contorta Dougl.	C	18–11	4 to –20	–50	3,000–1,000	70–200	350–700	1	6–33
Pinus flexilis James	C	16–12	–6 to –15	–50	2,000–1,000	50–100	450–1,000	1	3–20
Pinus monticola Dougl.	C	18–11	4 to –8	–30	2,500–1,500	100–200	500–1,000	2	12–45
Pinus ponderosa Laws.	C	22–15	–3 to –10	–40	>3,500–1,500	100–200	190–600	1	6–42
Pseudotsuga menziesii (Mirb.) Franco	C	21–14	4 to –18	–52	3,500–2,000	75–250	400–6,000	2	15–45

[a]The taxa are grouped into broad floristic groups: A, Arctic; B, Boreal; NT, Northern Temperate; G, Grassland; C, Cordilleran; P, Pacific; ST, Southern Temperate; and T, Temperate.

[b]Shade tolerance classes are 1 low, 2, medium, 3 high.

Greenland material, they should be used in interpretation with caution as he finds "great variation, even within a single species" (p. 26). Table 3.1 provides some information on the climate at the range limits of these arctic taxa.

Modern regional pollen spectra

The purpose of this section is to provide the reader with a concise account of the characteristic pollen spectra of the main regions of Canada as part of the auxiliary information needed for the interpretation of fossil data in later chapters. Some readers will not be familiar with this large, growing North American body of data, so I have chosen to present what I judge to be representative selections in the form of the traditional, but immediately assimilable, bar diagrams (Figs. 3.16, 3.18 and 3.20). These diagrams display the modern pollen registered at 164 sites across Canada, comprising less than one-tenth of all modern pollen spectra now available. The compilation of this large data bank (which is held on a computer file in my laboratory at the University of Toronto as well as at several other institutions) was initiated by Webb (1974a and b) at least fifteen years ago, and has resulted in several expanded versions as the set of samples increased (Davis and Webb 1975; Webb and McAndrews 1976; Webb 1982; and Delcourt, Delcourt, and Webb 1984). The more than 2,000 samples come from a wide range of site types, the majority being moss polsters. A recent compilation of presettlement spectra from lake sediments in Minnesota and adjacent states by E.J. Cushing (published in summary

Figure 3.16. Modern percentage pollen spectra from small lake sites in the Western Interior of Canada judged to be representative of the main vegetation zones. Only the more frequent taxa are included. See Appendix for details of sources.

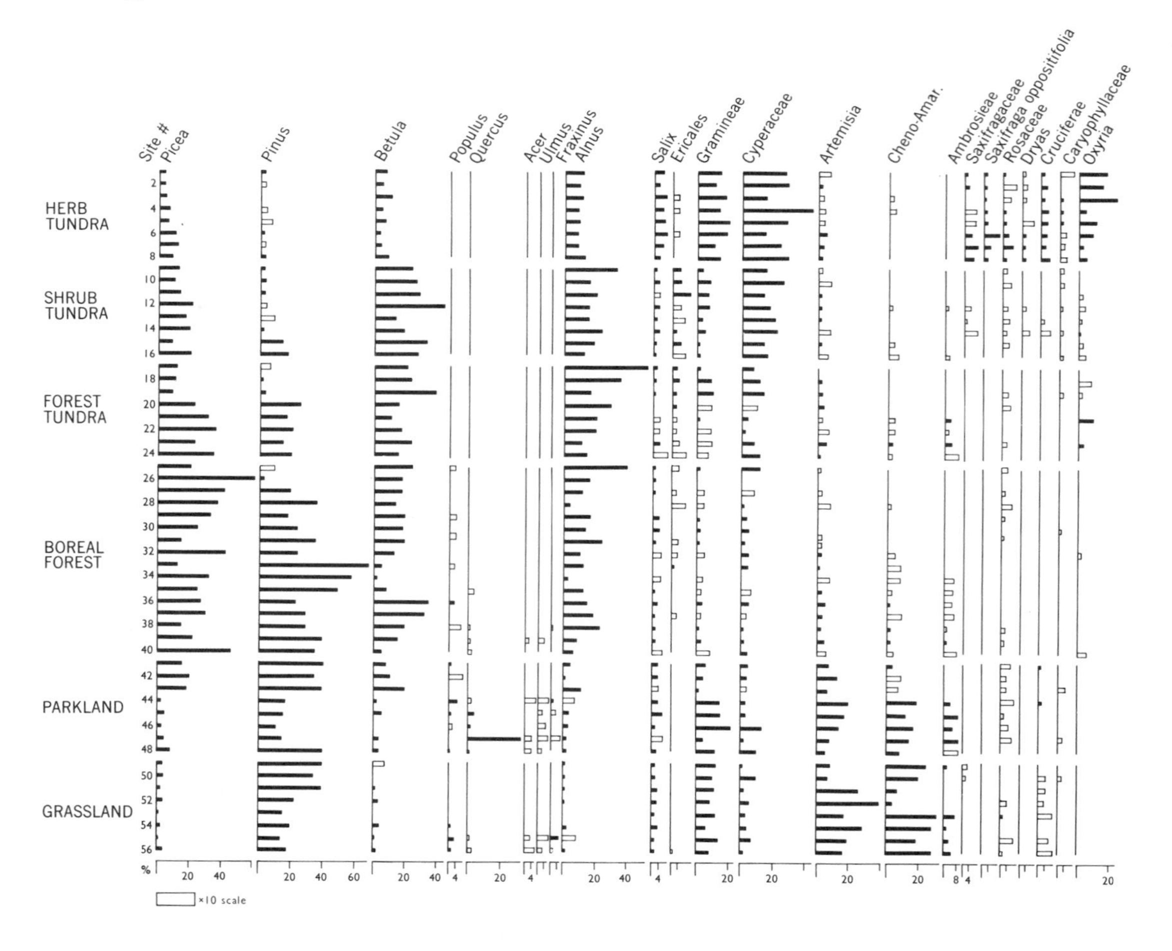

form by Jacobson and Grimm 1986) is an important addition of data of direct use in comparative studies.

Wherever possible, spectra from lakes are used because the great majority of sites in the subsequent review of the fossil record are small lake basins. Also, there is abundant evidence that moss polster samples are extremely variable in modern pollen registration, in part because of the large input of pollen from local plants (Ritchie 1974) and in part because polsters vary greatly in rain washout effects, bacterial decomposition, and retention attributes (Ibe 1983). Almost all the fossil records examined in Chapters 5 to 7 depict the pollen from a source area at the *regional* scale, as already noted in the Introduction in the brief discussion of the question of scale.

Figure 3.17. Localities of the percentage pollen spectra shown in Figure 3.16 for the Western Interior of Canada.

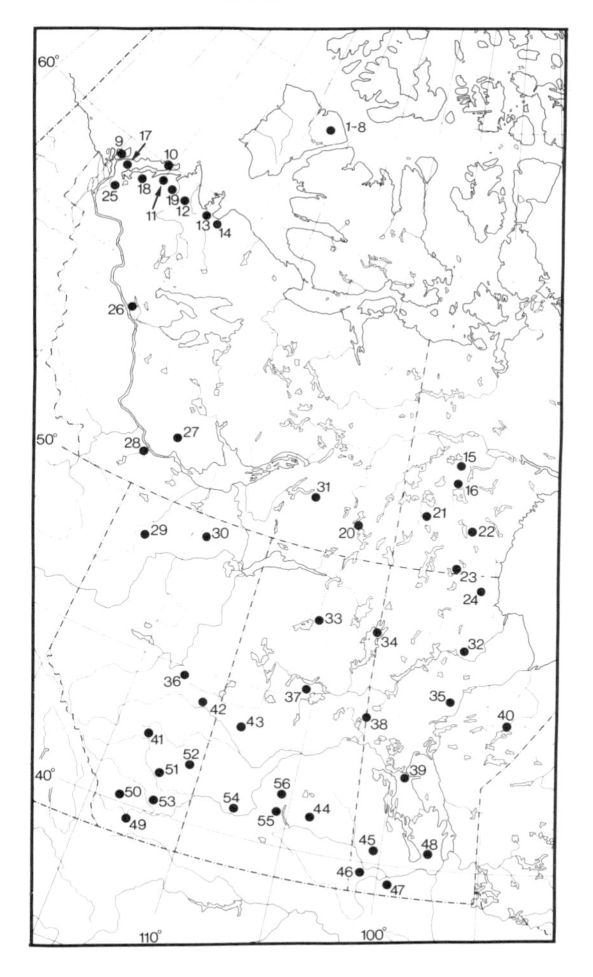

The southern Pacific and Cordilleran regions lack an adequate number of modern pollen spectra from lake sites because, in the case of the Cordilleran region, small lakes are very rare, and, in the case of the Pacific region, moss polster samples have been preferred because many of the early investigations were of mire deposits.

The localities of the modern spectra are shown in Figures 3.17, 3.19 and 3.21 by a solid dot with a number that also appears beside each bar diagram in Figures 3.16, 3.18, and 3.20. Three groups of modern spectra are presented here in broad transects that more or less encompass the five major regions of Canada.

The Western Interior

The first transect traverses the *Western Interior* (Figs. 3.16 and 3.17) and includes the subregions of the arctic, the subarctic, northern and southern sectors of the boreal forest west of Hudson Bay, and the parkland and grassland sectors of the subarid bioclimatic region. It draws on investigations published previously by Lichti-Federovich and Ritchie (1968), Mott (1969), Ritchie (1974), MacDonald and Ritchie (1986), and Ritchie, Hadden, and Gajewski (in press). The bar diagrams are self-explanatory and require little commentary. Herb tundra spectra are distinguished by low arboreal percentages (<25%) and high herb percentages of which certain taxa are endemic to arctic, or arctic–alpine regions (*Oxyria*, *Saxifraga*, *Dryas*). Others are important elements in the vegetation of high-arctic tundras (Cruciferae, Gramineae, Cyperaceae, Leguminosae, *Artemisia*). Mid-arctic spectra are very similar to high arctic, but the predominant shrub tundras of the low arctic are separable by the importance of Ericaceae and *Betula*. Spectra from the subarctic forest–tundra and northern boreal forest are difficult to separate, though they can be distinguished from adjacent regions to the north and to the south. *Alnus* and *Betula* have maximum percentages and *Picea* and Cyperaceae are also high. Central boreal forest sites are distinctive, being dominated by conifers (pine and spruce) with moderate values for birch (<10%), alder (<10%) and willow (<5%). The nonarboreal component is less than 5 percent. Southern boreal forest samples have consistent but low percentages of *Fraxinus*, *Ulmus*, and *Larix*. Parkland and grassland spectra are broadly similar to those from the high arctic in that grass, sedge and herb percentages are high (total 50 to 70%), whereas the tree percentage is 20 to 40. They can be distinguished by the higher percentages of *Artemisia* and Chenopodiaceae–Armaranthaceae, and the presence of certain indicator taxa, some of which are too infrequent to appear in Figure 3.16 (Ambrosieae, *Amorpha*, *Pe-*

Figure 3.18. Modern percentage pollen spectra from small lake sites in eastern Canada and Greenland judged to be representative of the main vegetation zones. Only the more frequent taxa are included. See Appendix for details of sources.

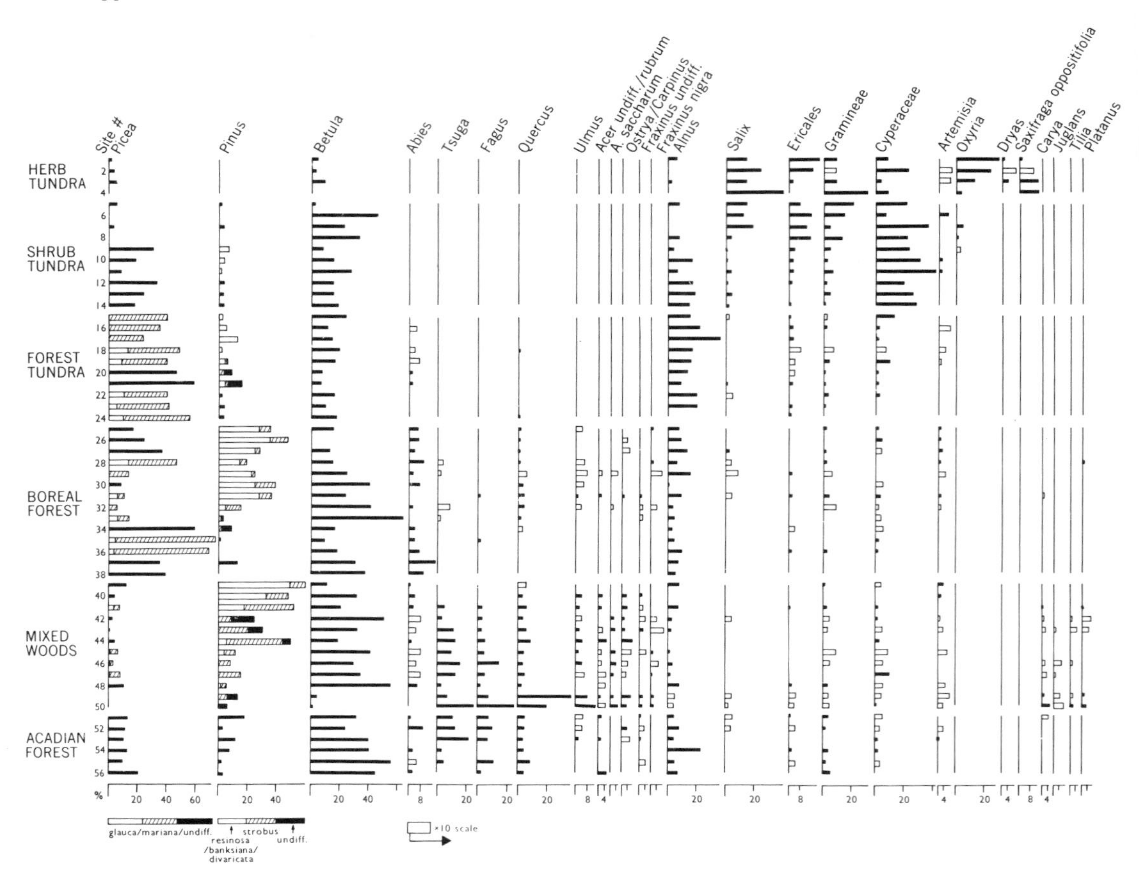

talostemum, *Sphaeralcea*, *Symphoricarpos*, and others).

The eastern plains transect

The second transect extends from northern Greenland and Ellesmere Island in the eastern high arctic, south through Baffin Island, and hence to northern Quebec, central Quebec and Labrador, and terminates in the temperate mixed forest and deciduous forest zones. The pollen spectra are arranged in a sequence from north to south (Figs. 3.18 and 3.19). This transect is simply an updated version of one published 20 years ago by Davis (1967).

The high-arctic herb tundra zone is poorly represented by lake sites, so I have included one from northwest Greenland (Fredskild 1969). The spectra are dominated by herb taxa, of which *Oxyria* (10 to 20%), *Saxifraga*, *Dryas*, Gramineae, Cyperaceae, Leguminosae, and Cruciferae are important, and arboreal and shrub percentages are low (<15%). Mid-arctic sites are also few, but they show similar spectra to the high arctic, with a higher proportion of *Salix* (30 to 40%) and Ericaceae (5%). Low-arctic spectra have higher shrub frequencies (*Alnus* <5%, *Betula* <5%) and high grass, sedge, and ericad percentages. The shrub tundra–forest tundra transition is difficult to detect palynologically, but Richard's (1981a) transect of lake sites in Ungava and Lamb's (1984) from Labrador provide most useful and thoroughly documented series of spectra, several of which are included in Figure 3.18. They show that the low-arctic shrub tundra on the eastern mainland of the continent has larger proportions of exotic taxa than sites on the arctic islands, largely as a function of distance from the source. The distinction between forest–tundra and low-arctic shrub tundra spectra is no clearer in northern Quebec (Richard 1981a, p. 72) than elsewhere in northern Canada (Fig. 3.16),

Figure 3.19. Localities of the percentage pollen spectra shown in Figure 3.18 for eastern Canada and Greenland.

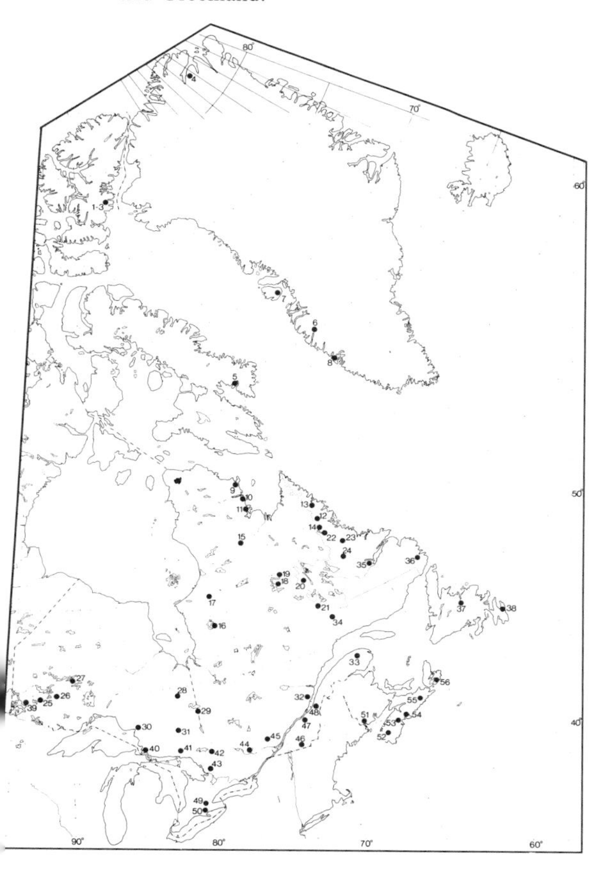

but Lamb's transect across the same boundaries in northern Labrador does provide some basis of differentiation. His tundra sites show consistently higher (>20%) sedge, willow (5%) and ericad (3%) components, while *Picea* has values lower than anywhere in boreal–subarctic Labrador. *Alnus* values decreased from the tundra (20%) through the forest tundra (10 to 15%) to the boreal forest (5 to 10%), and the Cyperaceae and herb taxa show similar decreasing trends. Boreal forest sites have high percentages of *Picea* (60%) and southern boreal sites show consistent percentages (5 to 10%) of the greatly underrepresented *Abies*. The various sectors of the temperate region are characterized by slight shifts in the proportions of *Pinus*, *Betula* and *Quercus* – overrepresented taxa; *Tsuga*, *Ulmus*, and *Fagus* – proportionally represented; and *Populus*, *Fraxinus*, *Acer rubrum*, and *Tilia* – underrepresented taxa.

The Pacific–Cordilleran transect

The Pacific–Cordilleran transect is the least complete, in two respects – there are large gaps in the record from much of north-central British Columbia, and the southern Cordilleran sector has few small lakes, particularly at low elevations. On the other hand, the modern pollen of the southern Cordilleran area has been investigated thoroughly by Mack and Bryant (1974), Mack, Bryant, and Pell (1978), and Hazell (1979) using moss polster and soil sample analyses. Polster samples are indicated by an asterisk in the set of bar diagrams from the Pacific–Cordilleran region, as shown in Figures 3.20 and 3.21.

The southern Pacific region is represented adequately in both modern and fossil lake sites only at lower elevations, in the coastal zone (the coastal western hemlock zone of Krajina, 1969), and the Puget lowland zone. Heusser (1969, 1978, 1985) and Hebda (1983a) have published many modern pollen spectra from polster samples in this zone, but I use only lake sites here for purposes of comparison with the fossil record, and, by contrast with the polster samples, the spectra chosen from the southern Pacific lake sites were those immediately below the settlement horizon. Presettlement spectra (Fig. 3.20, Nos. 1 to 8) are dominated by *Thuja plicata* (30 to 50%), *Pseudotsuga menziesii* (5 to 20%) and *Tsuga heterophylla* (20 to 15%), with variable amounts of *Pinus*, *Alnus*, *Picea*, and deciduous trees. A detailed investigation of aerial and fluvial pollen deposition in two small lakes (Marion and Surprise, also fossil pollen sites referred to in Chapter 7) has been completed by McLennan (1981) and published in part (McLennan and Mathewes 1984). A survey of two years of aerial deposition of pollen shows that, although the local vegetation at the two sites differs markedly, the pollen spectra are very similar, composed of roughly 90 percent regional sources. These data illustrate further the difficulties of pollen analysis in mountainous regions because of upslope transport.

The modern pollen sites in the southern Cordilleran region are a mixture of polster, soil, and lake mud samples (Hazell 1979; Mack and Bryant 1974; Mack et al. 1978; Mack, Bryant, and Pell 1984) to which I have added the upper spectra from selected fossil sites (Fig. 3.20). The sites are arranged in a north–south sequence; numbers 9 to 13 are in the Rocky Mountain foothills, 14 to 25 are at various elevations in the Rocky Mountain and Interior Mountain complexes, and 26 to 36 are from the arid interior zone. The steppe zone of the Columbia Basin, and the *Pinus ponderosa–Pseudotsuga*, *Abies grandis–Pseudotsuga* and a few other zonal associations are distinguishable (Fig. 3.20) and are reason-

Figure 3.20. Modern percentage pollen spectra from small lake, or polster/bog sites in the Pacific–Cordilleran region. See Appendix for details of sources.

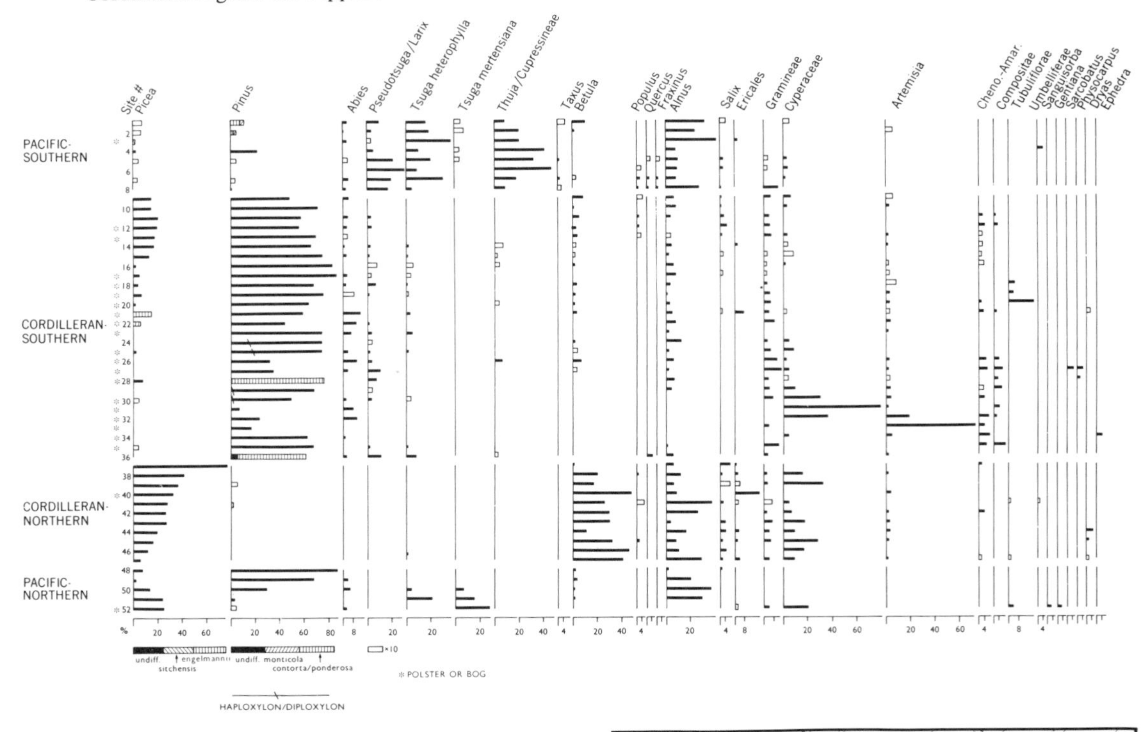

Figure 3.21. Localities of the percentage pollen spectra shown in Figure 3.20 for the Pacific–Cordilleran region.

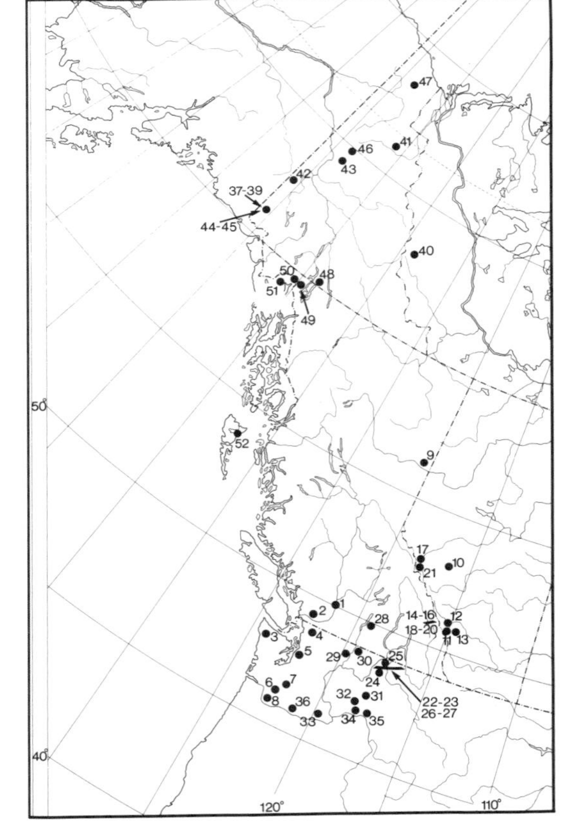

ably differentiated by the high percentages of *Pinus* (30 to 80%) with consistent percentages of *Abies* (2 to 10%), *Tsuga heterophylla* (1 to 5%), Gramineae (2 to 10%), *Artemisia* (1 to 2%) and *Pseudotsuga–Larix* (1 to 5%). Pollen assemblages from Northern Cordilleran sites vary with elevation, but all have high and variable proportions of *Picea*, *Alnus*, and *Betula* with *Salix*, Cyperaceae and Ericaceae increasing with elevation. Birks (1977) presents a detailed analysis of eleven spectra from lake muds and nineteen from polster samples in the forest, shrub–tundra, and tundra zones of the St. Elias Mountains, Yukon, and of fourteen lake and twenty-three polster samples from the Klutlan Glacier also in the Yukon Territory (Birks 1980). They show that the herb tundra (*Dryas*) and *Populus* woodlands produce distinctive assemblages, but that it is impossible to distinguish between the spruce forest and shrub–tundra samples. Rampton (1971) reached a similar conclusion from an analysis of surface samples (both polsters and pond muds) from the Snag–Klutlan area near the Antifreeze Pond site (Fig. 3.21, No. 42). Core-top spectra from sites in the northern Yukon are included to complete the set of Cordilleran samples, from Lateral Pond and Hanging Lake (Fig.

3.20, Nos. 41 and 47) and they show that a sensitive registration of the local vegetation zones in montane landscapes is not possible because of the large aerially transported *Alnus* and *Picea* components. Modern pollen spectra from lake sites have not been investigated in the northern Pacific zone, but Peteet (1986) provides a comparative study of seventy polster spectra and vegetation analyses from a range of communities in the Malaspina Glacier district of southeast Alaska. They vary with distance from *Pinus contorta* stands, elevation, and proximity to the Pacific coast. They are arranged along a gradient from the interior to the Pacific side of the coastal mountains and show inverse relationships of *Pinus* and *Tsuga* frequencies.

General comments on the modern pollen spectra

The southern Pacific–Cordilleran region is by far the most unpromising for conventional pollen analysis. In addition to the severe constraints imposed by the great orographic contrasts over short horizontal distances, several important tree genera and species cannot be identified with light microscopy. The Western Interior is floristically impoverished, but the species-poor forests are represented reasonably in the modern spectra with the notable exception of *Populus*. The difficulties of identifying pollen to genus and species of the arctic representatives of such large, important families as Leguminosae, Cruciferae, Rosaceae, and Ranunculaceae are compounded by their very low pollen production. The eastern temperate region is represented by the largest set of modern pollen spectra, but several important taxa are very underrepresented in samples (*Acer*, *Abies*, *Tilia*), and identification problems in others impose severe limitations on the palaeoecological information that can be derived (*Pinus*, *Quercus*).

4 Full-glacial refugia

Before tracing the vegetational history of Canada since deglaciation, it is necessary to have some conception of the nature of the plant communities that grew beyond the continental glaciers. It is evident from Figure 4.1 that two ice-free areas could have harboured the plant populations whose propagules spread into Canada. Most of the conterminous United States were ice-free, as were large tracts of lowland Alaska, the adjacent northern Yukon, and parts of the western arctic islands of Canada. The southern and northern areas are considered separately here.

The southern refugia

Several factors contribute to the great difficulty in deriving simple generalizations about the full-glacial vegetation south of the ice front, the most important being the sparse network of sites. The ice sheets expanded and contracted frequently, not always in association with climatic change (Mickelson et al. 1983) and the changes in frontal position were rarely synchronous along the length of the ice margin. The landscapes of the east-central United States were as varied and complex as they are today, with a major mountain axis (the Appalachians), a topographically complex interior continental region (Mickelson, Clayton, and Muller 1986), and extensive coastal plains. Depressions supporting lakes and mires are rare in unglaciated landscapes and this explains in part the scarcity of information on the full-glacial vegetation of the conterminous United States. The western montane region is considered here separately from the interior plains and eastern region.

Pacific–Cordilleran refugia

The record south of the Cordilleran ice complex is incomplete, but one or two sites are worthy of attention. An investigation of *Carp Lake* (Fig. 4.1, No. 1) by Barnosky (1985b) has yielded a detailed, well-dated, full-glacial sequence from the ponderosa pine–Douglas-fir interior region of the Columbia Basin, south of the maximum extent of the Fraser glaciation. The pollen and lithologic evidence for the full glacial is a nonarboreal assemblage dominated by *Artemisia* and Gramineae, and a matrix of silty clay with angular clasts. Barnosky suggests a steppe vegetation with some alpine elements prevailing in a cold, dry climate, and she concludes that "most of the Columbia Basin was too dry to support forests in either interglacial, glacial or postglacial time" (Barnosky 1985b, p. 120).

West of the Cascades, the *Fargher Lake* site (Fig. 4.1, No. 2) provides a dated pollen record spanning greater than 25,000 to 17,100 yr BP (Heusser 1983) and the assemblage is dominated by *Pinus contorta* type, *Picea*, and *Tsuga mertensiana* with a significant (25 to 35%) nonarboreal component. *T. heterophylla* and *Abies* are common earlier in the record. The *Battle Ground Lake* site in the Puget Lowland Trough (Fig. 4.1, No. 3) provides pollen and macrofossil evidence that the full-glacial vegetation there was a *Picea* cf. *engelmannii–Artemisia* park tundra at 20,000 to 17,000 yr BP, with macrofossil evidence for the presence of *Pseudotsuga*, *Picea sitchensis*, and *Alnus rubra* in a "slight warming" between 17,000 and 15,000 yr BP (Barnosky 1985a). This fine record (Fig. 4.2) has been amplified by a detailed investigation of the fossil bryophytes (Janssens and Barnosky 1985). Thirty-three taxa were found in the late-Pleistocene and early Holocene levels, falling into three ecological groups based on habitats – wetland species, upper-slope arctic–alpine taxa, and species of forest habitats. The records confirm and refine the vegetation reconstructions based on pollen and macrofossils –

Figure 4.1. The approximate position of the 18,000-yr-BP ice and the localities of the numbered fossil sites referred to in the chapter.

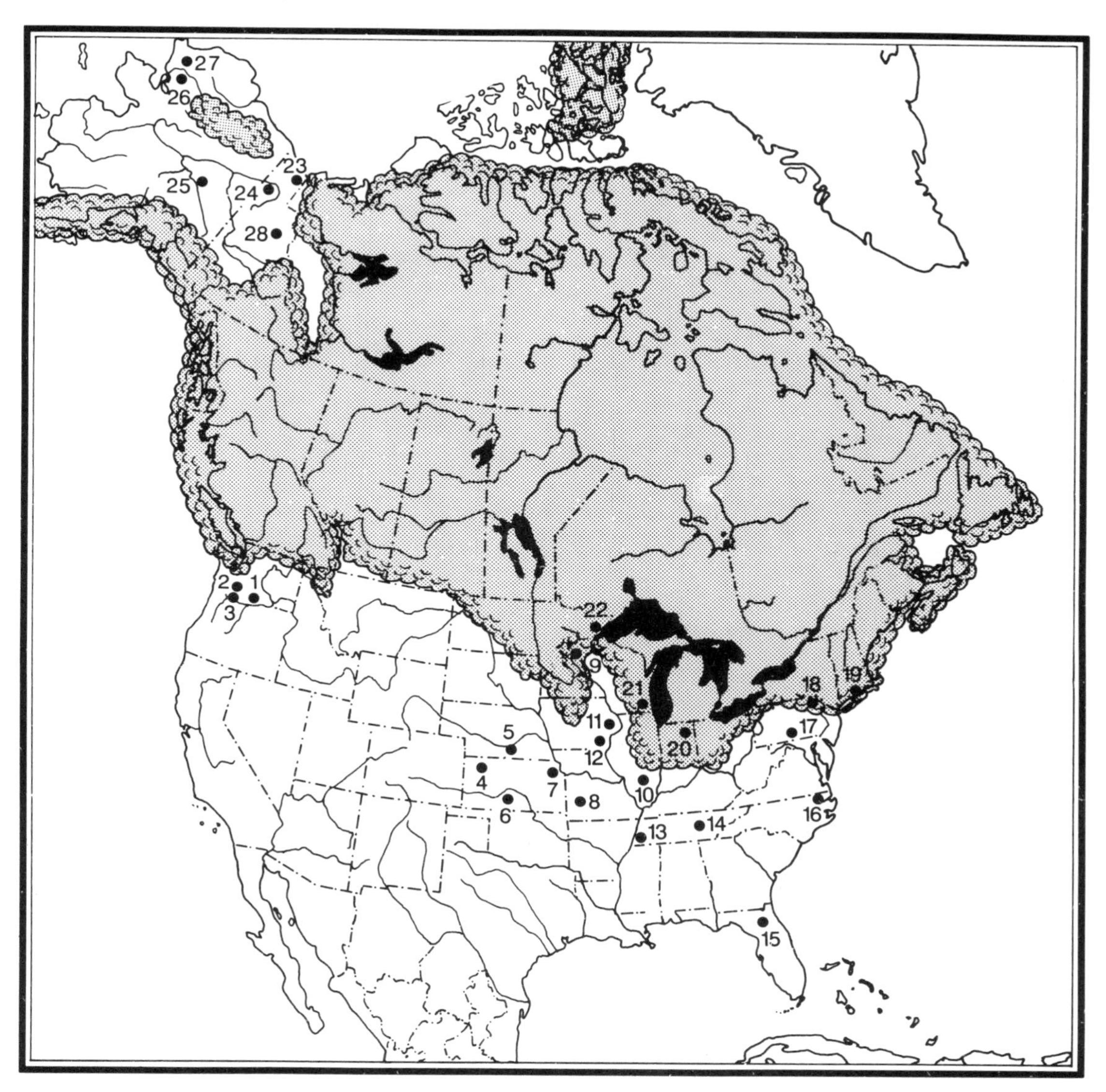

from 25,000 to 15,000 yr BP, a dry park–tundra assemblage; a diverse moss flora from 15,000 to 11,500 yr BP, suggesting a mosaic of forest and tundra; and temperate lowland taxa from 9,500 to 4,500 yr BP. Barnosky (1984) provides an excellent summary of the most recent full- and late-glacial records and the reader is recommended to examine that paper for full details. The reconstruction of the full-glacial vegetation of the Pacific Slope, Puget Trough, and Columbia Basin is shown in Figure 7.6. During an early stadial interval (the Evans Creek Stade, 20,000 yr BP), there was a *Tsuga mertensiana–Picea–Pinus* parkland on the Pacific slope of the Olympic Mountains, while the Puget Trough supported a *Picea–Pinus*–tundra parkland, and, as learned from the Carp Lake record, the Columbia Basin was occupied by periglacial steppe. Similar patterns are suggested for the Vashon stade. As Barnosky remarks (1984, p. 627), "Nothing is known yet about the ice age vegetation of Oregon and central and southeastern Washington".

Interior plains and eastern region

This section is brief because only a few additions to the record have appeared since the reviews by Watts (1983) and Baker and Waln (1985). Re-

Figure 4.2. A summary percentage pollen diagram based on the detailed diagram of the Battle Ground Lake, Washington, site (Fig. 4.1, No. 3) published by Barnosky (1985a).

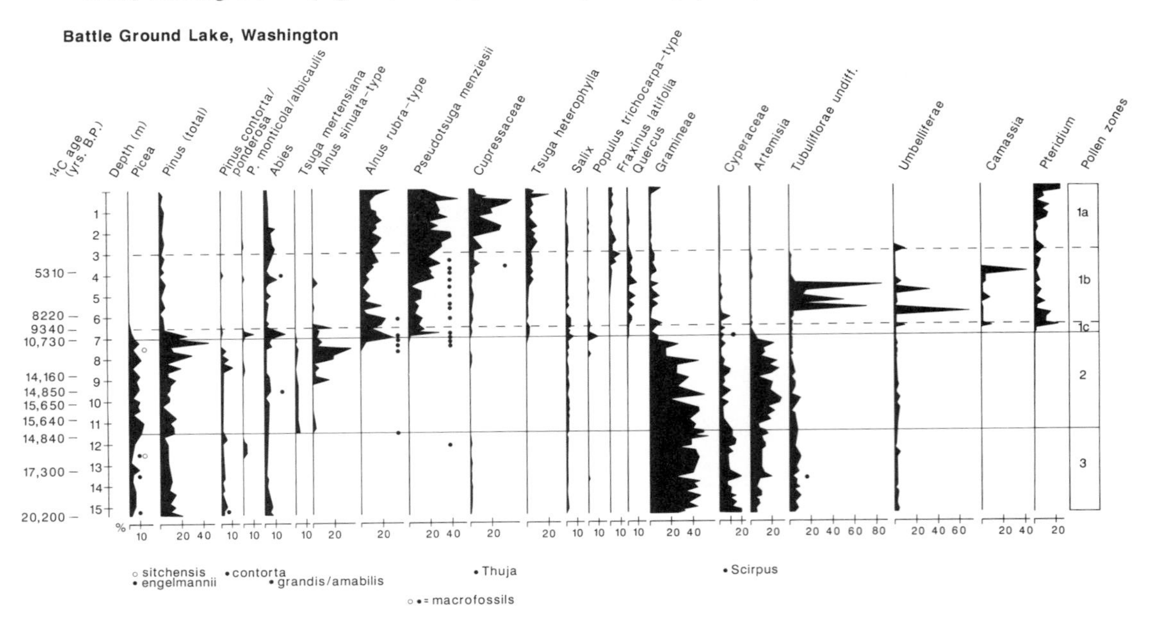

cently Dort et al. (1985) and Johnson et al. (1986) gave preliminary notice of significant new discoveries in the Great Plains. Sites in south-central Kansas and adjacent Nebraska have provided both plant and animal macrofossils from the radiocarbon interval 18,000 to 24,000 yr BP, and taxa of boreal–subarctic affinity (*Picea*, *Pinus*, *Populus*, *Mammuthus*) are common (Fig. 4.1, No. 6).

Two spring-fed marshes in northeastern Kansas (Fig. 4.1, No. 7) have yielded a rich pollen and macrofossil record of full-glacial vegetation (Grüger 1973). A *Picea*-dominated forest prevailed at these sites from 24,000 until roughly 14,000 yr BP. The macrofossil record indicates a rich associated flora of terrestrial and aquatic herbs with modern boreal–temperate distributions. Although the earlier and later sedimentary records at both sites are incomplete, it is probable that the (closed) spruce forest assemblage at 18,000 yr BP was preceded by a woodland with less spruce, some pine (30% maximum), moderate amounts of birch, alder, and willow, and a significant herb component, and was followed by a mixed spruce, oak, hornbeam, ash, and elm assemblage that was in turn replaced at about 10,000 yr BP by prairie.

Spring sites in the Ozark Plateau, southwestern Missouri, roughly 250 km southeast of the Kansas sites (Fig. 4.1, No. 8) have also provided pollen and macrofossil evidence for full-glacial vegetation (King 1973). The pollen record from the *Boney Spring* site is representative of this detailed investigation, supported by analyses of faunal remains from the same sediments. A spruce-dominated pollen zone begins at roughly 21,000 yr BP, preceded by a zone dominated by pine and herbs that King (1973, p. 559) interprets as an open parkland dominated by jackpine. The spruce zone, as at the Kansas sites, has a dominance of *Picea* pollen, interpreted as a closed spruce forest, which persisted until about 14,000 yr BP when a mixed assemblage of conifers (*Picea*, *Larix*) and deciduous angiosperms (*Quercus*, *Fraxinus*, *Ulmus*, *Ostrya*) assumes dominance.

The southwest limit of the full-glacial spruce forest is poorly defined because of the absence of sites (Wright 1981). The superb investigation by Birks (1976) of the *Wolf Creek* site in Minnesota (Fig. 4.1, No. 9) provides important data for our attempts to trace the origins of the Canadian flora. Detailed pollen and macrofossil analyses demonstrate that a floristically diverse, treeless, predominantly herbaceous vegetation occurred there between 20,000 and 14,000 yr BP (Fig. 4.3). Birks (1976, p. 411) points out that, while "the close correspondence between the modern and fossil influx rates suggests that the vegetation at Wolf Creek between 20,500 and 14,700 yr BP was a treeless tundra-like vegetation, . . . when the present-day ecological preferences of the taxa presumed to have been growing near the site at this time are combined with information about the topography near Wolf Creek, a tentative reconstruction of the past vegeta-

Figure 4.3. A summary percentage pollen diagram based on the detailed diagram of the Wolf Creek, Minnesota, site (Fig. 4.1, No. 9) published by Birks (1976).

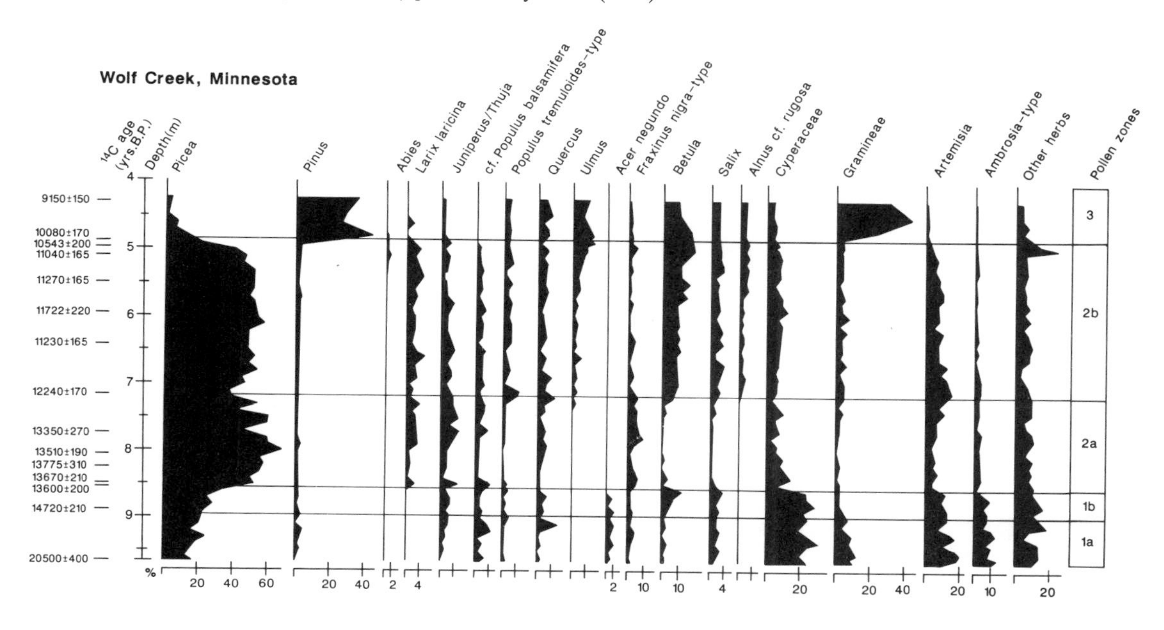

tion can be attempted." He goes on to reconstruct a mosaic of summit tundras, snowpatch herb tundras, local sheltered fern-rich habitats, grasslands with a mixture of prairie (*Amorpha*) and northern species (*Oxytropis*, *Hedysarum*), valley groves of *Larix*, *Populus balsamifera*, and *Fraxinus nigra* with boreal herbs and ferns, and rich calcareous fens. Of equal interest in this richly documented record is the occurrence of pollen of many tree taxa (*Acer*, *Juglans*, *Carya*, *Castanea*, *Celtis*, *Fagus*, *Ostrya–Carpinus*, *Pinus*, *Tilia*, and *Ulmus*). Birks (1976, p. 412) points out that while the low pollen frequencies of these taxa suggest they were not derived from the Wolf Creek region, the pollen "probably originates partly from trees in the boreal forest some 300–600 km to the south, and partly from trees in the mixed coniferous–deciduous and deciduous forest farther south."

In summary, although it is not known at present how representative the Wolf Creek record might be of a larger area south of the Great Lakes basin, it is important to note that the roughly 300 taxa of vascular plants and mosses recorded by Birks as either pollen or macrofossils, or both, include almost all the taxa recorded in all the pollen and macrofossil records published from late-Pleistocene and Holocene sites in the glaciated, continental part of North America. In other words, the vast majority of the species that comprise the pollen and macrofossil record of vegetational history for the past 15,000 years in continental Canada and adjacent United States were present immediately to the south of the ice sheets. However, only hazy notions exist of either their relative abundances or the vegetation patterns they formed. It is probable that a mosaic of communities – tundra, woodlands, forest of varied composition – occurred, varying regionally and locally in relation to the regional landform patterns (Mickelson, Clayton, and Muller 1986).

Findings from the *Conklin Quarry* site in southeastern Iowa (Fig. 4.1, No. 12), show that a full-glacial subarctic "treeline" assemblage prevailed. This multidisciplinary investigation of pond silts has contributed substantially to refining our understanding of full-glacial environments (Baker et al. 1986). Pollen, plant macrofossil, insect, small mammal, and mollusc data are presented, providing a detailed and coherent reconstruction. The pollen record (Fig. 4.4) is the least detailed because, the authors suggest, poor preservation appears to have been responsible for the absence of several likely taxa. Relatively high *Picea*, *Pinus* and Cyperaceae values were recorded (40 to 50% for the conifers and 30 to 40% sedge), exceeding any values near the modern treeline. The only indicator taxon noted was *Selaginella selaginoides*. On the other hand, the record of vascular plant and moss macrofossils is rich, consisting of taxa that are predominantly arctic–alpine in their modern ranges. The list includes *Betula glandulosa*, *Selaginella selaginoides*, *Saxifraga aizoides*, *Dryas integrifolia*, *Arenaria dawsonensis*, *A. stricta*, *Silene acaulis*, and *Vaccinium uliginosum*. In addition, nine taxa of mosses were

Figure 4.4. A summary percentage pollen diagram from the Conklin Quarry site in Iowa (Fig. 4.1, No. 12) based on Baker et al. (1986).

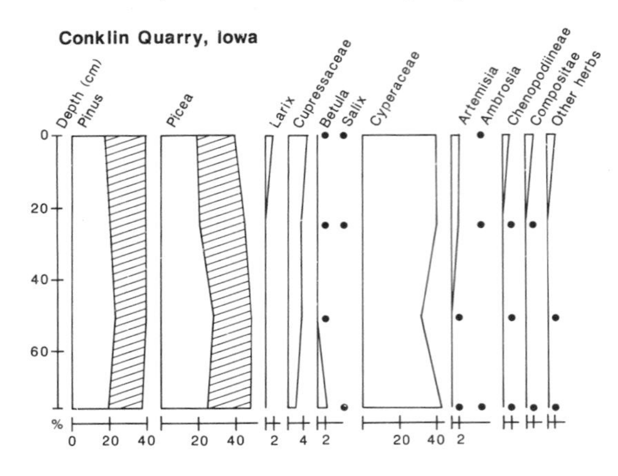

found, of which four have arctic–boreal ranges and two are reported for the first time as fossils in North America (published separately by Janssens and Baker 1984). All the mosses "indicate highly calcareous substrates and suggest open upland and wetland habitats" (Janssens and Baker 1984). Twenty-nine beetle species were found, all taxa that may be found today in the subarctic near the treeline. The authors note that the insect assemblage "has the most northern aspect of any fauna yet identified from (full-glacial sites in) the mid-continent." Similarly, five of the six taxa of small mammals are confined today to the forest–tundra transition zone. The molluscs "indicate an environment resembling present tundra–treeline situations, but with decidedly more available moisture than is now found in such environments." A remarkably rich and consistent reconstruction is possible from this fine record, and the authors conclude that the "dominant habitats . . . included open calcareous silty to sandy or gravelly upland sites, minerotrophic fens, ponds or streamside clayey to sandy shores, and shallow (possibly ephemeral), cold, clear-water ponds. Mean July temperatures were probably 10 to 12°C cooler than at present. The biota indicates that a *Picea–Larix* krummholz with extensive tundra openings was present in southeastern Iowa between 16,710 and 18,090 yr BP" (Janssens and Baker 1984, p. 2).

Farther east, a group of former lake sites on Illinoian till surfaces 60 km south of the Wisconsin ice limit has been investigated by Grüger (1972). These sites (Fig. 4.1, No. 10) have yielded rich pollen sequences, but the radiocarbon chronology is inadequate. Nevertheless, Wright (1981, p. 118) concludes that it is likely that the vegetation at 18,000 yr BP was a spruce forest type. Buried peat at the *Butler Farm* site (Fig. 4.1, No. 11) in eastern Iowa (Van Zant, Hallberg, and Baker 1980), while it dates from an interstade that immediately preceded the Wisconsin glacial, shows, in the upper samples, a clear trend toward dominance by *Picea* that the authors interpret as a closed spruce forest.

The nature of the full-glacial vegetation and flora of the southeastern region of the United States has been the central thrust of many investigations and one prominent, still unresolved, problem has been the location and composition of the "refugial" populations, of trees in particular, that subsequently provided the stocks to revegetate the northern areas as the ice receded. Watts (1979) reviews the topic thoroughly and states: "The location of refugia for deciduous trees has been keenly debated: were they in protected gorges in a mountainous area, or on exposed continental shelf, or far to the south of their present ranges?" A full-glacial record from northern Florida – the *Sheelar Lake* site (Fig. 4.1, No. 15) – investigated by Watts and Stuiver (1980) – provides clear evidence for the presence of *Ostrya–Carpinus*, *Ulmus*, *Fraxinus*, *Fagus*, *Acer saccharum*, and *Tilia* in the interval 18,500 to 13,500 yr BP, but sites farther north that might record refugial populations remain undiscovered. The detailed investigations of long records from the Carolina Bays of North Carolina (Fig. 4.1, No. 16) by Whitehead (1981) have shown that full-glacial assemblages were predominantly boreal, with abundant spruce and pine and only very small amounts of elements of the modern mixed deciduous hardwood–conifer forests of eastern North America until the late-glacial–Holocene transition.

The southern limit of the full-glacial spruce pollen zone is imprecisely known, but the *Nonconnah Creek* site, Tennessee (Fig. 4.1, No. 13) (Delcourt et al. 1980), and *Anderson Pond* (Fig. 4.1, No. 14), 400 km farther east in the same state (Delcourt 1979), are probably the southernmost sites with adequate documentation. The full glacial at sites in northwestern Georgia, 200 km south of Anderson Pond, is characterized by rich assemblages of both pollen and macrofossils, dominated by pine with small proportions of spruce, deciduous trees, and herbs (Watts 1970, 1983).

The northeastern sector of the region that lies beyond the Wisconsin ice has a very complex geomorphology, dominated by the Appalachian axis. The phytogeographical interest of this refugial area, and its late-Quaternary fossil record, have been elegantly and perceptively collated by Watts (1979) and reviewed by him more recently (Watts 1983). The *Crider's Pond* site in Pennsylvania (Fig. 4.1, No. 17) lies roughly 150 km beyond the limit of Wisconsin ice. Although its maximum age is 15,210 yr BP, it provides reliable evidence that the northern limit of

Figure 4.5. A summary percentage pollen diagram based on the detailed diagram of the Crider's Pond, Pennsylvania, site (Fig. 4.1, No. 17) published by Watts (1979).

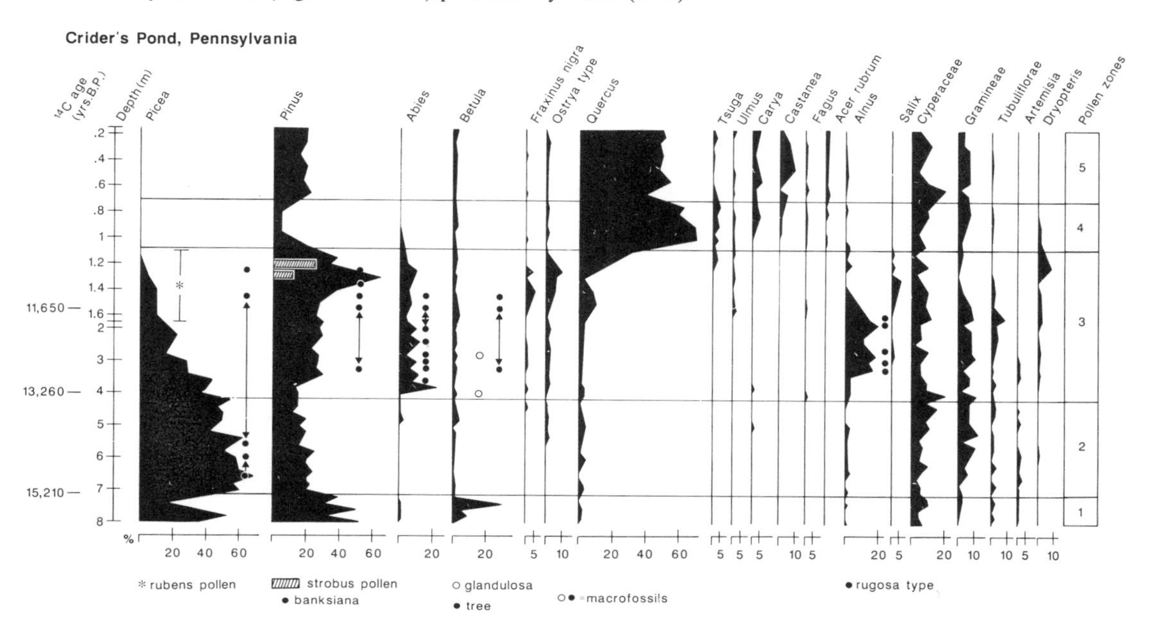

Picea lay south of the site during the full glacial, when a treeless vegetation with abundant dwarf birch pollen prevailed. Spruce colonized the area following 15,000 yr BP (Fig. 4.5).

In summary, both the geographical ranges of particular taxa and the structure and composition of the vegetation remain poorly documented for the southern, full-glacial region. The most detailed pollen and macrofossil records show that most taxa recorded in late-Pleistocene and Holocene records from continental Canada and adjacent United States were present during the full glacial but, with the exception of such dominant taxa as *Picea* and *Pinus*, we have only the vaguest conception of their ranges. In addition, it is likely that the most thoroughly documented pollen and macrofossil records comprise only about 15 to 25 percent of the total vascular plant flora of the surroundings of a site. Therefore, the probability is that the full-glacial whereabouts of a large fraction of the flora of Canada will remain unknown.

A few northern sites provide evidence for a treeless vegetation near the ice front but, if the Wolf Creek record can be assumed to be typical, we should, as Birks (1976) warns, hesitate to describe it as tundra because of the climatic implications of the term. The pollen record from central and southern localities provides a rough indication that the transition from *Picea* domination of the arboreal element to *Pinus* occurred at the latitude of south Kansas through Tennessee to North Carolina (Whitehead 1981). However, it is unlikely that the record is adequate to sustain such palaeovegetation maps as those prepared by Delcourt, Delcourt, and Webb (1983), though their confidence in such a procedure may be based on their remarkable assertion (p. 168) that "despite changes in distribution, and areal extent and fine-scale community composition, the majority of vegetation types today characteristic of boreal and temperate regions of eastern North America have persisted since the full glacial." It will be discovered that the late-Pleistocene and Holocene pollen record for continental Canada gives evidence for somewhat different conclusions.

Between the time of the maximum extent of the ice cover, which varied between 21,000 and 14,000 yr BP, but was centred on 18,000, and 12,700, yr BP, which represents the end of a disintegration phase (Denton and Hughes 1981), very little, if any, of the land areas of what is now Canada was exposed (Fig. 4.1). However, changes occurred in the plant cover south of the Canadian border, and these should be examined briefly before dealing with Canadian sites and the post-12,700-yr-BP record.

The *Tannersville Bog* site in northeastern Pennsylvania (Fig. 4.1, No. 18) is located a short distance inside the area of Wisconsinan ice. The bog developed in a moraine-dammed lake and yielded a pollen and macrofossil record dating from roughly 14,000 yr BP (Watts 1979). The pollen record has a basal zone that Watts interprets as "a sparsely vegetated, treeless landscape with pioneer herbs, among which sedges may have been especially prominent," although tree percentages (50% pine and 20% spruce) are quite high. However, pollen accumulation rates are low, and Watts suggests that trees were either absent or very rare. At 13,300 yr BP, spruce increases in both percentage and accumulation rate, associated with juniper, alder, and poplar, and macrofossils of these taxa strengthen the interpretation of this second zone as patchy spruce forest.

A detailed analysis of the *Rogers Lake*, Connecticut, deposit (Fig. 4.1, No. 19), yielded a familiar, now classical record of pollen accumulation rates from this site just inside the glacial limit (Davis 1969). The area was ice-free by 15,000 and remained treeless until about 12,300 yr BP when a sharp rise occurred in the spruce accumulation rates, although spruce was common only 400 km to the southwest in Pennsylvania by 13,300 yr BP (Watts 1979).

A similar treeless episode preceding the spruce rise occurred at sites in New Hampshire and south-central Maine, reviewed by Davis and Jacobson (1985) and Davis (1983). The *Moulton Pond* site near the coast in southern Maine was isolated on an island by rising sea level at 13,000 yr BP, and this local environment is adduced to account for the low pollen accumulation rates, interpreted in terms of "tundra" vegetation persisting until about 10,000 yr BP (Davis et al. 1975).

The treeless vegetation that occupied most of glaciated New England following deglaciation (14,000 to 12,000 yr BP and later) has been the topic of some speculation. Pollen analysis has not been used very effectively, so far, to elucidate the nature of the vegetation, but studies of macrofossils, in conjunction with pollen analysis, have proved to be illuminating, although they are based on sediments that are slightly younger than 12,000 yr BP. The *Columbia Bridge* site in Vermont, analyzed by Miller and Thompson (1979), revealed a rich assemblage of plants preserved in laminated, glaciolacustrine muds aged 11,500 yr BP. While most of the macrofossils were of nonarboreal taxa, spruce and balsam poplar did occur at the site, associated with a varied assemblage of shrubs, herbs, and mosses, some arctic–alpines, many boreal–subarctic taxa, and a few with highly disjunct modern distributions. Miller and Thompson (1979, p. 209) conclude that this assemblage offers little scope for palaeoclimatic reconstruction, but rather "suggests that the absence of competition from trees and soil immaturity may have allowed heterogeneous associations of species to coexist." It is of interest to note that of the 63 taxa of vascular plants and mosses recorded by Miller and Thompson (1979) and Miller (1980) from this site, more than half also occur in the full-glacial assemblage at Wolf Creek, Minnesota.

A recent investigation in Maine by Davis and Kuhns (personal communication) of the pollen and macrofossils of a 13,000-year-old section of lake sediment, the *Chase Lake* site, demonstrates that *Populus balsamifera* was probably present locally in an otherwise treeless landscape at roughly 13,000 to 12,000 yr BP (Figs. 8.1a and 8.1b). Pollen concentrations were very low ($<10{,}000 \cdot cm^{-3}$), dominated by Cyperaceae, Gramineae, and *Artemisia*, with macrofossils of *Dryas integrifolia*, Ericaceae, *Arenaria*, Gramineae, and *Rubus*. The recent study by Tolonen and Tolonen (1984) of the pollen and macrofossils of four sites in coastal Maine further illustrates the value of detailed macrofossil analyses of vascular plant and bryophyte remains in vegetation reconstruction. They show that an open mosaic of heterogeneous vegetational elements occurred about 11,000 yr BP, shortly after deglaciation, consisting of a "pioneer" vegetation of mosses (but not lichens), herbs (sedges, *Ambrosia*, *Artemisia*, other Compositae, Chenopodiaceae), dwarf shrubs (*Salix*, *Betula glandulosa*), and some trees (*Picea mariana*, *P. glauca*).

Farther to the west, in northern Indiana, freshly deglaciated landscapes were exposed by 15,000 yr BP. Williams (1974) provides detailed percentage and PAR data from three radiocarbon dated cores recovered from *Pretty Lake* (Fig. 4.1, No. 20). A brief treeless episode, from 14,300 to 13,800 yr BP, dominated by sedge pollen and interpreted by the author as "tundra" is followed by *Picea*-dominated spectra until 11,000 yr BP.

A site west of Lake Michigan, *Hook Lake Bog* in Dane County, Wisconsin (Fig. 4.1, No. 21), similarly positioned in relation to the outer limit of Wisconsin ice, has yielded a long pollen record with a very closely controlled radiocarbon chronology (Winkler 1985a). Two pollen assemblage zones are recognized for the interval 16,000 to 12,500 yr BP. A lower zone dominated by *Picea* (70%) but with PAR (2,500 to 5,000 grains $\cdot$ cm^{-2} $\cdot$ yr^{-1}) that is interpreted as an open spruce parkland, followed (13,500 to 12,500 yr BP.) by a *Picea*-dominated assemblage with significant amounts of *Fraxinus nigra*, *Ostrya–Carpinus*, *Salix* and *Alnus*, reconstructed as a closed conifer–deciduous forest.

The remaining site to be considered on the freshly exposed fringe area just beyond the Canadian border is the *Kylen Lake* site in northeast Minnesota (Birks 1981a) (Fig. 4.1, No. 22). Birks (p. 341) recognizes a subzone from 15,850 to 13,600 yr BP with low pollen accumulation rates and a predominance of nonarboreal taxa, interpreted as "a discontinuous cover of stunted grasses, sedges and dwarf-willows" with a variety of herbs characteristic today of "open, well-drained, base-rich mineral soils. . . . " The next zone, from 13,600 to 12,000 yr BP shows increased pollen accumulation rates with the development of more extensive, closed vegetation cover resembling modern shrub–tundra in physiognomy and species composition (particularly *Juniperus*, *Betula glandulosa*, and various heaths). *Picea* macrofossils do not occur until later, at 12,000 yr BP, suggesting a progressive spread of spruce forest northward from Wolf Creek, beginning at about 13,600 yr BP (Birks 1976).

The Beringian refugia

One of the earliest phytogeographical theories concerning North American refugia was Hultén's (1937) germinal proposition that the unglaciated montane landscapes of Alaska–Yukon and eastern Siberia (Beringia) provided a range of habitats suitable for plants to survive a glacial period and from which a significant proportion of the modern northern floras was subsequently derived.

A few additions have been made to the botanical record of the full glacial of Beringia since recent reviews (Matthews 1982; Heusser 1983; Ritchie 1984a; Ager and Brubaker 1985). The record from the earlier investigations of continuous records of primary sediment from Imuruk Lake showed that the full-glacial interval was characterized by nonarboreal pollen assemblages (Colinvaux 1967; Shackleton 1982).

The pollen record for the full glacial (15,000 to 30,000 yr BP) from the Yukon lowlands near Fairbanks, Alaska (Fig. 4.1, No. 25), lacks any evidence of arboreal elements (Matthews 1974; Edwards et al. personal communication) and indicates a herb tundra vegetation. Recent findings from northwestern Alaska from cores of reliably dated primary lake sediment provide further evidence that the vegetation of the full glacial (39,000 to 14,000 yr BP) was a treeless, "meadow-like tundra" in lowlands (Anderson 1985, p. 315). She also notes, however, that at the *Squirrel Lake* site (Fig. 4.1, No. 26), low PARs of *Betula* and *Populus* might indicate the presence of these taxa in favourable, stream-margin sites in the lowland valleys of the region.

Evidence from the *Old Crow*, Yukon, region (Fig. 4.1, No. 24) and from *Hungry Creek* area (Fig. 4.1, No. 28) , reported by Lichti-Federovich (1974) and Hughes et al. (1981), respectively, indicates that a spruce forest or woodland probably occurred regionally at approximately 37,000 yr BP, but so far no reliable data have been reported that would indicate that spruce and associated boreal elements survived the full-glacial period from 25,000 to 15,000 yr BP. *Picea* did not reach the Yukon–Alaska region until 9,500 yr BP, although climatic conditions favourable to its growth probably prevailed in the area from at least 12,000 yr BP.

A detailed pollen analysis of a continuous deposit of primary sediment from the tundra zone of the northern Yukon (the *Hanging Lake* site, Fig. 4.1, No. 23, reported by Cwynar 1982) demonstrated that a significant number of arctic–subarctic taxa were present in the area during the final stages of the full glacial, but that trees were absent until the early Holocene.

In summary, eastern Beringia undoubtedly served as an important full-glacial refugium, and I examine in Chapter 8 the interesting question of what the respective contributions of the two refugial areas might have been to the flora of Canada.

5 Eastern Canada – fossil record and reconstruction

Introduction

This chapter assembles the fossil record from sites in eastern Canada, including the eastern arctic. The western limit of the region in the south coincides roughly with the western boundary of the temperate mixed-forest zone, while to the north it is set arbitrarily by the 90th longitude, which runs along the southern Manitoba–Ontario border, along the west coast of Hudson Bay, north to the west coast of Ellesmere Island. Site locations mentioned in this chapter are shown on Figure 5.1.

The chapter is organized into three main sections, the first dealing with the period from 12,700 to 10,000 yr BP (the late glacial) and the second from 10,000 yr BP to the present (the Holocene). Within each section, the fossil sites are grouped geographically into Quebec–Maritime and Great Lakes sectors. The third section is devoted to the record from the eastern arctic.

The procedure is to present the pollen and plant macrofossil records from selected sites in the region, followed by a reconstruction of the past vegetation. By keeping separate the data and the interpretation, readers will be able to develop their own interpretation that might differ from what is offered here.

The reader can best grasp the temporal and spatial framework of the material to be examined here by referring to Figure 2.2, where the approximate positions of the Laurentide Ice Sheet are shown for 12,700, 10,000 and 8,000 yr BP. At 12,700 yr BP only parts of southern Ontario, coastal Newfoundland, and some of the Queen Elizabeth Islands were ice-free. By 10,000 yr BP, the ice margin was north of Lake Superior at the Chapleau moraine (Saarnisto 1974) and extended eastward from near Lake Timiskaming, passing south of Lac St. Jean to form the north shore of a reduced Champlain Sea north of Prince Edward Island (Prest 1970) (Fig. 2.2). Farther east, large portions of Newfoundland were ice-free by 10,000 yr BP as was a wide tract of coastal Labrador to north of 55°N. The position of the 10,000-yr-BP ice margin, as shown in Figure 2.2, is highly generalized because of the presence of proglacial lakes, various uncertainties about the local extent of

Figure 5.1. A map of eastern Canada and adjacent United States showing the locations of the fossil sites referred to in the chapter.

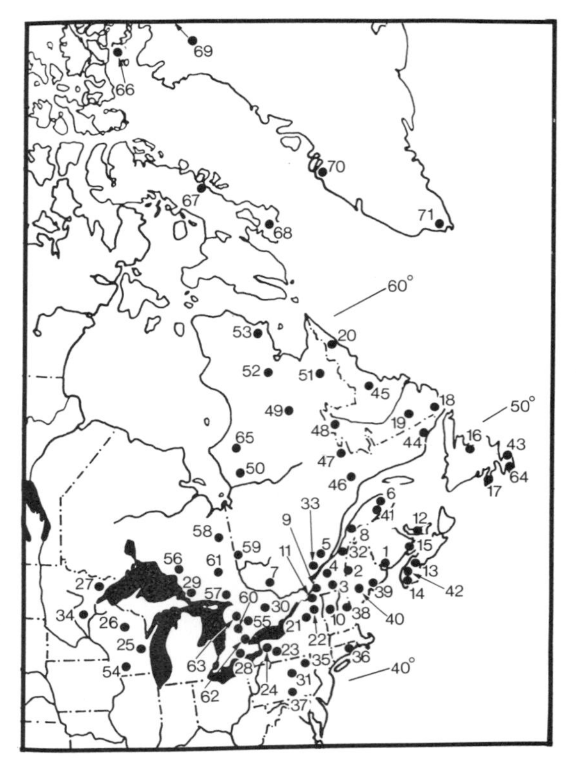

Figure 5.2. Summary percentage pollen diagrams for the sites indicated. In these, as in all subsequent composite pollen diagrams, the pollen percentages are plotted against a radiocarbon age scale. In preparing the diagrams from originals drawn on a depth scale, uniform sedimentation rate was assumed between dated levels and, where necessary, pollen curves were stretched or compressed to fit a constant time scale. The pollen sum in all diagrams includes all terrestrial taxa. Note occasional percentage scale changes used to improve visibility of changes in taxa with low percentage values (e.g., *Abies*, *Acer*). The details of the original investigations are given in the text, including the literature citations; locations are on Figure 5.1.

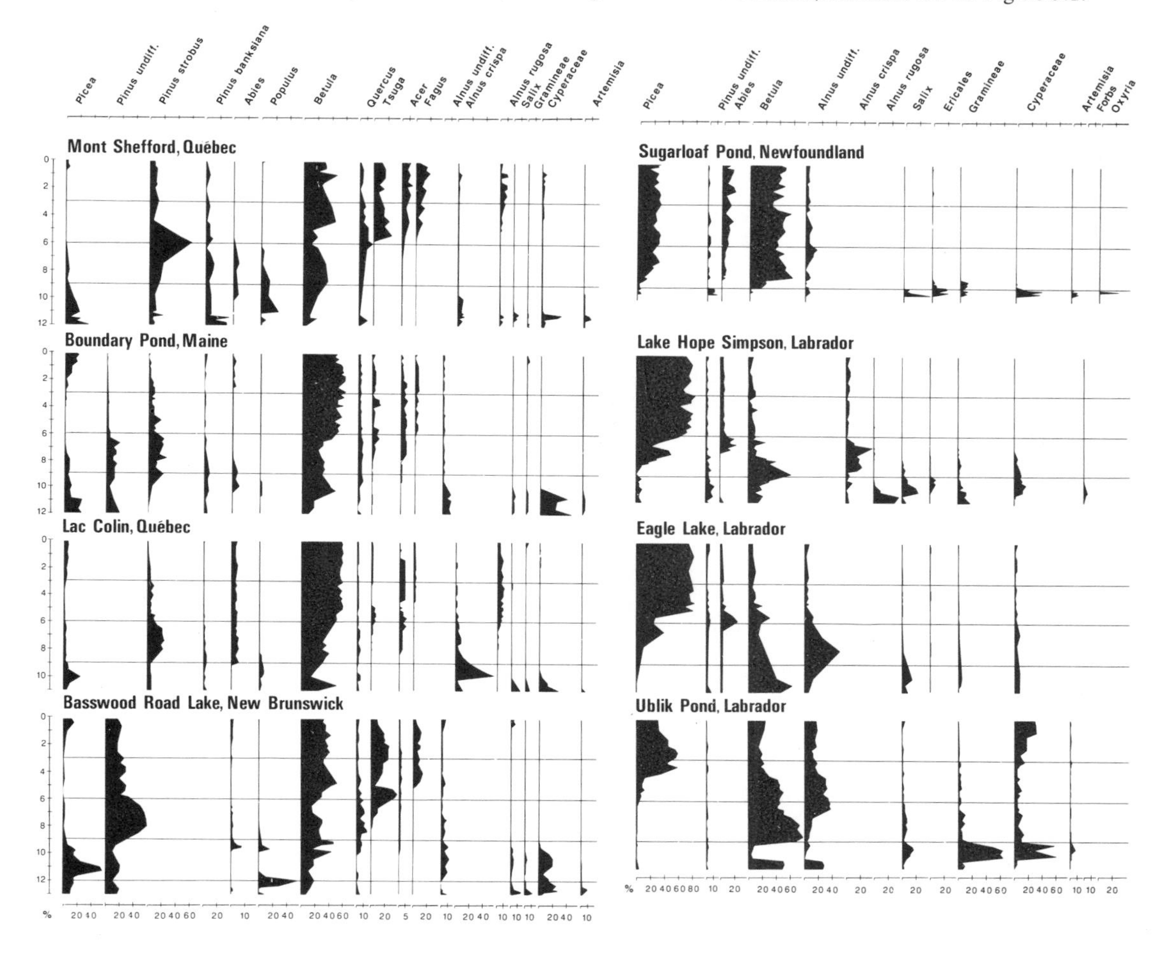

glaciers, and an incomplete understanding of the history of sea-level change, as discussed in Chapter 2.

The late glacial – 12,700 to 10,000 yr BP

Southern Quebec and New Brunswick

Basswood Road Lake (Fig. 5.1, No. 1), New Brunswick, a small (4 ha) lake in a tectonic basin at the edge of the New Brunswick Highlands section of the Appalachian Physiographic region yielded over 6 m of polliniferous sediment, investigated in detail by Mott (1975). A basal date of 12,600 ± 600 and similar values from other cores suggest that the site was exposed at about that time by the retreat of a maritime lobe of the main Laurentide Ice Sheet. This lobe disintegrated during the ensuing 2,000 years and by 10,000 yr BP was represented in the vicinity by two small ice caps on the Gaspé Peninsula. The site is in a region of considerable vegetational diversity. The immediate forests are dominated by *Acer saccharum*, *Fagus grandifolia*, *Quercus rubra*, and *Pinus strobus*, with local dominance of *Picea rubens*, *Abies balsamea*, *Tsuga canadensis*, and *Fraxinus nigra*.

Six well-spaced radiocarbon dates in the sediment provide a secure chronology, making possible the determination of PARs. From 13,000 until 12,500 yr BP, very low PAR values were recorded. The percentage diagram (Fig. 5.2) shows roughly equal proportions of spruce, pine, birch, willow, grass and sedge pollen. At 11,500 yr BP, rapid changes occur with a sharp rise in *Populus* and *Betula* percentages, followed by a steep rise in *Picea* to values greater than 60 percent, with decreases in both *Populus* and *Betula*. The spruce-dominated spectra persist until 9,500 yr BP.

At *Boundary Pond*, Maine (Fig. 5.1, No 2), 4 m of lake sediment were recovered from this site, which lies just inside the northeastern Maine border with Quebec (Mott 1977). The oldest radiocarbon age was 11,200 ± 200 at 389 to 392 cm in the sediment. At that time *Picea* dominated the pollen spectra associated with birch, alder, grass, sedge, and willow in similar frequencies to those at the Basswood Road site for the same time interval (Fig. 5.2).

Mont Shefford, Quebec (Fig. 5.1, No. 3), a small bog-filled depression at 280 m above sea level on the Mont Shefford bedrock island in the St. Lawrence lowlands, yielded 6.4 m of polliniferous sediment from which a very detailed percentage pollen diagram was prepared (Richard 1978). Ten radiocarbon dates provide a very secure chronology. From 12,000 to 11,000 yr BP, very low pollen concentrations were recorded, with relatively high proportions of herbs associated with dwarf birch, juniper, and willow. The period from 11,100 to 10,000 yr BP has high values of *Picea*, associated with *Populus*, *Pinus*, and *Juniperus* (Fig. 5.2).

The *Lac Marcotte* site (Fig. 5.1, No.4) is a small shallow pond at the southeast edge of the Precambrian Shield, 503 m above sea level, roughly 100 km downstream from Quebec City, 35 km inland from the north shore of the St. Lawrence River. Labelle and Richard (1981) recovered 3.72 m of polliniferous sediment and present a detailed percentage diagram with five radiocarbon dates, shown here in summary form (Fig. 5.3). The period of our immediate interest is characterized by an initial assemblage, from 11,100 to about 10,000 yr BP, composed of such arctic taxa as *Polygonum viviparum*, *Oxyria digyna*, *Saxifraga* type *oppositifolia*, *Armeria maritima*, and *Dryas*, associated with high frequencies of grass, sedge, and composite pollen types; a younger phase has *Salix*, *Artemisia*, and *Betula glandulosa* dominance. At about 10,000 yr BP, *Populus* reaches high values (30%), associated with *Salix* and *Alnus crispa*, with a later increase of *Picea* (distinguished palynologically as *P. mariana*).

An earlier investigation by Richard and Poulin (1976) at the nearby *Lac Mimi* site (Fig. 5.1, No. 5) yielded a similar record, with an initial pollen assemblage, from 11,000 to 10,000 yr BP, dominated by grass and sedge, with occurrences of *Polygonum viviparum*, *Oxyria digyna*, *Saxifraga* type *oppositifolia*, and *Koenigia islandica*. Following a brief phase with abundant *Salix*, *Artemisia*, and *Betula glandulosa*, *Populus* reaches high values at about 10,000 yr BP, associated with *Shepherdia canadensis*, *Salix*, and *Alnus crispa*.

From Mont-Saint-Pierre, Gaspé, Labelle and Richard (1984) report the pollen and macrofossils from a small pond (*Lac Turcotte*, Fig. 5.1, No. 6) at 457 m above sea level on a plateau surface, 8 km inland from the south shore of the St. Lawrence estuary. From the beginning of sedimentation, estimated at 11,000 until 10,360 yr BP, nonarboreal pollen taxa in low concentrations (500 to 7,000 grains · cm^{-3}) predominate, consisting of Gramineae, Cyperaceae, *Artemisia*, *Salix*, and *Juniperus*, and associated with macrofossils of *Diapensia*, *Dryas*, *Vaccinium*, and *Salix herbacea*. Tree taxa, particularly *Picea*, increase at 10,360 yr BP, associated with pollen and macrofossils of *Larix*, *Populus*, and *Abies* (Fig. 5.3).

Ramsay Lake, Quebec, in the Gatineau Park section of the Precambrian Shield uplands, about 37 km north northwest of Ottawa (Fig. 5.1, No. 7), is small (9 ha), 200 m above sea level, and situated immediately to the west and above the maximum extension of the Champlain Sea. Mott and Farley-Gill (1981) report on a detailed pollen analysis of 9 m of radiocarbon-dated sediment summarized in Figure 5.4. Their analysis includes estimates of PARs for the main taxa. Herb pollen dominate the levels from the beginning of sedimentation, estimated at earlier than 11,000 to about 10,500 yr BP, when a prominent peak of *Populus* occurs, followed at 10,200 yr BP by *Picea* dominance. Another site, *Pink Lake*, 29 km southeast of Ramsay Lake, yielded a very similar pollen record.

In addition to these primary sites that appear to be representative of this subregion, pollen records showing similar trends for this time interval have been recorded in southern Quebec by Richard (1971, 1973a) near Quebec City, by Richard (1973b, 1975b) in the Laurentide Park area, by Richard (1975a) in the St. Lawrence Lowlands, and by Savoie and Richard (1979) for sites in the Sainte Agathe region between the north shore of the Champlain Sea and the 10,000-yr-BP ice-front margin.

Informative, thorough investigations of plant macrofossils and associated microfossils, provide additional data essential to any attempt to reconstruct late-glacial plant communities. The first, by Mott, Anderson, and Matthews (1981), describes and illustrates beautifully the pollen, plant macrofossils, and insect remains in radiocarbon-dated organic deposits at two sites near the margin of the Champlain Sea. The *St. Eugene* site (Fig. 5.1, No. 8) is located south of the St. Lawrence River, and consists of a bed of organic debris that yielded a radiocarbon determination of 11,050 ± 130 yr BP. The single sample analyzed for pollen produced a

Figure 5.3. Summary percentage pollen diagrams for the sites indicated. Details as in the legend to Figure 5.2.

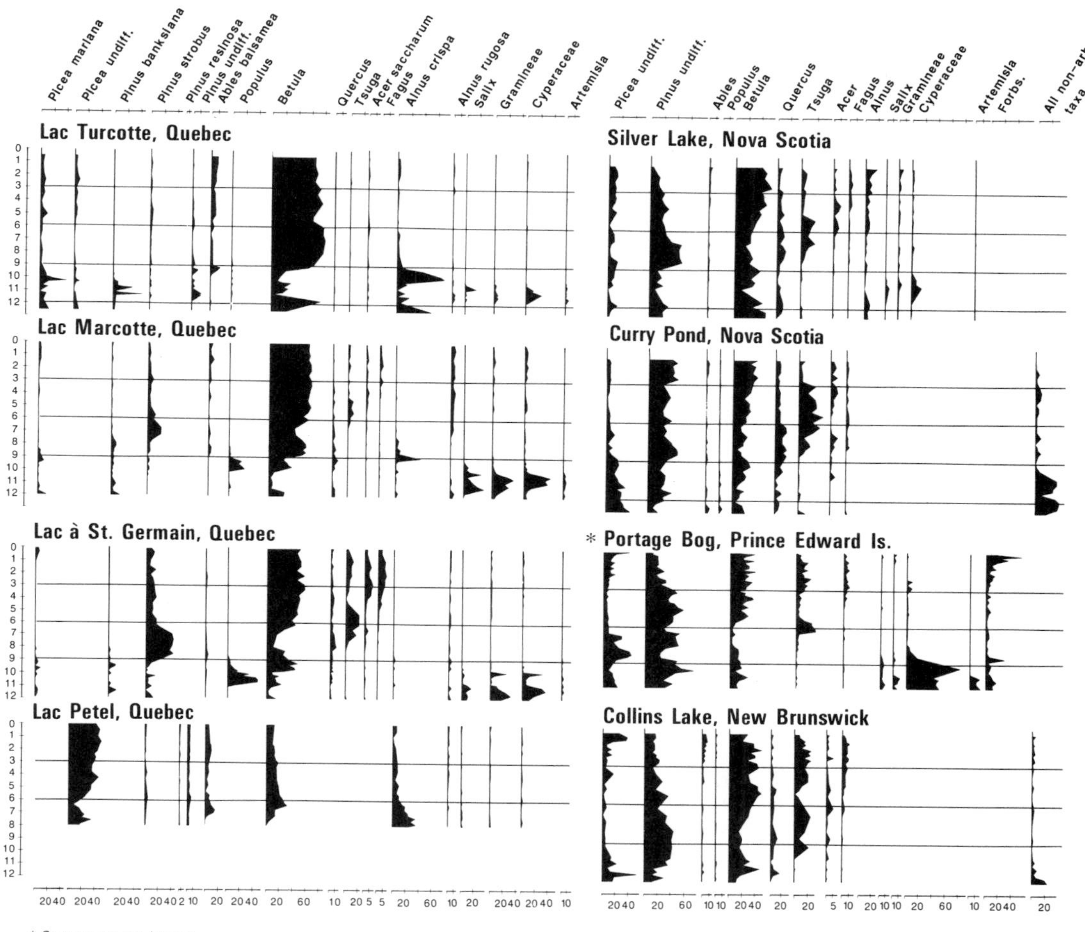

spectrum with negligible amounts of tree pollen, 60 percent grass, 9 percent *Oxyria digyna*, and about 15 percent other herb taxa. The plant macrofossil analysis yielded a large number of specimens grouped into 42 taxa of vascular plants and mosses (Table 5.1). Common taxa are *Salix herbacea*, *Dryas integrifolia*, *Oxyria digyna*, *Armeria maritima*, *Carex maritima*, *C. bigelowii*, and *C. nardina*.

The *Mont St. Hilaire* (Fig. 5.1, No. 9) site near Montreal consists of a unit of terrestrial organic debris that had been washed into the margins of the Champlain Sea and buried. A radiocarbon-dated wood sample gave an age of 10,100 yr BP. A single pollen sample gave a spectrum with 13.5 percent spruce, 10 percent pine, 12 percent birch, 23 percent oak, and negligible amounts of nonarboreal taxa. The plant macrofossils comprise 29 taxa of vascular plants and 16 of mosses (Table 5.1). By contrast with the St. Eugene site, the most common taxa are trees (*Picea mariana*, *Larix laricina*, *Betula* species) and the herb *Dryas* cf. *drummondii*.

An exposure of laminated lacustrine and alluvial sediment at *Columbia Bridge* in the Connecticut Valley, New Hampshire (Fig. 5.1, No. 10), provides the second investigation of pollen and macrofossils of relevance to our analysis. Miller and Thompson (1979) report on the pollen and vascular plant macrofossils. Two radiocarbon dates from wood, 11,390 and 11,540 yr BP, provide an estimate of the age of the deposit. The pollen diagram (Miller and Thompson 1979, Fig. 2) shows 20 to 40 percent *Picea*, 20 percent *Pinus*, 5 to 10 percent *Juniperus*, 1 to 2 percent *Fraxinus nigra*, 20 percent Cyperaceae, and small amounts of *Alnus*, *Salix*, *Artemisia*, and Compositae. The vascular plant fossil remains consist of a few trees (*Picea*, *Populus balsamifera*), shrubs (*Shepherdia canadensis*, *Elaeagnus commutata*, *Betula* cf. *glandulosa*, *Salix* (three taxa), *Juni-*

Figure 5.4. Summary percentage pollen diagrams for the sites indicated. Details as in the legend to Figure 5.2.

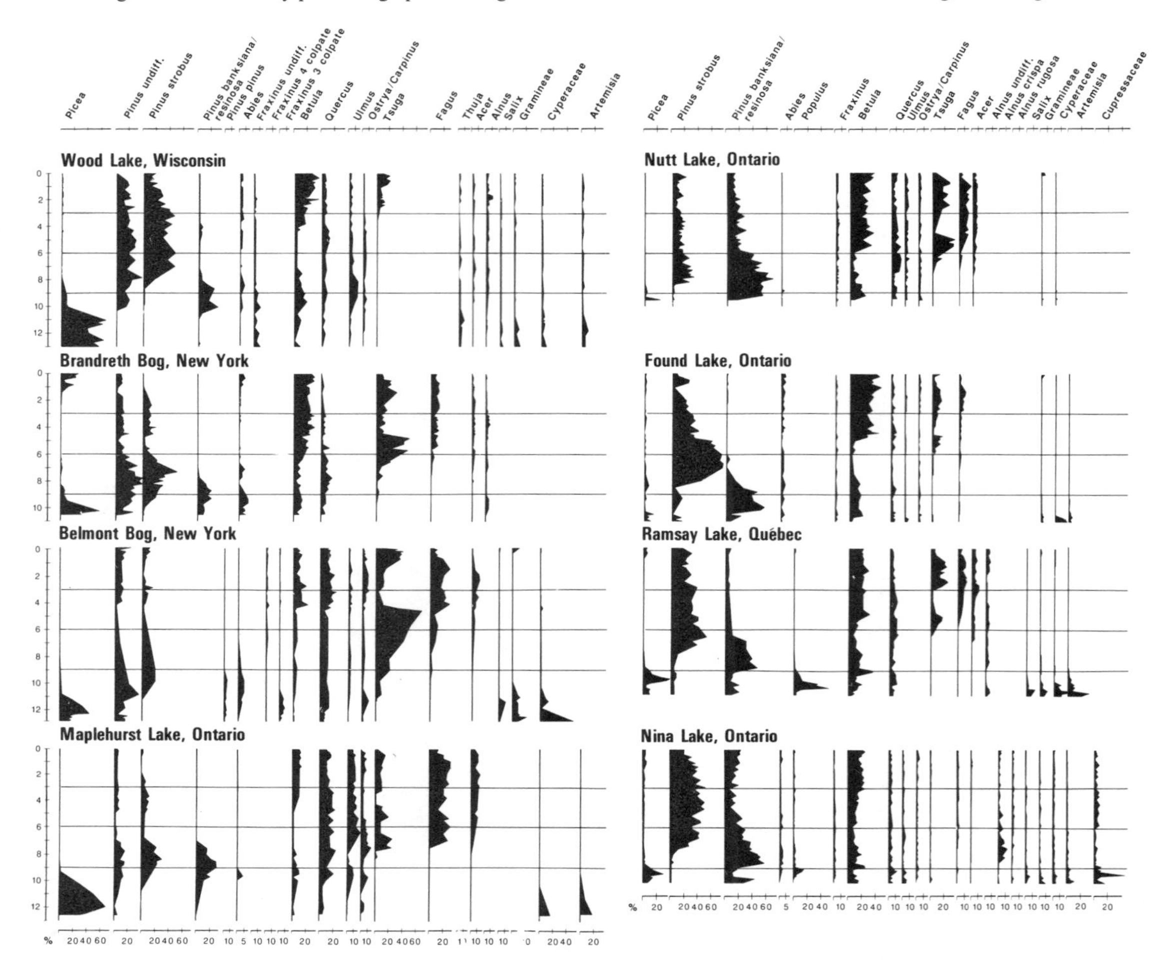

perus communis, three ericad genera (*Ledum*, *Vaccinium*, *Arctostaphylos*)), and the herbaceous taxa (*Geum*, *Dryas drummondii*, *D. integrifolia*, *Potentilla*, *Sibbaldia procumbens*, *Parnassia*, *Saxifraga aizoides*, *Ranunculus cymbalaria*, *Silene acaulis*, *Arenaria*, *Polygonum*, *Oxyria*, *Carex* (two species), *Potamogeton* (two species), *Woodsia ilvensis*, *Selaginella selaginoides*, and *Equisetum*). A large moss record (Table 5.1) was reported by Miller (1980). Many of these taxa have considerable indicator value, as discussed below.

Delage et al. (1985) describe macrofossils from two radiocarbon-dated organic layers excavated at the *Hinchinbrook* site (Fig. 5.1, No. 11), southwest of Montreal near the Quebec–New York boundary. The sediments are described as sands and silts with intercalated organic horizons, deposited by the Champlain Sea, dated at 10,480 yr BP. The relatively small number of macrofossil taxa (six) are arctic–boreal species (*Dryas integrifolia*, *Salix herbacea*, *Vaccinium uliginosum*, *Cerastium alpinum*) or undifferentiated genera (*Carex*, *Equisetum*). Pollen spectra from nine sampled levels showed a predominance of NAP, including *Oxyria* and *Artemisia*. The authors conclude that, while the local vegetation was probably treeless, the occurrence of grains of several trees (*Picea*, *Pinus*, *Betula*, *Quercus*, *Tsuga*, *Populus*, *Castanea*, *Ulmus*) is accountable by distant transport from the surrounding uplands (Shefford and Albion hills), where afforestation had begun by 11,000 yr BP. Finally, Tolonen and Tolonen (1984) report on an investigation of four coastal sites in Maine, including pollen analysis and vascular plant and moss macrofossil records for sediments dating from around 10,500 yr BP.

Maritime Canada

By 10,000 yr BP, Prince Edward Island, Nova Scotia, most of Newfoundland, and a broad tract of southeast Labrador were ice-free (Fig. 2.2, based on

Table 5.1. *Late-glacial macrofossils from southern Quebec and adjacent Vermont*

	Columbia	St. Eugene	St. Hilaire
VASCULAR PLANTS			
Androsace sp.			1
Arctostaphylos sp.			1
Arctostaphylos uva-ursi	11		
Arenaria cf. *dawsonensis*	1		
cf. *Armeria maritima*		11	
Armeria maritima ssp. *labradorica*		28	
Betula cf. *glandulosa*	1		
Betula cf. *lutea*			1
Betula papyrifera		1	
Betula sp.			6
Bidens sp.			1
Carex aquatilis	5		3
Carex bipartita	1		
Carex cf. *atlantica*			2
Carex cf. *bigelowii*		47	
Carex cf. *crinita*			3
Carex cf. *nardina*		81	
Carex cf. *trisperma*			2
Carex flava		3	
Carex maritima		157	
Carex rostrata			3
Carex spp.		267	3
Cerastium cf. *alpinum*		36	
Chenopodium spp.			2
Compositae	2		
Cornus sp.			1
Cornus stolonifera			1
Draba sp.	1		
Dryas cf. *drummondii*			33
Dryas drummondii	many		5
Dryas integrifolia	many	42	
Elaeagnus commutata	5		
Eleocharis acicularis			1
Epilobium sp.			1
Fragaria cf. *virginiana*		2	
Fragaria sp.		4	
Fragaria vesca		1	
Geum sp.	2		
Hippuris vulgaris			1
Juncus spp.		2	9
Juniperus communis	42		
Larix laricina			18
Ledum groenlandicum	11		
Lycopus americanus			1
Myrica gale			1
Najas flexilis		1	
Oxyria digyna	1	66	
Parnassia cf. *kotzebuei*	2		
Picea cf. *mariana*			37
Picea sp.	11		
Polygonum ramosissimum	4		
Populus balsamifera	many		

Table 5.1. (*continued*)

	Columbia	St. Eugene	St. Hilaire
Potamogeton cf. *gramineus*			1
Potamogeton filiformis	27	9	
Potamogeton pusillus	1		
Potamogeton sp.		1	2
Potentilla cf. *anserina*			4
Potentilla palustris		1	
Potentilla spp.	7	189	
Ranunculus cf. *trichophyllus* var. *eradicatus*		1	
Ranunculus cymbalaria	2		
Ranunculus pedatifidus/R. sulphureus		40	
Ranunculus sp.		1	2
Rubus sp.			2
Rumex cf. *triangulivalvis*			1
Salix cf. *argyrocarpa* or *pellita*	20		
Salix cf. *uva-ursi*	41		
Salix herbacea		151	
Salix sp.		51	1
Salix vestita	17		
cf. *Saxifraga* sp.		1	
Saxifraga aizoides	1		
Scirpus paludosus			1
Selaginella selaginoides	2		
Shepherdia canadensis	7		
Sibbaldia procumbens	2	2	
Silene acaulis var. *exscapa*	1	41	
Sparganium cf. *chlorocarpum*		1	
Taraxacum cf. *pumilum*		6	
Vaccinium uliginosum var. *alpinum*	many		
Viburnum sp.			1
Woodsia ilvensis	1		
BRYOPHYTES			
Aulacomnium palustre		30	
Aulacomnium turgidum		39	
Bryum pseudotriquetrum		16	
Bryum pseudotriquetrum f. *neodamense*			2
Calliergon sarmentosum		2	
Campylium polygamum			2
Drepanocladus revolvens		1	
Drepanocladus uncinatus			5
Fissidens taxifolius			2
Hylocomium splendens		2	1
Philonotis fontana			2
Pogonatum alpinum		78	
Scorpidium scorpioides		1	
Tomenthypnum nitens		2	2
Tortella tortuosa			5
Tortula mucronifolia			6

Source: Based on Mott, Anderson, and Matthews (1981) and Miller and Thompson (1979).

Prest 1970; Grant 1977; and King 1986). Anderson (1980) has investigated four mire sites on Prince Edward Island, one of which (*Portage Bog*, Fig. 5.1, No. 12) is summarized here as being representative (Fig. 5.3). The basal levels yielded spectra with moderate frequencies of shrub birch pollen and high percentages of herb taxa and sedges. The author explains the 10 percent spruce and 15 to 20 percent pine values in terms of dispersal from the south.

The pollen record of Nova Scotia for the period immediately following deglaciation was established initially by Livingstone (1968) in a detailed investigation of seven lake sites, but the number of radiocarbon dates available at that time was too small to establish the early chronology precisely. The *Silver Lake* diagram (Fig. 5.3; Fig. 5.1, No. 13) shows that the earliest pollen spectra reflect a dominance of herb pollen with *Salix* and *Shepherdia canadensis*, broadly similar to the lowest assemblage at Portage Bog, Prince Edward Island. Green (1981 and in press) has published percentage and PAR diagrams for one site in Nova Scotia (*Curry Pond*, Fig. 5.1, No. 14) and one nearby in New Brunswick (*Collins Lake*, Fig. 5.1, No. 15), and they indicate that a pollen assemblage with very low accumulation rates and nonarboreal taxa predominated until approximately 11,500, after which spruce, birch, and pine increased (Fig. 5.3). The age of the oldest sediment was determined by linear extrapolation from the lowest available radiocarbon dates. Railton (1972) provides unpublished pollen diagrams for sites in southwestern Nova Scotia but the age of the earliest sediments could not be determined accurately.

Although deglaciation of Newfoundland began before 10,000 yr BP, the only complete pollen records, which will be examined below in the section on the Holocene, begin at about 9,300 yr BP (Macpherson 1982). However, Macpherson and Anderson (1985) and Anderson (1983) provide preliminary analyses of late-Wisconsin sediments from two sites, one in north-central Newfoundland near *Notre Dame Bay* (Fig. 5.1, No. 16) and the other in the south on the *Burin Peninsula* (Fig. 5.1, No. 17). Macpherson and Anderson (1985) identify three pollen assemblages, associated with three sediment types, bracketed by two radiocarbon dates – 13,200 ± 300 and 10,500 ± 140. The dominant taxa in the zones are Gramineae, Cyperaceae, *Salix*, *Myrica*, *Betula*, *Oxyria digyna*, and other herbs. A tripartite pattern is reported also for a site on the Burin Peninsula by these authors, and they make the interesting suggestion that the sediment sequence and associated pollen spectra, interpreted in terms of a warming at 13,200 to 11,000 yr BP, a cooler episode, and a warming again at 10,500, could be correlated with the palaeoclimatic reconstructions proposed by Ruddiman and McIntyre (1981) for the adjacent Atlantic sedimentary record.

The southeastern corner of Labrador was ice-free by 11,000 yr BP (Fig. 2.2) and a recent pollen record by Engstrom and Hansen (1985) documents the early vegetation history. Their *Lake Hope Simpson* site (Fig. 5.1, No. 18) provides a representative record (Fig. 5.2). Between 10,500 and 9,500 yr BP, a predominantly nonarboreal assemblage was recorded, with low PAR values (100 to 900 grains · $cm^{-2} \cdot yr^{-1}$), dominated by *Salix*, Gramineae, Cyperaceae and various herbaceous taxa. *Alnus rugosa* and *Betula* make up 20 to 40 percent and 20 percent, respectively, of the spectra, but their PAR values are very low. A similar record was reported by Lamb (1980) for three sites in the same general region. His *Eagle Lake* site (Fig. 5.1, No. 19) shows that at about 10,500 yr BP, the first pollen spectra to be registered were dominated by dwarf birch, willow, and sedge, associated with grass and herbs (Fig. 5.2).

While the glacial retreat models show an ice-free coastal fringe along the Atlantic seaboard of Labrador and northern Quebec, only one site so far has produced a reliably dated pollen assemblage for the period that ended at 10,000 yr BP. It is the *Ublik* site (Fig. 5.1, No. 20) on the northeast Labrador coast, and Short and Nichols (1977) report a basal assemblage with high relative frequencies of Gramineae, Cyperaceae, *Salix*, and herb pollen (Fig. 5.2).

The Great Lakes Basin

It is appropriate here, before reviewing the data from Canadian sites in the Great Lakes Basin, to collate the pertinent evidence from sites lying immediately south of the international border. From Chapter 2, it can be recalled that many asynchronous advances and recessions of various ice lobes and associated glacial lakes occurred throughout the Great Lakes Basin (Prest 1970; Mickelson et al. 1983).

Overpeck (1985) has recently reported the results of pollen analysis of bog sediments at the *Brandreth* site in northern New York (Fig. 5.1, No. 21), and his results provide an important intermediate site between the investigations in New England and southeast Quebec and those to the west in the vicinity of Lake Ontario. Organic sedimentation began at 10,500 yr BP, and the earliest pollen assemblage, ending at 9,600, was dominated by *Picea*, associated with *Abies*, *Pinus*, and *Betula* (Fig. 5.4).

A combined pollen and plant macrofossil investigation of six sites in the *High Peaks* region of the Adirondack Mountains, northeast New York (Fig. 5.1, No. 22), provides detailed documentation along an elevational gradient that a fairly uniform, open, white-spruce woodland prevailed from 12,000 to

10,500 yr BP. One site produced evidence that herb–tundra prevailed before 12,000 yr BP. Closed forests of *Abies*, *Betula papyrifera*, and *Picea* spread at all elevations between 10,000 and 9,500 yr BP (Jackson 1983, 1986).

The *Belmont Bog* site in western New York (Fig. 5.1, No. 23) has yielded a continuous pollen record from 16,000 yr BP to the present (Spear and Miller 1976), and the period of our interest is well bracketed by radiocarbon dates of 12,565 ± 115 and 10,100 ± 100. The pollen spectra, both in the percentage and PAR diagrams, are dominated by *Picea* which reaches its maximum percentages (40%) in this interval. It is associated with low frequencies of *Pinus* and up to 10 percent of *Quercus*, *Fraxinus nigra*, *Carpinus–Ostrya*, *Salix*, Gramineae and Cyperaceae. The undated sediment below the level of the 12,565-yr-BP date gave pollen spectra with high (70%) values for nonarboreal taxa, but included apparently anomalous taxa (e.g., *Tsuga*) ascribed to sampling contamination (Fig. 5.4).

Farther to the northwest in New York, at the *Lockport* site (Fig. 5.1, No. 24), Miller (1973b) has reported on the pollen and plant macrofossils in an organic bed associated with glacial Lake Iroquois sediments, dated at 12,100 ± 100. The pollen spectra are similar to those from Belmont Bog with up to 50 percent *Picea*, and smaller amounts of *Fraxinus nigra*, *Carpinus–Ostrya*, *Quercus*, and nonarboreal types. Twenty-three taxa of vascular plants and 30 mosses were identified, the majority (80%) of which have modern ranges throughout the boreal region.

In the south-central portion of the Basin, on the west shore of Lake Michigan (Fig. 5.1, No. 25), the well-studied *Two Creeks Forest Bed* exposure is dated at 11,850 yr BP. The Two Creeks results are drawn upon in an attempt to reconstruct the past vegetation, but it should be noted here that the site has yielded pollen (West 1961), plant macrofossils (Miller 1976), and beetle remains (Morgan and Morgan 1979), and the collective assemblage is a spruce-dominated community with many taxa of vascular plants, mosses, and beetles whose modern ranges and ecology are boreal.

Roughly 200 km to the north, several sites have been studied in the conifer–hardwood zone of northern Wisconsin and adjacent northern Michigan, reviewed recently by Webb, Cushing, and Wright (1983). The pollen stratigraphy is similar at all sites and I have selected the *Wood Lake* site (Fig. 5.1, No. 26), reported by Heide (1984), because it provides the most complete record, with six well-spaced radiocarbon dates and both percentage and PAR data, and has also a record from a nearby small hollow that enables more detailed interpretation. From 13,000 to 10,000 yr BP, *Picea* (50 to 60%) dominated the spectra, associated with *Larix*, *Thuja*, *Fraxinus*, *Populus*, *Artemisia*, Gramineae, and Cyperaceae (Fig. 5.4).

Finally, the *Kylen Lake* sequence (Birks 1981a), already considered in an older context, provides a rich source of information on the pollen and macrofossils deposited at a site only 100 km from the southern border of western Ontario (Fig. 5.1, No. 27). From 12,000 to 10,700 yr BP, *Betula* (type *glandulosa*) and *Picea* (both *glauca* and *mariana* are common as pollen and macrofossils) predominated, associated with *Larix*, *Shepherdia canadensis*, *Populus balsamifera*, *Juniperus*, *Salix*, and a diverse assemblage of herb taxa. The period 10,700 to 10,400 yr BP is transitional with increases in *Picea* and *Pinus* (mainly *banksiana*) associated with *Larix*, *Populus*, *Abies*, *Fraxinus nigra*, *Quercus*, and *Ulmus*.

The *Maplehurst Lake*, Ontario, site (Fig. 5.1, No. 28) is one of the most complete pollen records for southern Ontario, with five radiocarbon dates and detailed pollen counts (Mott and Farley-Gill, 1978). It lies near Woodstock, in an interlobate zone that became ice-free about 12,500 yr BP. The pollen diagram (Fig. 5.4), shows that, during the interval from 12,500 to 10,500 yr BP, *Picea* dominated the spectra (70%) with small proportions (5%) of other tree taxa and relatively high frequencies (5 to 10%) of Gramineae, *Artemisia*, and Cyperaceae. At about 9,600 yr BP, the *Picea* zone was replaced by a *Pinus* zone.

The *Gage Street* site, 20 km from Maplehurst Lake, provides pollen, macrofossil, and insect data for the period between 13,000 and 6,000 yr BP, and contributes significantly to an understanding of the vegetation and environment that might have prevailed (Schwert et al. 1985). A pollen assemblage between 13,000 and 10,000 yr BP is dominated by *Picea*, including macrofossils of both *P. glauca* and *P. mariana*, associated with *Larix* (pollen and macrofossils), and relatively high percentages of *Juniperus*, *Populus*, *Salix*, *Artemisia*, and Cyperaceae. Macrofossils of *Cornus stolonifera* and *Thuja occidentalis* occur in this zone, along with *Dryas*, *Vaccinium uliginosum*, and *Empetrum*. The Coleoptera in this zone are "dominantly boreal forest inhabitants." The authors note that "no obligate tundra or tree-line insects were found" (Schwert et al. 1985, p. 208).

Saarnisto (1974) has reported the pollen stratigraphy of lake sites to the east of Lake Superior, near Sault Sainte Marie, of which the *Upper Twin Lake* is representative (Fig. 5.1, No. 29). The lake was near the ice margin, and Saarnisto (1974, p. 330) estimates that the area was deglaciated by 11,000 yr BP, and Upper Twin Lake emerged from a large glacial lake (Lake Minong; Prest 1970) at about 10,800. There was a short-lived (500 years) episode at this site when the pollen spectra were characterized by high percentages of nonarboreal pollen (>30%), dominated by Cyperaceae, *Artemisia* and Ambrosieae, and *Salix*, *Juniperus*, and *Alnus crispa* were common. The percentages for *Picea* and for *Pinus* type *banksiana* increase during this period, from 10 and 30 percent, respectively, to 20 and 40 percent, respectively.

The *Found Lake* site (Fig. 5.1, No. 30), investigated by Boyko and McAndrews (McAndrews 1981) is a small lake in bedrock that yielded a 4.5-m core of sediment. The pollen diagram has seven radiocarbon dates. The lowest levels, dated at 10,400 yr BP, have high pollen percentages of *Betula glandulosa*, *Juniperus*, *Artemisia*, and Cyperaceae (Fig. 5.4).

Vegetation reconstruction

The earliest deglaciation occurred at 13,000 yr BP, exposing two areas of eastern Canada to colonization by plants from the south. The Laurentide Ice margin at that time was at the Paris moraine in southern Ontario (Fig. 2.2) although there remain various conflicting views among geologists about the precise chronology and extent of the Late Wisconsin glacial record in the Great Lakes region. In any case, it appears to be accepted that southern Ontario became ice-free relatively early – between 13,000 and 12,000 yr BP – when lobes of ice to the east and west extended into New York and Michigan, respectively. Parts of southern New Brunswick, between the isostatically raised sea level and the ice front, were also exposed by 13,000 yr BP (Gadd, McDonald and Shilts 1972).

It has already been noted (Chapter 4) that when these Canadian landscapes were first emerging, the areas to the south were fully vegetated and the pollen and macrofossil records provide a reasonable portrayal of the patterns. East of the Appalachians, at latitude 40°N, a boreal forest assemblage dominated by *Pinus banksiana*, *Abies*, arboreal *Betula*, and *Fraxinus nigra* had replaced an earlier boreal vegetation dominated by *Picea glauca* associated with shrub *Betula* (Watts 1983). This spruce community extended north to about 42°N (northern Pennsylvania) by 13,000 yr BP, but to the north a treeless vegetation prevailed, variously described as tundra or herb–tundra. The nature of the treeless vegetation that occurred at several sites between the 13,000-yr-BP ice margin and the woodland or forest vegetation to the south is poorly known, but its floristic composition is different from the treeless vegetation that fringed the ice margin at localities beyond its full-glacial, maximum extent (Watts 1979, p. 458). He notes that at the *Longswamp*, Pennsylvania, site (Fig. 5.1, No. 31), the "tundra vegetation was present at 200 m elevation 40 km from the ice front. The tundra was rich in grasses and contained ericaceous shrubs and dwarf birch." By contrast, the first communities recorded at late-glacial sites farther north in the glaciated zone are dominated by *Salix*, sedges, grasses, *Oxyria*, *Artemisia*, and other herbs. An increase in *Picea* PAR at sites in Connecticut and Massachusetts, and later in New Hampshire and New Brunswick, reflects the time-transgressive northward spread of spruce into eastern Canada (Davis 1983).

The Basswood Road Lake site (Mott 1975) reveals a very similar pollen assemblage, representing the initial plant communities to occupy the landscape. Little can be added to Mott's interpretation of the data. He points out that the 10 percent *Picea* and 20 percent *Pinus* components are probably of distant origin as the total PAR of these spectra is very low. Shortly after, at about 12,600 yr BP, the NAP element decreases and a prominent *Populus* peak indicates that poplar spread on to the landscape to be followed within 1,000 years by *Picea*, reaching a maximum at 11,300 yr BP. The diversity of these communities increased by the addition of *Larix*, *Abies*, *Fraxinus* and *Carpinus–Ostrya* to form a conifer forest that prevailed until 10,000 yr BP.

Elsewhere in this region, a roughly similar pattern developed, illustrated by the Boundary Pond and *Lac Colin* sites (Mott 1977; Fig. 5.1, No. 32) and the Mont Shefford site (Richard 1978; Fig. 5.2), and it is likely that most of the Appalachian Highlands of southern Quebec and adjacent New Brunswick were dominated by this closed spruce–fir–birch forest with *Larix* and *Fraxinus nigra* on the lowlands, between 11,500 and 10,000 yr BP. The coniferous forests did not extend eastwards far beyond Lac Colin at 10,000 yr BP. The St. Eugene site, with its rich macrofossil record, documents the presence of a treeless assemblage with many herb taxa characteristic of open habitats – *Salix herbacea*, *Dryas*, *Sibbaldia procumbens*, *Oxyria digyna*, *Silene acaulis*, *Cerastium* cf. *alpinum*, *Ranunculus pedatifidus*, *Armeria maritima*, *Saxifraga*, and others. The insect assemblage confirms the botanical indications of an open vegetation with a considerable diversity of habitats, and interestingly, "some of the insects imply that (at 10,500 yr BP) treeline was no further than a few tens of kilometers south of the site" (Mott, Anderson, and Matthews 1981, p. 161). The Lac Colin and Boundary Pond sites provide corroboration of this interpretation.

The tract of land between the Champlain Sea and the Laurentide Ice margin, from the Ramsay Lake site at the western extremity (Mott and Farley-Gill 1981) through the Sainte Agathe region sites (Savoie and Richard 1979) to the Lac Marcotte (Labelle and Richard 1981) and Mimi sites (Richard and Poulin 1976), remained treeless until shortly after 10,000 yr BP when spruce and poplar and, a few centuries later, fir and birch expanded into the area. Between deglaciation and 10,000 yr BP, two basic treeless assemblages prevailed, and no doubt their pattern and extent were dependent on such local factors as microclimate and substratum, for which the record at present is sparse (Richard 1977b). Labelle and Richard (1981 p. 350 ff.) distinguish stages in the development of the initial vegetation at the Lac Marcotte and adjacent sites, based on very detailed pollen identification. The first community they recognize is a sparse herbaceous, probably discontinuous vegetation with abundant Gramineae, Cyperaceae, *Artemisia* and *Oxyria*, and subsequent succession involves *Salix herbacea*, *Saxifraga oppositifolia*, *Dryas integrifolia*, *Oxytropis*, *Polygonum viviparum*, and *Armeria*. Later development toward a shrub cover involves the apparent closing of the vegetation by the spread of *Betula glandulosa*, *Myrica gale*, *Juniperus*, *Shepherdia canadensis*, and *Salix* (Fig. 5.3). A very similar vegetation appears to have prevailed at the *Lac à Saint-Germain* site (Fig. 5.1, No. 33) near Sainte Agathe (Savoie and Richard 1979), and it is clear that many species recorded at the rich macrofossil site near Eugene, Quebec (Mott, Anderson, and Matthews 1981), were common also at these sites on the north shore of the Champlain Sea, between 11,000 and 10,000 yr BP. It should be noted that the Columbia Bridge record (Miller and Thompson 1979; Miller 1980), referred to earlier, dates from the same time (11,500 yr BP) and was associated with a local proglacial lake that was roughly contemporaneous with the Champlain Sea less than 100 km to the north. The vascular plants and mosses recorded include a significant number of taxa also found at St. Eugene, Quebec (Table 5.1), and the general conclusions of Miller and Thompson (1979, pp. 208–9) for the Columbia Bridge records probably have applicability in the St. Lawrence Lowlands region, in that "although the landscape . . . was predominantly treeless, a diverse flora of mosses, herbs, woody plants of herblike habit, and shrubs was established . . . both calcicolous and noncalcicolous plants are represented. This suggests that the absence of competition from trees and soil immaturity may have allowed heterogeneous associations of species to exist."

To the east of the Champlain Sea, the records from Mont Saint Pierre in Gaspé (Labelle and Richard 1984), Nova Scotia (Livingstone 1968; Green 1981), Newfoundland (Macpherson and Anderson 1985; Anderson 1983), and Labrador (Engstrom and Hansen 1985; Short and Nichols 1977) show that the deglaciated landscape that lay to the south and east of the ice sheet was treeless. The vegetation was registered in the pollen records by a small group of taxa noted also at all other late-glacial records in the region, summarized appropriately by Engstrom and Hansen (1985, p. 547) as follows: "high pollen frequencies of *Salix* (10–30%), Gramineae (10–35%), Cyperaceae (10–15%), and entomophilous herbaceous taxa including Rosaceae, Ranunculaceae, Leguminosae, Tubuliflorae, Caryophyllaceae, and *Epilobium*." Later, they offer a useful interpretation as follows: "The vegetation represented by this pollen assemblage was probably open tundra dominated by sedges, grasses and other herbs such as *Epilobium*, Caryophyllaceae, and Leguminosae, along with dwarf willows as the dominant shrub type. Initially birch and alder coverage was probably sparse, as suggested by their relatively poor representation among less prolific pollen producers such as willows and herbs and by their low pollen influx relative to subsequent zones. The modern sedge–shrub tundra of the drift-covered belts of north-central Labrador–Ungava (Hare 1959) may serve as a rough analogue to this early postglacial vegetation, although numerical comparisons with modern pollen spectra by Lamb (1984) demonstrate that these early fossil assemblages of the southeast are unlike any found in Labrador today. In particular *Salix* and Gramineae pollen do not exceed 5 percent in any of Lamb's modern tundra assemblages, while *Picea* pollen, which never exceeds 10 percent in zone I of this study, ranges from 20–40 percent in recent tundra spectra."

In summary, during the late-glacial period in the region centred on the St. Lawrence estuary and gulf, a predominantly spruce forest with significant amounts of *Abies*, *Betula* and *Larix* predominated by 10,000 yr BP on the landscapes of southern Quebec and southern New Brunswick to about the latitude of the Boundary Pond site of Mott (1977). To the south, small but increasing proportions of *Quercus*, *Acer*, and *Pinus* occurred. The northern section, between these mixed coniferous forests and the Champlain Sea, had a spruce-dominated complex, with few other trees except *Populus*. Treeless communities prevailed north of the Champlain Sea and at all sites from east of the position of Quebec City to the Atlantic seaboard, including Gaspé, Nova Scotia, Newfoundland, and Labrador.

To the west of the Appalachian massif, the earliest vegetation cover in eastern Canada occurred in southern Ontario, as registered at the Maplehurst Lake and Gage Street sites (Mott and Farley-Gill 1978; Schwert et al. 1985). The insect (Coleoptera) and plant (pollen, macrofossil) records, as happens frequently, provide palaeoecological indications that both amplify and contradict each other. On the one hand, the beetle species from the oldest sediments (13,000 yr BP) are predominantly boreal forest taxa, and "no obligate tundra or treeline insects were found." On the other hand, the coincidence of *Picea* with high *Salix*, *Artemisia* and Cyperaceae pollen, along with *Dryas*, *Vaccinium*, and *Empetrum* macrofossils might prompt the traditional "open subarctic parkland" interpretation, as indeed the authors suggest. After 12,500 yr BP, however, the record of both insects and plants preclude any interpretation other than that of a closed conifer forest very similar to the modern midboreal zone, with macrofossils of *Cornus stolonifera* and *Thuja occidentalis* associated with beetle species of exclusively modern boreal affinities. The moss record is equally significant. "The rich fen bryophytes include *Calliergon trifarium*, *C. subsarmentosum*, *Drepanocladus vernicosus*, *D. exannulatus*, and *Scorpidium scorpidioides*. By 11,000 yr BP, the site resembled a typical small lake or pond in the present-day boreal forest, with floating leaves of water lilies and pond weeds covering the water surface" (Schwert et al. 1985, pp. 212–13). In fact, the rich fen mosses noted above are characteristic of calcareous open fen habitats, and it is probable that such mires occurred locally at the Gage Street site along with closed white-spruce forests. These moss taxa have been recorded also at *Wolf Creek*, Minnesota (Fig. 5.1, No. 34) (Birks 1976), Two Creeks, Wisconsin (Miller 1980), and Lockport, New York (Miller 1973b), sites and Miller (1980 p. 386 ff.) suggests that "rich fen communities appear to have been a conspicuous feature of the lowland landscape in the Upper Great Lakes–New England region where fens of this type are still found, although their area is much less now than formerly." The beetles recovered from the Two Creeks site were made up almost entirely of taxa that today are confined to the central boreal forest (Morgan and Morgan 1979).

By the time the Laurentide Ice had receded to just north of Lake Superior, all sites in south-central Ontario were registering a *Picea*-dominated assemblage except the Upper Twin Lake site near Sault Sainte Marie, where a treeless assemblage was recorded from 10,650 to 9,940 yr BP that is ascribed to a poorly defined "tundra vegetation" (Saarnisto 1974).

The Holocene – 10,000 yr BP to the present

Southern Quebec and New Brunswick

It will be useful to outline the record of past vegetation for the late-glacial to Holocene transition at sites in the adjacent United States. Several records of continuous pollen percentages and PAR from the glaciated region to the south of Quebec and New Brunswick, spanning the past eleven millennia, provide an adequate portrayal of the record. The *Tannersville Bog* site in Pennsylvania (Watts 1979; Fig. 5.1, No. 35) and the *Rogers Lake* site in Connecticut (Davis 1983; Fig. 5.1, No. 36) are so similar that they can be considered together. Both show a decrease in *Picea* (percentage and PAR, though the latter values indicate a much more gradual decline than the percentages) and an increase in *Pinus*, *Abies*, *Betula*, and *Alnus* at 10,000 yr BP. Very shortly afterwards, *Quercus* and *Betula* and later *Tsuga* increase markedly at Rogers Lake, but the *Betula* curve quickly decreases and at Tannersville it shows only a gradual decrease. Macrofossil analyses at Tannersville, however, suggest that, while the dominant *Betula* species prior to 10,000 yr BP was *B. papyrifera*, *B. populifolia* occupied the site from about 9,800 yr BP. Similar significant changes occur earlier in the pollen record of sites farther south in deglaciated areas – namely from a *Picea* assemblage through a stage of *Pinus*, *Abies*, *Alnus*, *Betula* to a *Pinus strobus*, *Quercus* and *Tsuga* assemblage by 10,000 yr BP, for example, at *Crider's Pond* in south central Pennsylvania (Watts 1979; Fig. 5.1, No. 37). A *Quercus* maximum and a steep decline in the *Pinus* curve occur at both the Rogers Lake and Tannersville sites at 8,500 yr BP, after which *Fagus* increases, and between 8,500 and 4,700 yr BP *Quercus*, *Tsuga*, and *Fagus* dominate the spectra. A *Tsuga* decline at 5,000 yr BP is followed by increases in *Pinus* and *Betula*. The *Tsuga* pollen frequency increases gradually at 4,000 yr BP accompanied by an increase in birch and a decline in *Quercus*.

The Brandreth site in the Adirondacks of north central New York (Overpeck 1985), the *Mirror Lake* site in new Hampshire (Likens and Davis 1975), the *Moulton Pond* site (Davis et al. 1975) and *Chase Lake* sites in Maine (Davis and Kuhns personal communication) provide a record of change from areas closer to the Canadian border (Fig. 5.1, Nos. 38–40). Both Brandreth Bog and Mirror Lake are in areas of low granitic mountains, on the west and east sides, respectively, of the southern terminus of the Appalachians, while Moulton Pond is a coastal site in northern Maine surrounded by low-relief topography. All four sites have a spruce-dominated assemblage at 10,000 yr BP, with associated *Populus*, *Abies*, and *Alnus* at Moulton Pond, Chase Lake, and Mirror Lake, but not at Brandreth Bog. A spruce decline, followed by a *Pinus* increase to more than 60 percent, occurs at all sites at 9,500 yr BP. The sequence of *Pinus* species is *P. banksiana* and/or *resinosa* followed by *P. strobus*, whose pollen frequencies quickly reach 50 percent. *Betula* and *Quercus* increase at about 9,000 yr BP, and *Acer*, *Tsuga*, and *Fagus* increase at slightly different times between 8,000 and 7,000 yr BP. *Tsuga* and *Betula*, associated with *Quercus*, *Fagus*, and *Acer*, dominate from 7,000 to 5,000 yr BP, when the large-scale synchronous decline in *Tsuga* pollen frequencies occurs. *Fagus*, *Acer*, *Betula* and later *Tsuga* dominate the spectra until late in the Holocene when decreases in *Quercus* and *Pinus* and increases in *Picea* and *Abies* occur. Though the percentages are 25 to 60 percent, the peak PAR of *Picea* at Moulton Pond is very low. I discuss in Chapter 8 the interpretation offered by Davis et al. (1975) in the broader context of vegetation reconstruction. The Chase Lake site is examined more fully later also, as it includes one of the most detailed and informative macrofossil records for the eastern region (summary diagram shown in Fig. 8.2). Detailed pollen and macrofossil analyses from four additional sites, recently completed by the Davis and Jacobson group (personal communication 1985), have greatly enriched and consolidated the record. They repeat, with only minor variations, the basic sequence displayed here (Fig. 5.2 and Fig. 8.1) for the Basswood Road, New Brunswick, and Chase Lake, Maine, sites.

Finally, the detailed macrofossil analyses of late-glacial and Holocene sediments from six sites in the High Peaks region of the Adirondack Mountains, New York, provide an important contribution to understanding the regional vegetational history (Jackson 1983, 1986). Following the development of closed spruce–fir–birch forests between 10,500 and 9,500 yr BP, white pine, and later hemlock and yellow birch, spread into the region. The expansion of hemlock and yellow birch resulted in substantial declines of low-elevation populations of spruce, fir, paper birch, and later white pine. The presence in the early to middle Holocene of significant quantities of hemlock and yellow birch 150 m above their present limit is interpreted by Jackson (1983) as indicating "a warmer, possibly drier climate in the region."

At Basswood Road Lake, the *Picea*-dominated assemblage that terminated at 9,500 yr BP was replaced by a short-lived *Abies* phase with higher fir percentages than in modern samples, and then by *Pinus* (40 to 60%), *Quercus* (10 to 20%) and *Betula* (30%). The pine pollen is the *Pinus strobus* type, replacing the Diploxylon type that prevailed in older levels. At 6,500 yr BP, *Tsuga* frequencies increase sharply from less than 10 percent to more than 30 percent, accompanied by increases in birch and decreases in pine. *Fagus* increases from less than 1 percent to 10 percent at 5,100 yr BP and hemlock declines. Thereafter, the hemlock frequencies increase slowly, and finally, at 1,000 yr BP, *Picea* and *Abies* increase (Fig. 5.2).

At the Boundary Pond site there is a transition at 10,000 yr BP from the Picea–Betula–Alnus–NAP pollen spectra to those dominated by *Pinus* (Diploxylon type) and *P. strobus* associated with *Abies*, *Quercus* and *Betula*. *Acer* reaches its Holocene maximum at 7,700 yr BP, and *Tsuga* and *Fagus* increase successively at about 6,000. The *Tsuga* decline occurs here at 5,700, although, at adjacent sites, Mott (1977) records it closer to the common date of 5,000 yr BP. Slight changes in the relative amounts of the main taxa occur until 1,390 yr BP when *Picea* and *Abies* increase (Fig. 5.2).

The Mont Shefford site shows a Holocene sequence similar to that of Basswood Road Lake with a delay of 500 to 1,000 yr in the arrival of several of the main tree taxa. *Abies* and *Pinus* (*strobus*) increase at 9,000 yr BP, whereas *Quercus* increases a few centuries later at 8,700. These taxa, along with *Betula* and *Ulmus*, dominate the spectra until 5,800 yr BP when *Tsuga* increases from less than 5 to more than 20 percent and *Pinus* decreases. By 4,400 yr BP, *Pinus* has declined to 10 percent and *Fagus* and *Acer* have increased. There is no clear evidence of an abrupt *Tsuga* decline at this site. Between 4,000 yr BP and the present day, the only percentage changes are a gradual decline of *Quercus* to less than 5 percent, an increase in *Alnus rugosa*, and slight increases in *Fagus* and *Tsuga* (Fig. 5.2).

The Lac Colin site of Mott (1977) lies near the boundary between the boreal forest and the mixed conifer–hardwood zones and, therefore, should provide a sensitive registration of the history of some important taxa. At 10,000 yr BP, a *Picea*–NAP assemblage was replaced by a short-lived peak of *Alnus crispa* associated with *Populus*, followed at about 9,000 yr BP by an abrupt increase in *Abies*, and increases in *Pinus strobus*, *Pinus* Diploxylon type, *Betula* (arboreal), and *Quercus*. *Acer* appears at about 7,000 yr BP and maintains its characteristically low frequencies, with a break at about 5,000 yr BP, to the present day. *Tsuga* shows a continuous registration for the first time at about 6,000 yr BP and *Fagus* at 4,000 yr BP. There is a *Tsuga* decline to sporadic occurrences of individual grains until 3,300 yr BP when a continuous curve resumes. *Alnus rugosa* shows an increase at 6,300 yr BP. *Picea* increases steadily from 3,000 to 0 yr BP (Fig. 5.2).

The Lac Marcotte site (Labelle and Richard 1981) is situated just within the boreal (spruce–fir) zone, but resembles the Lac Colin site in the general pattern of pollen frequencies (Fig. 5.3). It lies near the transition between mixed forest and boreal forest. The present vegetation includes abundant *Abies*, *Picea*, *Betula papyrifera* and *Populus*, with *Acer*, *Betula alleghaniensis*, *Tsuga* and *Fagus* on better sites. *Tilia*, *Juglans*, and *Quercus rubra* are at their northern limit here. At 9,900 yr BP, *Populus*, *Juniperus*, and shortly afterwards *Picea* (cf. *mariana*) and *Quercus* increase. They decline at 9,000 yr BP, when arboreal *Betula* rises to 60 percent and *Abies* rises to 5 percent, and later, at 7,500 yr BP, *Pinus strobus* increases from 5 to 25 percent. At 6,000 yr BP, the *Tsuga* curve rises, and later, at very small percentages, continuous curves of *Fagus* and *Acer* begin. Similar results, with less chronological precision, were reported by Richard (1971) from two sites in the precincts of Quebec City and by Richard and Poulin (1976) from the Lac Mimi site. At about 10,000 yr BP, a short-lived peak of *Populus* occurs, associated with *Alnus crispa*. A brief peak of *Picea* (type *mariana*) follows, with increases to 5 percent of *Quercus* and *Pinus strobus*. *Abies* increases at 9,500 yr BP as *Picea* decreases, and *Betula* (arboreal, assumed by the authors to be *B. papyrifera*) increases markedly to 60 percent. Only minor changes occur later. There is a very small peak (to 4%) in *Tsuga* at roughly 6,500 yr BP, an increase of *Acer* shortly after to values between 2 and 3 percent, and finally *Fagus* increases from occasional grains to a continuous curve of 2 to 5 percent, at roughly 4,000 yr BP.

The Lac Turcotte site (Labelle and Richard 1984) shows that, following a nonarboreal phase prior to 10,000 yr BP, an arboreal assemblage dominated by *Picea* (*mariana*), *Populus*, *Abies*, and *Betula papyrifera* persisted from 10,000 to 9,000, marked also by a short-lived maximum (70%) of *Alnus crispa*. From 9,000 yr BP, *Betula* (65%) dominates the spectra, and only *Abies*, *Picea*, and *Pinus* (both *strobus* and cf. *banksiana*) register percentages greater than one. *Fraxinus nigra*, *Ulmus*, *Quercus*, *Acer*, *Fagus*, and *Tsuga* show discontinuous curves, but none exceeds 1 percent at any level (Fig. 5.3). Macrofossils of *Betula papyrifera*, *Picea*, *Abies balsamea*, and *Larix* occur scattered in the sediment between 9,000 yr BP and the present day. It is of significance to note that a nearby valley site (*Lac à Léonard*, Fig. 5.1, No. 41), in contrast with the Lac Turcotte site, which is located on the upland plateau, shows percentage curves for *Pinus strobus*, *Fraxinus nigra*, *Acer saccharum*, and *Ulmus* that are continuous from roughly 7,000 yr BP to the present day.

The Ramsay Lake, Quebec, record provides percentage and PAR data that give a detailed record of the pollen spectra for the past 10,000 yr. The *Picea*-dominated spectra that became established by 10,200 yr BP, with their associated *Populus* maxima, declined at about 9,500 yr BP when first *Pinus banksiana* then *P. strobus* increased to between 30 and 40 percent, with 5 to 10 percent *Betula* and *Quercus*. *Tsuga* and, at very low frequencies, *Acer* enter the record at 6,500 yr BP, followed very shortly by *Fagus* at frequencies between 2 and 5 percent. The *Tsuga* decline and recovery are registered at roughly 5,000 and 3,000 yr BP, respectively, but only slight increases in *Betula* and *Fagus* comprise the remaining changes in the spectra (Fig. 5.4).

The Lac à Saint-Germain site near the St. Narcisse moraine, above the level of the Champlain Sea, is one of three deposits in the area investigated by Savoie and Richard (1979). The site is in a bedrock depression in the mixed deciduous–conifer zone, at present dominated by forests of *Abies balsamea*, *Acer saccharum*, and *Betula alleghaniensis*, with *Pinus strobus*, *Fagus grandifolia*, and *Tsuga*. Following a brief period prior to 10,420 yr BP with abundant NAP, *Populus*, *Picea*, and *Juniperus* dominate the spectra from 10,400 to 10,000 yr BP, if one radiocarbon age at 6.5 m that is out of sequence is overlooked. At about 10,000 yr BP, *Picea* and *Populus* decline to negligible amounts, and *Pinus* cf. *banksiana*, *Betula* (arboreal), *Pinus strobus*, and *Quercus* dominate until 7,000 yr BP when *Tsuga* increases markedly followed by gradual increases of *Acer* (cf. *saccharum*) and *Fagus*. The *Tsuga* decline occurs at about 4,000 yr BP followed by a gradual increase. *Fagus* reaches a maximum (10%) about 3,000 yr BP, and the only other change is a slight increase of *Picea* at an undated level near the top of the section (Fig. 5.3).

The Maritimes, Labrador, and Northern Quebec

The Portage Bog site (Fig. 5.3) is representative of the four sites that Anderson (1980) investigated on Prince Edward Island. The island is within the mixed conifer–hardwood forest region, dominated by communities of *Picea* (*P. glauca* and *P. rubens*), *Abies*, *Acer*, *Fagus*, *Tsuga*, *Pinus strobus*, and *Betula lutea*.

Following a short-lived NAP assemblage dominated by dwarf birch and herbs (*Artemisia*, Gramineae, Cyperaceae), *Picea* and *Pinus* increase at about 9,500 yr BP. Later increases of *Pinus strobus*, *Tsuga* at 7,000 yr BP, *Betula*, and *Fagus* complete the record.

The Silver Lake site (Fig. 5.3) is chosen here to represent the record from Nova Scotia, mainly because it includes three radiocarbon dates. The reader might find it of interest to compare the record with the PAR diagram for the *Everitt Lake* site (Fig. 5.1, No. 42) reported by Green (1981). The earliest reliable pollen spectra at Silver Lake are dominated by *Salix* and Cyperaceae, followed at 10,000 yr BP by increases in spruce and pine; then *Abies*, *Quercus*, and *Betula* increase at about 9,000, followed by *Tsuga* at 7,500. *Acer*, *Fraxinus*, and *Fagus* all increase gradually between 8,000 and 3,000. The PAR diagram at Everitt Lake shows no evidence of a recovery of *Tsuga* following the sharp decline at 5000 yr BP. It is interesting to note a steep decline there in the *Betula* PAR curve at 4,000 yr BP, which has the effect of increasing the percentage representation of *Tsuga* and other taxa at that level, implying that at Everitt Lake, as at Moulton Pond, the hemlock recovery is an artefact of a percentage diagram. The Holocene pollen stratigraphy of Nova Scotia has been reviewed recently by Green (in press) and he includes a previously unpublished diagram from Curry Pond, a 22-ha lake in southwestern Nova Scotia (Fig. 5.3). In its general outlines, it is similar to the Silver Lake record. However, Green (in press) offers an interestingly original interpretation of the low occurrences of tree pollen and suggests that all the tree taxa present today except *Fagus* had arrived by about 10,000 yr BP. Subsequent changes were partly responses to a highly varied landscape that controlled the extent of fire and partly competition effects between early immigrants (*Picea*, *Pinus*) and later immigrants. The problem is examined in Chapter 8.

The *Sugarloaf Pond* site (Fig. 5.1, No. 43), one of two on the Avalon Peninsula reported by Macpherson (1982), provides the most detailed pollen record for Newfoundland and amplifies the early investigations of Terasmae (1963). The site is in the boreal forest zone, and upland surfaces support an open parkland and krummholz that reflect the subarctic climate. The dominant trees are fir, spruce, larch, and white birch. A 5.5-m core yielded a basal assemblage from 10,000 to 9,300 yr BP dominated by *Salix*, Cyperaceae, and various herbs, followed by an increase in shrub birch and ericads, and later by increases in the percentages of tree birch, spruce, and fir. *Populus* shows small but continuous registration in the early Holocene. No significant changes occur in the spectra between 7,000 yr BP and the present (Fig. 5.2).

The combination of pollen, macrofossil, and geochemical investigation of four sites in the southeastern corner of Labrador by Engstrom and Hansen (1985) marks an important advance in our understanding of the nature and chronology of vegetation development in that region. Their Lake Hope Simpson site is representative of the study and the region. It is a 65-ha lake in a bedrock plateau, at 295 m above sea level, surrounded by boreal forest vegetation. *Picea mariana* is the dominant tree, with *Abies* and *Picea glauca* on rich mesic soils, and stands of *Betula papyrifera* and *Populus tremuloides* on recently burned areas. The pollen record, following the initial nonarboreal assemblage at 10,500 to 9,500 yr BP, consists of spectra dominated by shrub birch (45 to 60%), ericads and *Alnus crispa* (25 to 30%), followed at about 8,000 yr BP by a rise in *Picea*, predominantly *P. glauca* (distinguished palynologically, but recorded also as abundant macrofossils in the original investigation), and later by an increase to 20 percent *Abies*. Following the fir increase, *Alnus* declines and later *Picea glauca* decreases markedly, whereas *P. mariana* increases from less than 10 percent at 7,000 to 50 percent by 6,000 yr BP. The pollen assemblage remains stable from 6,000 yr BP to the top of the sediment column. The percentage diagram, shown in summary form here, differs only slightly from the PAR frequencies (Fig. 5.2).

Lamb (1980) published percentage and PAR diagrams for three lake sites in southeastern Labrador; one (*Whitney's Gulch*, Fig. 5.1, No. 44) lies on the north shore of the Gulf of St. Lawrence 80 km southwest of Lake Hope Simpson, and the others are a similar distance to the northeast near the Eagle River (*Paradise Lake* and Eagle Lake). The Eagle Lake site is described here. A core of sediment from Eagle Lake, a large (8 × 1 km) body of water, yielded a 10,500-year-old record. The site is in the boreal forest, and open stands of *Picea mariana* with extensive lichen cover dominate, associated with local stands of *Abies* and *Betula papyrifera*. The pollen sequence is broadly similar to the Lake Hope Simpson record, with dominance of *Betula* (shrub) with *Salix* and NAP from 10,500 to 9,000 yr BP, when *Alnus crispa* increases at 8,000 followed by increases in *Abies* and *Picea* at 6,000 (Fig. 5.2). Lamb (1980, p. 126) suggests that *P. glauca* predominates until about 5,000 yr BP when *P. mariana* assumes dominance. A similar sequence, less precisely dated, was demonstrated by Jordan (1975) for sites in the Lake Melville region to the immediate northeast of Lamb's Eagle River sites.

The record northwards from Hamilton Inlet along the Atlantic seaboard of Labrador and adjacent Nouveau Québec is relatively detailed. Four lake cores were recovered by Lamb (1985) in the forest–tundra zone centred on 50°N in Labrador, and the pollen record from one site (*Gravel Ridge*, Fig. 5.1, No. 45) is presented in some detail in his monograph, and summarized in Figure 5.5. The site is near the altitudinal limit of trees, on the Nain-George Plateau, an upland of Archaean bedrock. A 3.7-m core yielded a simple sequence of pollen spectra beginning with a basal assemblage (roughly 6,470 to 6,000 yr BP) dominated by *Alnus crispa* (60%) with *Salix*, Gramineae, Cyperaceae, and *Betula* (20%). At 5,500 yr BP, *Picea* (both *P. glauca* and *P. mariana*) increases to 65 percent while *Alnus crispa* decreases and *Betula* remains at 20 percent. *Abies* shows a minor maximum to 3 percent between 5,500 and 4,500 yr BP. Increases in NAP taxa and a decrease of *Picea* are the only significant changes in the latest part of the record, dating from only a few centuries ago.

Mott (1976), Richard, Larouche, and Bouchard (1982) and King (1984) have, independently, investigated sites that form a transect from near Sept-Iles on the north shore of the St. Lawrence Estuary, Quebec, to the Kaniapiscau Plateau, Quebec. These sites are at the centre of the Labrador Sector of the Laurentide Ice Sheet, one of the last regions in eastern North America to be freed of continental ice. The transect passes through the boreal forest zone in the south dominated by closed stands of *Picea mariana*, with *Betula papyrifera* and *Populus tremuloides*, and local populations of *Picea glauca*, *Abies*, and *Pinus banksiana*, to the northern

Figure 5.5. Summary percentage pollen diagrams for the sites indicated. Details as in the legend to Figure 5.2.

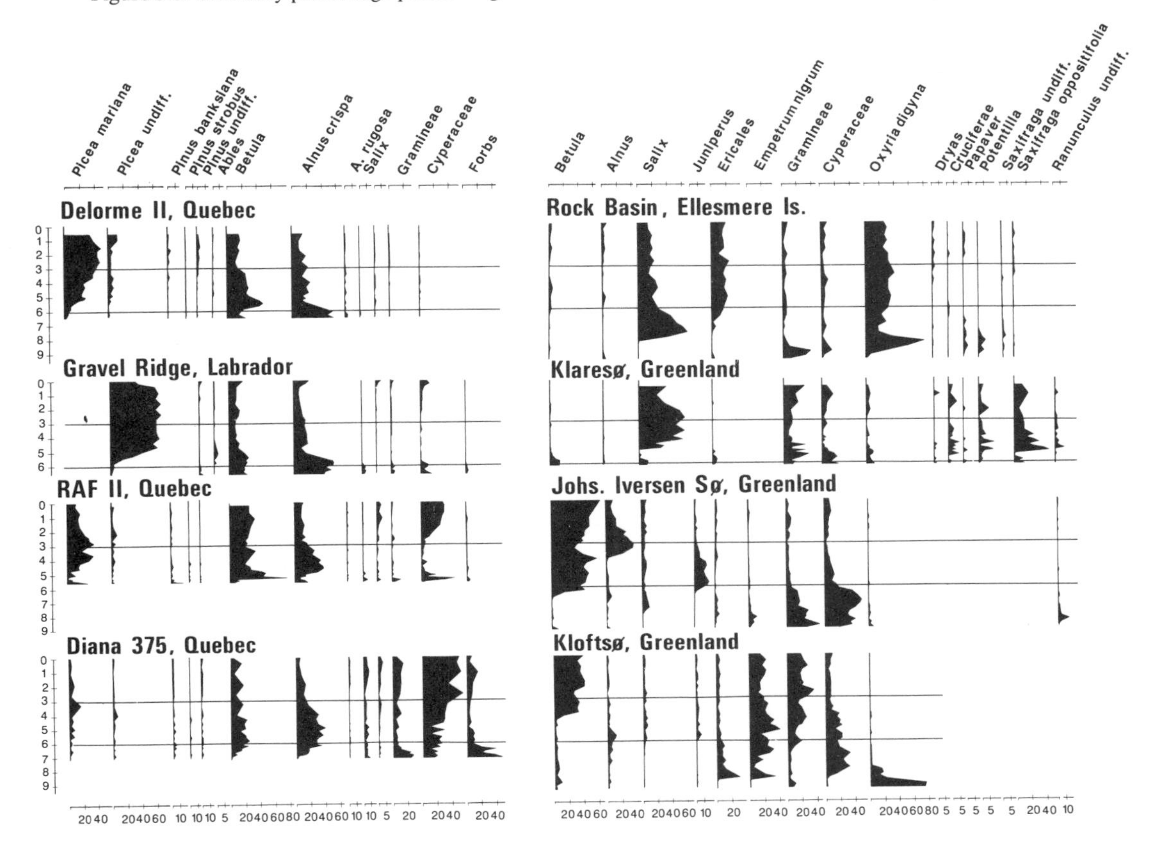

subarctic woodlands dominated by open stands of *Picea mariana* with *Cladina* lichen cover and very localized stands of *P. glauca* and *Abies*.

The *Lac Petel* site (King 1984; Fig. 5.1, No. 46) provides an 8,000-year percentage and PAR record (Fig. 5.3). *Alnus crispa*, *Betula*, and *Salix*, associated with *Picea glauca* in small amounts made up the initial pollen assemblage after deglaciation, followed by an increase in *Abies* at 7,000 yr BP and later, at 5,500 yr BP, an increase of *Picea* (chiefly *P. mariana*) to greater than 50 percent. *Betula* values remain high from 6,500 yr BP to the present day. *Pinus banksiana* frequencies increase at 7,000 yr BP to very low percentages (<1%), but PAR high enough to indicate presence at the site. They remain constant throughout the succeeding record. Mott's (1976) percentage and PAR records from a few kilometres to the south are almost identical to the Lac Petel sequence.

At *Lac Gros* (Fig. 5.1, No. 47), 180 km to the north, King (1984) records a 7,000-year sequence of pollen percentages and PAR with an initial, short-lived assemblage dominated by shrubs (*Betula*, *Salix*, and *Alnus crispa*) followed by an expansion of *Picea* at 6,500 yr BP with only a slight subsequent increase in *Abies*. No significant changes occur after 6,000 yr BP.

The *Coghill Lake* site in Labrador (Fig. 5.1, No. 48), roughly 150 km north of the previous site, has a basal age of 6,200 yr BP, and King (1984) reports the following pollen sequence: an initial dominance of *Alnus crispa* (30 to 40%) associated with *Betula* followed by an increase of *Picea*, at 5,000 yr BP, from less than 10 percent to more than 50 percent, with reduced proportions of *Alnus* (10%) and *Betula* (10%). A detailed analysis of pollen and macrofossils at four sites on the Kaniapiscau Plateau of Nouveau Québec was reported by Richard, Larouche, and Bouchard (1982) and the *Delorme II* site (Fig. 5.1, No. 49) is summarized to represent the record for this region (Fig. 5.5). A 2.6-m sequence with five radiocarbon dates, very detailed percentage and PAR pollen diagrams, and supporting macrofossil data comprise a superb record. The major features of the pollen diagram are similar to those from the Coghill Lake site. From 6,320 to 5,500 yr BP, *Alnus crispa* dominates (>50%) associated with *Larix* (macrofossils) and *Betula*, with small quantities of *Picea* (*mariana*). *P. mariana* increases steadily from 10 to 30 percent at 5,000 yr BP and dominates the spectra throughout the rest of the sequence, associated with both tree and shrub birch, and alder. *Abies* is represented only by scattered grains and the maximum *Pinus* frequencies do not exceed 10 percent.

Farther west, in the subarctic woodland zone of Nouveau Québec, large areas remain unknown with the exception of an investigation of five sites in the middle catchment of *La Grande Rivière* (Richard 1979; Fig. 5.1, No. 50). From 7,000 yr BP, *Pinus* (cf. *banksiana*) (20%), *Picea mariana* (10%) and various NAP taxa predominate, but the very low PAR values lead Richard (1979, p. 107) to suggest that only the NAP element represented local inputs. After a short-lived peak of *Populus* cf. *tremuloides* at 6,500 yr BP, *Alnus crispa* and *Picea mariana* increase to 30 percent each, associated with *Betula* (both tree and shrub forms). The only significant later change in the spectra is an increase of *Pinus banksiana* at 3,000 yr BP to 20 percent.

The Labrador–Ungava region has been investigated by two major research groups with the result that a relatively dense network of sites is available, most with detailed radiocarbon chronologies. The Ublik Pond site is one of several investigated by Short and Nichols (1977) and it provides a detailed record from a coastal tundra site that was free of ice by 10,000 yr BP. Following an initial phase when Gramineae, Cyperaceae, and *Salix* predominated, *Betula* (shrub) increased markedly at 9,000 yr BP in both percentage and PAR with decreases in herb taxa. *Alnus* then increased at 6,700 yr BP to 30 percent, and finally *Picea* increased from less than 10 to more than 50 percent at about 4,400 yr BP. Spruce values declined again after 2,000 yr BP (Fig. 5.2).

The same authors provide a record from an inland site, the *Pyramid Hills* site east of the George River in the forest–tundra zone (Fig. 5.1, No. 51) and, although a few of the radiocarbon dates appear out of sequence, the basic pollen configuration is probably reliable. The first organic sediments date from 7,000 yr BP and contain low pollen concentrations dominated by Gramineae and Cyperaceae, followed by increases at 6,500 of both shrub birch and *Alnus crispa*. *Picea* increases gradually from 10 percent at 5,000 to more than 30 percent at 2,500 yr BP and then decreases slightly to the present day.

A major investigation of seven lake sites and three mires on the east side of Ungava Bay, supplemented by a survey of modern pollen spectra in the area, resulted in an important, fully illustrated monograph by Richard (1981a). The *RAF II* site (Fig. 5.1, No. 52) illustrates the sequence recorded in the southern sector. The site is located in the catchment basin of the Rivière aux Feuilles, on the boundary between the shrub tundra zone and the forest–tundra, which also delimits the northern limit of *Larix*, and the northern limit of *Picea* krummholz. The first pollen was registered at 5,000 yr BP following the recession of a proglacial lake, and Gramineae, Cyperaceae, and *Betula glandulosa* predominated briefly. Dwarf birch then increased, followed at 4,900 by a sharp rise in both the percentage and PAR values of *Alnus crispa*. *Picea mariana* increased steadily from 4,500 yr BP, reaching its maximum (60%) by 3,000 yr BP, after which it decreased and dwarf birch and sedge increased to the present day (Fig. 5.5). The second group of sites is to the north in the herb tundra zone, in the vicinity of Diana Bay at 60°N. The *Diana 375* site (Fig. 5.1, No. 53) is representative, a small round lake in a glacial cirque depression. One and one-half metres of sediment yielded three radiocarbon dates and a detailed percentage and PAR diagram. The initial pollen registration is dominated by herb taxa, with Ranunculaceae (20%), Cyperaceae (20%), Gramineae (25%), and *Oxyria digyna* the most important elements. Between 6,500 and 6,000 yr BP, a transition occurs and the frequencies of these herb taxa, except Cyperaceae, decline while those of *Betula glandulosa* and *Alnus crispa* increase. From 6,000 to 3,500 yr BP, *Alnus* and *Betula* dominate the spectra associated with Cyperaceae and *Salix*, which remains constant at 7 percent throughout the profile. *Alnus* and, to a lesser extent, *Betula* decrease after 3,000 yr BP while Cyperaceae, Gramineae, and Ericaceae increase (Fig. 5.5).

The Great Lakes

Several sites in the lower Great Lakes area of the United States provide a detailed record of the vegetation that prevailed during the transition from the late glacial to the Holocene. At both the *Hook Lake* (Winkler 1985a; Fig. 5.1, No. 54) and Wood Lake (Heide 1984) sites in Wisconsin, rapid changes in the pollen record are registered. At Hook Lake, in southern Wisconsin, Diploxylon pine (*Pinus banksiana/resinosa*) and *Betula* increased at 10,500 yr BP following decreases in *Picea* and *Fraxinus nigra*. Between 10,000 and 9,000 yr BP, deciduous tree species (*Ulmus*, *Quercus*, *Carya*, *Juglans*, *Acer*, and *Tilia*) increase and *Pinus strobus* becomes conspicuous. At the north-central Wood Lake site, *Pinus banksiana/resinosa*, *Abies*, *Betula*, *Ostrya–Carpinus*, and *Ulmus* increased as *Picea* decreased at about 10,000 yr BP (Fig. 5.4). Later, at about 8,000 yr BP, *Pinus strobus* and *Acer saccharum* increased. *Tsuga* was absent from both sites until much later in the Holocene. Farther to the northwest, at Kylen Lake, Minnesota, the period 10,700 to 9,250 yr BP was dominated by *Picea* and *Pinus banksiana*, later (9,250 to 8,400) replaced by *Pinus resinosa* and *Be-*

tula papyrifera (Birks 1981a). By contrast, the Belmont Bog site in northwestern New York shows a gradual decline in spruce in the late glacial, an increase in *Pinus strobus* at 11,000 yr BP, a maximum at 10,000 yr BP, followed immediately by increases in *Tsuga*, *Acer saccharum*, and, later, *Fagus*.

At the Maplehurst Lake site in southern Ontario (Fig. 5.4), the late-glacial *Picea*-dominated assemblage is replaced partially at about 10,500 yr BP by increases in *Pinus banksiana* and *Quercus*, each to 25 percent. *Pinus strobus* increases rapidly at about 9,000 yr BP to 30 percent (Mott and Farley-Gill 1978). By then, *Picea* has decreased to negligible proportions. *Carpinus–Ostrya* and *Acer* show gradual increases at about 8,000 and 7,500 yr BP. *Tsuga* and, shortly afterwards, both *Ulmus* and *Fagus* show rapid increases. *Fagus* (30%) is the chief component of the spectra after *Tsuga* declines at 6,000 yr BP, and a few other significant changes occur until near the top of the sequence when Ambrosieae and Gramineae increase.

Bennett (in press, personal communication) has analyzed in great detail four sites in south-central Ontario in a transect from the deciduous forest zone to the northern limit of *Fagus*. One of these diagrams, from *Nutt Lake* (Fig. 5.1, No. 55) in the mixed-forest zone, is shown in summary form (Fig. 5.4). A 5.2-m core was recovered from the centre of the small (7.9 ha) relatively deep (8.2 m) lake, ^{14}C dated at ten levels, and analyzed at close intervals to provide the most precise Holocene pollen record available from south-central Ontario. Following a decline in *Picea* and *Populus* at 10,000 yr BP, *Pinus strobus* increases to 20 to 30 percent at 9,000, associated with *Betula*, *Quercus*, *Ulmus*, *Ostrya–Carpinus*, and *Fraxinus*. *Acer*, *Tsuga*, and *Fagus* enter the record at 8,000, 7,400 and 7,000 yr BP, respectively, and all increase rapidly. A steep decline in hemlock at 5,000 yr BP is accompanied by an increase in beech. Hemlock increases steadily to its former amounts by 2,000 yr BP. At 800 yr BP, both beech and hemlock decrease and *Betula* increases.

The Found Lake site (McAndrews 1981) provides a pollen record from a site on the Shield with several radiocarbon dates, but only a simplified diagram has been published so far. *Picea*, *Betula*, and NAP taxa dominate until 10,400 yr BP when *Pinus banksiana/resinosa* increases to 60 percent, associated with *Quercus*. *Pinus strobus* increases to 70 percent at 7,500, and *Tsuga* increases at about 5,500 yr BP. *Acer* and *Fagus* increase slowly to low percentages, and *Tsuga* decreases abruptly at 4,640 yr BP, preceded by a steep rise in *Betula*. *Tsuga* increases at 3,800 yr BP, after which *Pinus* decreases and *Fagus* increases, until near the top of the section there is a small peak in pine (Fig. 5.4).

The Upper Twin Lake site, reported by Saarnisto (1974), shows a shrub–NAP assemblage from 10,650 to 9,940 yr BP when *Picea* increases to 40 percent and *Pinus banksiana* to 20 percent. *Betula* (arboreal) increases from less than 10 percent to more than 40 percent at 8,760 with slight increases in *Quercus*, *Ulmus*, *Abies*, and *Fagus*. An undated increase in *Pinus strobus* to 20 percent completes the record, estimated at 8,000 yr BP. The *Alfies Lake* site (Fig. 5.1, No. 56) of Saarnisto (1975) is the only locality for which a complete Holocene record is available from these investigations – they were designed to study Late Wisconsin and early Holocene shoreline displacement. The herb zone is missing from this site, and the record begins at 9,210 yr BP with a *Picea* assemblage, followed at 8,500 yr BP by a *Betula–Abies*-dominated zone. Both *Pinus strobus* and *P. banksiana* increase at 7,000 yr BP. Small but continuous percentages of *Tsuga* and *Acer* begin at about 6,500 yr BP and *Pinus* decreases. *Betula* increases to more than 50 percent and *Fagus* shows an almost continuous record. *Picea* increases at an undated level near the top of the sequence.

Liu (1982) has completed a detailed investigation of a transect of sites in northern Ontario. His *Nina Lake* site (Fig. 5.1, No. 57) provides a well-dated, detailed record for the northern part of the mixed conifer–hardwood forest zone. *Pinus banksiana*, *Betula papyrifera*, *Populus*, and *Abies* are common today with lesser amounts of *Acer*, *Pinus strobus* and *Picea glauca*. The record begins at 9,500 yr BP with successive short-lived peaks of *Pinus banksiana*, *Juniperus*, *Picea glauca*, *Populus*, and *Betula*. At 8,000 yr BP, *Pinus banksiana* increases to more than 40 percent with 20 percent *Betula*. At 7,000 yr BP, *Pinus strobus* increases to 50 percent while *Betula* and *Pinus banksiana* decrease, although birch increases again from about 400 yr BP to the present. *Tsuga*, *Fagus* and *Acer* produce continuous records, but with very low percentages (<5%) from about 7,000 yr BP to the present (Fig. 5.4).

Liu's (1982) northernmost site, *Crates Lake* (Fig. 5.1, No. 58), provides an illuminating comparison with Nina Lake. It is a small (3.4 ha) kettle lake in the Cochrane till plain, within the area of proglacial Lake Barlow-Ojibway. The surrounding vegetation is boreal forest, dominated by *Picea glauca*, *P. mariana*, *Betula papyrifera*, and *Populus balsamifera*. Local stands of *Fraxinus nigra* and *Thuja* occur in swamps. Unfortunately, some anomalous radiocarbon dates limit the precision with which a chronology can be established, but the pollen diagram is interesting in any event. After a short-lived assemblage with high frequencies of *Artemisia*, Cyperaceae, and Gramineae, *Picea*, *Populus*, *Pinus banksiana*, and *Betula* increase. Later, at 7,000 yr BP, *Pinus strobus* increases to 20 percent and then, before 3,000 yr BP, decreases to insignificant frequencies while *Picea* (both *P. glauca* and *P. mariana*) increases steadily to 40 percent and *Abies* increases to 15 percent. *Quercus*, *Ulmus*, and *Carpinus–Ostrya* have continuous curves throughout the section, but with very low percentages, whereas *Tsuga* and *Fagus* are represented discontinuously at frequencies less than 1 percent.

Richard (1980a), provides two detailed pollen sequences from near Lake Abitibi on the Quebec–Ontario boundary at about latitude 48°30′. The sites are very similar in pollen stratigraphy and chronology, and the *Lac Yelle* site (Fig. 5.1, No. 59) is summarized here. It lies in the boreal forest zone, with a predominance of *Picea mariana*, *Pinus banksiana*, *Abies*, *Thuja*, *Larix*, and *Betula papyrifera*, but it is noteworthy that several elements of the mixed conifer–hardwood zone occur very locally, close to their northern limit: *Pinus strobus*, *P. resinosa*, *Fraxinus nigra*, *Ulmus americana*, *Acer rubrum*, *A. saccharum*, *Tsuga*, and *Betula alleghaniensis*. Lac Yelle is 3 ha, in an area of Precambrian bedrock uplands with extensive lowland clays deposited by proglacial Lake Ojibway. A unit of inorganic silty clay deposited before 8,900 yr BP yielded pollen spectra dominated by *Pinus* (50%), with a mixture of *Picea*, *Quercus*, *Juniperus*, *Alnus*, and miscellaneous taxa in very low frequencies (*Carya*, *Tilia*, *Juglans*). The pollen concentration is less than 10,000 grains · cm^{-3}, suggesting that the assemblage consisted of either redeposited or long-distance elements. At about 8,900 yr BP, organic sediments yielded higher concentrations (150,000 grains · cm^{-3}) and *Populus*, *Pinus banksiana*, and later *P. strobus* (40%) dominate with *Betula*. *Pinus strobus* decreases by 5,600 yr BP to 20 percent, and few significant changes occur throughout the upper part of the profile.

Vegetation reconstruction

I will make use of data on modern pollen representation and collations of autecological information on the major taxa, both summarized in Chapter 3. The interpretations of the original investigations are referred to in detail, although in some cases it might seem appropriate to propose alternatives. Sites are grouped according to the two major biomes of the region – the southeastern temperate forest region and the northern boreal–subarctic zone (Fig. 2.3a). The former is a relatively complex region, floristically and physiographically rich, made up of the northern Great Lakes Basin, the St. Lawrence River lowlands, and adjacent southern Quebec, New Brunswick, and Nova Scotia. The region has undergone major changes in vegetation from 10,000 yr BP to the present, and the pollen frequencies of the dozen or so major tree taxa have changed significantly during this period. Some changes were asynchronous from site to site, implying time-transgressive events, while others were synchronous over wide areas, suggesting a response in species abundance to a factor of relatively widespread occurrence. By contrast, the boreal–subarctic zone in the eastern half of Canada is relatively simple, dominated by very few major taxa of transcontinental range, and the physiography, dominated by the Canadian Shield region, is relatively uniform.

Each of the two major regions is divided into subregions that have a characteristic vegetation and physiography.

Southeastern temperate forest region

Appalachian uplands

The uplands south of the St. Lawrence, made up of the Appalachian physiographic region, where the topography is controlled by bedrock and where relief is considerable, can be considered as a unit, in part, because the final product of Holocene vegetation change is more or less uniform throughout the region, and, in part, because of similarities in the pollen sequences. The Boundary Pond (Mott 1977), *Albion* (Richard 1975b), Basswood Road Lake (Mott 1975), and Mont Shefford (Richard 1978) sites lie in this region and form the basis of the following interpretation. By 10,000 yr BP, a forest assemblage dominated by *Abies*, *Picea glauca*, *Betula papyrifera*, *Populus*, and *Quercus* (cf. *rubra*) occupied mesic sites. *Larix laricina*, *Fraxinus nigra*, and *Picea mariana* were the major species in poorly drained habitats, and *Ulmus*, *Ostrya*, and *Acer saccharum* were minor components of the vegetation. As Richard (1978 and 1981a) points out in his perceptive discussions of interpretive problems in southern Quebec, the reconstructed vegetation shows little direct correspondence to the pollen stratigraphic patterns, chiefly because *Pinus* and *Betula* are overrepresented pollen taxa whereas *Acer saccharum* and *Abies balsamea* are greatly underrepresented. He shows (1981a, Fig. 2) that *Acer* and *Abies* make up about 4 and 10 percent, respectively, of modern pollen spectra from the vegetation zones where they dominate the modern forests. It can be assumed that mesic sites supported a mosaic pattern with *Betula* and *Populus* taking the role of gap-filling, rapid-growing, and relatively short-lived species while *Abies*, *Picea glauca*, and, to a lesser extent, *Quercus rubra* were longer-lived, shade-tolerant taxa that entered the later stage succession. The change in composition of these primarily coniferous forests on upland sites between 11,000 and 9,500 yr BP appears to have been the result of a slightly slower spreading from the south of *Abies* and *Quercus* than of *Picea*.

At about 9,000 yr BP, the total pollen concentration at most sites in this region increased from 100,000 to over 200,000 grains · cm^{-3} and the species composition of the forests changed significantly. *Acer saccharum* expanded steadily, apparently replacing in part *Abies* and almost entirely *Picea glauca* on mesic sites. *Pinus strobus* and *Quercus* increased in abundance, occupying xeric sites, such as rocky scarps and outcrops in the uplands and local sand deposits in the lowlands. It is also likely that the white pine and oak, along with *Betula papyrifera*, played a role in the patch dynamics of the dominant maple forest on mesic sites. *Fraxinus* (cf. *americana*), *Ostrya*, *Carpinus*, *Tsuga*, and *Ulmus* were minor elements in the upland forest while *Fraxinus nigra*, *Acer rubrum*, *Ulmus*, and *Larix* prevailed in poorly drained habitats.

The question of the presence in the landscape of such important taxa as *Tsuga* and *Fagus* when their pollen representation is less than 1 percent (discussed in Chapter 8) will not be resolved until detailed macrofossil evidence is at hand for several Canadian sites as at Chase Lake, Maine (Fig. 8.1b). One interpretation of the continuous registration of very low proportions (<1%) of *Tsuga* and *Fagus* is that the trees were present in low density, between 9,000 and 7,000 yr BP (Bennett 1985). During the ensuing few millennia, from 7,000 to roughly 4,000 yr BP, the primary changes in these forest communities involved the expansion, either by spreading from the south or by increases in existing small populations, of several dominant taxa – *Betula alleghaniensis*, distinguished by Richard (1978) on the basis of pollen-size frequencies; *Tsuga canadensis*, whose percentage pollen curve at all the sites in this region shows a relatively rapid increase at about 6,000 yr BP; and finally *Fagus*, *Tilia*, *Carya*, and *Juglans* increase at about 4,500 to 5,000 yr BP although the pattern of occurrence of *Tilia*, *Carya*, and *Juglans* cannot be ascertained precisely because of their sparse representation in the pollen data. Similarly, some sites (Basswood Road Lake) show a continuous *Fagus* curve from about 7,000 yr BP, suggesting presence of the tree at that time, while others (Mont Shefford) lack this feature. At the same time as *Tsuga* and *Fagus* pollen percentages increase, those of *Pinus* and *Quercus* decrease, and it is likely that the decreases were due in part to a real reduction in the abundances of oak and pine in the face of competition from *Tsuga*, and slightly later from *Fagus*. Hemlock has many ecological and life cycle attributes that promote dominance in forest communities. It tolerates a wide range of soils from xeric to moist; it has the greatest longevity of all eastern North American tree species; it can regenerate under a wide range of conditions; it can fill its own gaps in a forest without any intermediate successional species; and it achieves optimal height and shading efficiency. *Acer saccharum* increased along with *Fagus* and, at about 5,000 yr BP (4,500 at Mont Shefford), *Tsuga* values decline while *Fagus* and, at some sites, *Betula* increase. A beech–maple–yellow birch community dominated all mesic sites at this time with very local white pine and oak on xeric or successional sites and *Fraxinus*, *Ulmus*, *Acer rubrum*, and *Alnus rugosa* dominating swamps and swales. However, hemlock remained an important member of the regional forests, as its pollen representations in both percentage and PAR show, and was never less than it is at modern sites in the Great Lakes–St. Lawrence region, where *Tsuga* is very common. After 7,000 yr BP, the four major tree species of beech, maple, hemlock, and yellow birch remained abundant in the landscape to the present day, with shifts in their relative amounts. Three (beech, sugar maple, and hemlock) have very similar reproductive strategies and ecological amplitudes, being long-lived with slow to moderate growth rates, shade-tolerant, and having the capacity to replace themselves in succession. Yellow birch has a shorter life span, a more rapid growth rate and is a prolific producer of mobile seeds. As a result, it plays a more important role in intermediate successional stages after disturbance or gap formation. In this interpretation of the record, it is suggested that the important dominants (hemlock, maple, beech, and yellow birch) were present in the landscape of the entire region surrounding the St. Lawrence lowlands, from Ontario to the longitude of Quebec City, by at least 8,000 yr BP. The subsequent variations in species abundance throughout the rest of the Holocene were a function of any one, or combinations, of changing climate, changing fire frequencies, and interspecific competition.

An alternative view is that the initial steep rises in the frequency curves of the main tree taxa depict the arrival of the species and therefore that the landscapes under consideration here were progressively enriched after 9,000 yr BP by the successive arrivals of fir, white pine, maple, hemlock, and oak (Davis 1981). This question is reexamined in Chapter 8 when reconstructions of past environments are attempted.

Later, at 3,000 yr BP, some sites show an increase in *Tsuga* frequencies and decreases in pine and oak, suggesting changes in the relative abundance of these trees in the regional vegetation. In the Appalachian uplands, but not at the Mont Shefford site, an increase in *Picea* and *Abies* on high-elevation surfaces occurred late in the record, at roughly 1,500 yr BP.

Acadian subregion

Similar patterns of change occurred in Prince Edward Island and Nova Scotia, although the precision of the chronology, the density of sites, and the detail of the palynology are inadequate for firm conclusions. However, the sequences described above for the sites in Prince Edward Island (Anderson 1980) and Nova Scotia (Livingstone 1968; Green 1981, in press) show that following a coniferous forest phase dominated by spruce, fir, jackpine and/or red pine and white birch, oak, and white pine expanded along with *Acer*, and by 8,000 yr BP, these taxa, as well as small populations of *Ostrya–Carpinus*, *Fraxinus*, *Ulmus*, *Tsuga*, and possibly *Fagus*, made up the forest core. Subsequent changes involved changes in species abundances, particularly among the taxa that dominate the later stages of succession – maple, yellow birch, beech and hemlock. Green (1982) interprets his PAR and charcoal record in terms of a "combination of climatic changes, species migrations, competition and fire . . . " (p. 38). In a later contribution (Green in press), he suggests an alternative interpretation to what is suggested above, as noted earlier on p. 82, and the difficult question of the registration of plants at very low densities in local pollen spectra is addressed later in Chapter 8. The observation made many years ago by Livingstone (1968, p. 121) about many of the Nova Scotia records, that the changes in pollen frequencies recorded in the diagrams from southern Quebec, New Brunswick, and the Maritimes "are not greater than the differences between the Great Lakes–St. Lawrence Forest and the Acadian Forest," can be extended here.

St. Lawrence lowlands and the adjacent Laurentian scarp

The sites at Ramsay Lake, Lac à St. Germain, and Lac Marcotte, summarized on pages 80 to 81, make a coherent grouping and are representative of the larger data set from additional sites in the region reported by Richard (1971, 1973a, 1973b, 1973c, 1975a, 1975b, 1977a). The sites all occur in a lowland region, much of which was occupied by the Champlain Sea, and all are close to the northern margin of a major physiographic scarp – the southern limit of the Canadian Shield. The northern edge of the region coincides with the northern extent of the Great Lakes–St. Lawrence forest region (Rowe 1972). The northern limits of beech and hemlock occur in this region, while sugar maple, white pine, and yellow birch occur farther north but in progressively diminishing abundance (Figs. 1.3 and 3.11).

An identical pattern can be seen in the pollen records along the length of the lowlands subsequent to 10,000 yr BP when a *Picea–Populus–Betula* forest occurred at the west end near Ramsay Lake and a *Populus* and very short-lived spruce–fir forest at Lac à St. Germain and Lac Marcotte. Most of the species of the modern forests had arrived by 9,000 yr BP. Toward the eastern extremity of the lowlands, fir and yellow birch forests dominated mesic sites from 8,400 to 7,500 yr BP, and toward the west, a mixed forest with white pine, oak, hornbeam, ironwood, and maple, but with little or no spruce and fir, prevailed. Hemlock was present, and possibly beech, from 8,000 yr BP, but in small isolated populations. At about 6,500 yr BP, hemlock expanded at all sites along with yellow birch and maple. Later, beech expanded (at 5,500 yr BP) and the regional vegetation throughout the lowlands changed very little in the subsequent time.

Gaspé and South Laurentians

Although Labelle and Richard (1984) emphasize that their pollen and macrofossil study of a valley and a plateau site on the northern edge of the Gaspé Peninsula is very preliminary, this investigation is of critical significance because the sites lie near the northeastern limits of a few tree species characteristic of the Great Lakes–St. Lawrence forest zone. *Acer saccharum*, *Betula alleghaniensis*, *Ulmus americana*, *Fraxinus nigra*, and *Pinus strobus* spread to the valley site (Lac à Léonard) in the early Holocene, following a boreal forest phase at 9,300 yr BP dominated by *Picea*, *Abies*, *Larix* and *Betula papyrifera*. *Betula alleghaniensis* arrived early, followed by *Pinus strobus* at 6,500 yr BP. *Fraxinus nigra*, *Ulmus americana*, and *Acer saccharum* spread into the area at about 5,000 yr BP, and the composition and structure of the vegetation has remained constant to the present day. Labelle and Richard (1984, p. 272) ascribe the very low (<0.5%) discontinuous frequencies of *Tsuga*, *Tilia*, *Fagus*, *Juglans*, and *Ostrya* to long-distance transport by wind. By contrast, the first forest vegetation established around the plateau site (Lac Turcotte) was a Picea–Abies–*Betula papyrifera* association with local stands of *Larix* and *Populus*. Apart from the establishment of very scattered patches of *Betula alleghaniensis* and *Pinus strobus* at 5,500 and 6,500 yr BP, respectively, the upland vegetation has not changed throughout the Holocene.

The records from Lac Marcotte (Labelle and Richard 1981) and Lac Mimi (Richard and Poulin 1976) are similar to that from the Gaspé valley site. Following the initial treeless vegetation, *Populus* and *Picea* formed the first tree cover in the area at about 10,000 yr BP. At 9,500 yr BP, *Abies* and *Betula papyrifera* expanded and a spruce–fir–birch forest was established, preceded by a short phase of abundant *Alnus crispa* that the authors interpret as a response to cooler climate. Very local stands of *Pinus strobus* occurred on well-drained soils from about 7,000 yr BP, and *Ulmus* and *Fraxinus nigra* spread into lowland habitats. *Acer saccharum* and later *Fagus* spread into the area in very low population densities.

South-central Ontario subregion

It was concluded earlier in this chapter that between 12,500 and roughly 10,000 yr BP, a closed forest dominated by *Picea* occupied mesic sites. At roughly

10,500 yr BP, the abundance of *Picea* decreased rapidly and a mixed forest developed, consisting of local stands of *Pinus banksiana* and *Pinus strobus* on better-drained sites, *Abies*, *Ulmus*, *Fraxinus nigra*, and *Larix* on moist to hydric sites, and *Betula*, *Quercus*, *Carpinus–Ostrya*, and *Acer* (probably *saccharum*) on mesic sites. *Picea* was present in low frequencies, chiefly *P. mariana* on poorly drained sites. *Fagus*, *Tsuga*, and *Acer* were all registered as continuous curves at Belmont Bog in northwest New York by at least 10,000 yr BP. In other words, it is probable that the full complement of forest species found today in southern Ontario was present early in the Holocene, and subsequent changes in pollen frequencies reflect changes in the relative abundances of the major species. The above assemblage changed at 7,600 yr BP when *Tsuga* expanded to population sizes greater than at any subsequent time at any site in Eastern Canada. It can be assumed that *Tsuga* is represented in percentage pollen diagrams in approximately equal proportion to its representation in the landscape, so it must have been at least a codominant member of the forest. The increase at sites in the vicinity of Lake Ontario was more or less synchronous, suggesting an environmentally forced response rather than a migration effect. The Maplehurst Lake (Fig. 5.4), Gage Street (Schwert et al. 1985), *Edward Lake* (Fig. 5.1, No. 60) (McAndrews 1981), and Belmont Bog (Spear and Miller 1976) sites all show a major expansion of *Tsuga* at 7,500 yr BP, preceded by a decrease in *Pinus strobus* and *P. banksiana* to virtual absence in the landscape.

Shortly after, at 6,000 yr BP at Maplehurst and Edward Lake, *Tsuga* percentages decline and *Fagus*, *Ulmus*, and *Acer* expand. Whatever the causal factor (this is discussed in Chapter 8), a predominantly deciduous forest prevailed throughout south-central Ontario dominated by *Acer saccharum* and *Fagus* with *Fraxinus*, *Tilia*, *Carpinus–Ostrya*, *Ulmus*, *Quercus*, and *Tsuga* locally abundant on particular soil types.

However, the pollen record of *Tsuga* prior to its major increase, when it occurs in some diagrams as a continuous curve at low percentages, appears inconsistent between sites (e.g., Belmont Bog and Nutt Lake) and I consider the problem later in a more general context.

The Alfies Lake site of Saarnisto (1975) and Liu's (1982) Nina Lake site provide the two northernmost records of vegetation change from the mixed forest zone. The initial vegetation was a mosaic of shrub tundra, white spruce, and poplar groves, and a scrub of *Shepherdia* and *Juniperus*. At about 9,500 yr BP, forests spread on to all mesic sites. *Picea glauca* dominated, associated with *Populus* and minor amounts of *Ulmus*, *Pinus*, and *Quercus*. At about 9,200 yr BP, *Larix*, *Betula papyrifera*, and later *Pinus banksiana* expanded and *Picea* decreased. The increase in *Alnus* at this time and the *Pinus banksiana* maximum probably indicate the occurrence of fire in the forests. *Pinus strobus* increased significantly at 7,400 yr BP. *Tsuga* and *Acer* spread into the area and, in the case of *Acer*, became significant components of the forest. The *Tsuga* pollen evidence is strengthened by the discovery of a macrofossil in a level dated at 7,500 yr BP. *Fagus* was present in small populations at about 6000 yr BP. At 6,000 yr BP, *Thuja* (macrofossil evidence to substantiate the otherwise equivocal pollen data) and *Betula alleghaniensis* expanded to replace white pine on mesic sites. From 4,000 yr BP to the present, *Picea*, *Pinus banksiana*, and *Betula papyrifera* increased. *Tsuga*, *Acer saccharum*, and *Fagus* became rare or absent. The uplands were dominated, as they are today, by *Pinus banksiana*, *Betula*, *Abies*, and *Populus*. The Nina record is similar to the sequence at *Jack Lake*, roughly 100 km to the north (Fig. 5.1, No. 61), in that both show a mid-Holocene (7,300 to 3,000 yr BP) expansion northward of the mixed conifer–hardwood forest communities, beyond their present-day limits (Liu and Lam 1985).

The effects of Indian agriculture on the maple–beech forest have been demonstrated skillfully by Boyko (1973) at the *Crawford Lake* site near Toronto (Fig. 5.1, No. 62), where annual sediment laminae made possible a precise chronology of pollen change (McAndrews and Boyko-Diakanow in press). The effects of three hundred years of Indian farming, from 1360 to 1660 AD, were to alter the forest by cutting and burning, thereby reducing the abundance of maple and beech. Abandoned fields were then reoccupied by successional species – red oak, poplar, and white pine. A similar explanation seems reasonable for the beech and hemlock curves at Nutt Lake (Fig. 5.4). A recent detailed analysis of sediments from two small lakes less than 2 km apart, adjacent to Georgian Bay, Lake Huron, has produced an informative reconstruction of the influence of Indian and European cultural activities on the local erosion (Burden, Norris, and McAndrews 1986) and vegetational history (Burden, McAndrews and Norris 1986) during the past six centuries (Fig. 5.1, No. 63). Several Indian village sites from the Huron occupation have been excavated near the lakes. Maple, beech, hemlock, and oak forests were cleared by Huron Indians between 1450 and 1650 AD in order to plant maize (*Zea*). A pattern of clearing, farming, and abandonment of fields is indicated in the pollen record by decreases in the forest dominants and increases in herbs, grasses, and bracken (*Pteridium*), the last named a reliable indicator of local abandonment. A general abandonment occurred about 1650 when the Huron Indians were displaced and increases in pine and oak pollen indicate local reforestation. Forest clearing by European farmers followed at 1850, indicated by sharp declines in arboreal pollen and increases in *Ambrosia*, *Rumex*, and *Plantago*. The *Ambrosia* rise is a well-established pollen stratigraphic marker in southeastern Canada, dated precisely by Boyko (1973) from the annually laminated sediments at Crawford Lake at 1846 to 1851. An interesting secondary outcome of the investigations by Burden, McAndrews, and Norris (1986) was the observation that the pollen sedimentation patterns of the two lakes differed markedly, apparently controlled by their different morphometric characteristics. Both inter- and intralake variations were observed in the sedimentation of several important pollen taxa.

Boreal region

Newfoundland

The details of the deglaciation of Newfoundland remain unknown, but it is accepted that the latest glaciation was represented by an ice cap that covered the island incompletely, and it has been suggested that the Burin Peninsula was ice-free in the full glacial (Brookes 1972; Grant 1977; Tucker and McCann 1980). The results reported by Anderson (1983) from a lake site in the Burin Peninsula and by Macpherson and Anderson (1985) from the Notre Dame site in the northeast provide evidence of the earliest vegetation to colonize the island. Their results indicate that from more than 13,200 until about 10,000 yr BP treeless vegetation of varying composition dominated the ice-free uplands, and shortly afterwards trees arrived.

The Sugarloaf Pond record from the Avalon Peninsula and the *Hawke Hills* site (Fig. 5.1, No. 64) from south of Conception Bay (Macpherson 1982) provide the basis for the following summary of the Holocene vegetational history of southern Newfoundland. The basal assemblage, with *Salix*, sedges, grasses, and such herb types as *Oxyria* and *Androsace*, led Macpherson to suggest a tundra vegetation roughly analogous to modern communities in central Baffin Island, and she compares the subsequent shrub pollen zone to the modern treeless heaths found on uplands in the Avalon Peninsula. *Picea*, arboreal *Betula*, *Abies*, and *Populus* were present in the area by 8,300 yr BP. Macpherson (1982, p. 190) suggests that the persistence of high percentages of shrub birch indicates that the community remained open for 3,000 years. The high frequency of charcoal reflects the frequent fires, favouring the persistence of extensive alder and birch communities. PAR values reach maximum values at 5,000 yr BP, and Macpherson (1982) proposes a closure of forest canopy at that time. The Hawke Hills site, inland south of Conception Bay, shows a reduction of PAR after 3,000 yr BP with increases in the relative amounts of dwarf birch, whereas the Sugarloaf site shows no change to the present day.

Quebec–Labrador

The Quebec–Labrador Peninsula was one of the last areas to be totally freed of Laurentide Ice, by about 6,000 yr BP. The description of the vegetational history is based on the records from sites in that area. The summary pollen diagrams in Figures 5.2 and 5.3, are useful points of reference while reading the following account.

A 100-km wide tract in the southeast was deglaciated by 10,000 yr BP (Fig. 2.2). A herbaceous tundra vegetation dominated mesic sites, with dwarf willow as the dominant shrub. After 9,500 yr BP, a dwarf shrub tundra replaced these initial herb tundras, for example, at Whitney's Gulch (Lamb 1980) and at Lake Hope Simpson (Engstrom and Hansen 1985). The shrub tundra was very similar to modern communities in Labrador–Ungava. As the ice receded from the south coast of Quebec, in the Sept-Iles region, the first plant cover to occupy the narrow deglaciated strip between the sea and the ice front was a shrub tundra with dwarf birch, *Salix*, and *Alnus crispa* (Mott 1976; King 1984). The subsequent changes record the arrival and expansion of tree species. As Engstrom and Hansen (1985, p. 550) note: "Sediments deposited between 8,000 and 6,000 yr BP record the transformation of shrub–tundra to conifer forest in southeastern Labrador. White spruce was the first tree to invade the region, arriving as early as 8,000 yr BP." Generally, the white-spruce communities were associated with nonarboreal elements, particularly dwarf birch and herbs, suggesting that spruce groves occupied lower slopes and valley floors with tundra on the uplands. At about 7,000 yr BP, *Abies* spread into southeast Labrador and southern Quebec, probably from Newfoundland in the former situation and from farther west in Quebec in the case of the Sept-Iles region. The ensuing increase in PAR of trees and decreases in alder, shrub birch, and herbs is interpreted by most investigators (e.g., Engstrom and Hansen 1985; King 1986) as a spread of closed spruce–fir forest on the landscape and a sharp reduction in the extent of tundra.

Abies, as the dominant tree, and *Picea glauca* formed the conifer forests that occupied most of the inland landscapes of southeast Labrador between 7,000 and 6,000 yr BP. Both Lamb (1980) and Engstrom and Hansen (1985) record a major change in forest composition at 6,500 yr BP when *Picea mariana* gradually replaced both *P. glauca* and *Abies*. From 6,000 yr BP to the present, *Picea mariana* dominated the pollen spectra (70 to 85%) and little change is observed in any of the several sequences from the region during the latter half of the Holocene. The modern vegetation of southeastern

Labrador – a closed forest of black spruce with associated balsam fir and white birch on mesic sites, *Larix* in mire habitats, open woodland or shrub tundra on exposed summits and along the coasts – has been in place for the last six millennia.

Farther north, various predictable changes occurred in the fossil record, both in timing and composition. King's (1984) transect of sites in the central region illustrates aspects of the variations. North of the modern boundary between the taiga and the forest–tundra, at about latitude 53°N, the initial *Picea* expansion occurred shortly after deglaciation (at 5,500 yr BP). *Abies* was never a significant tree in the landscape at this latitude (Coghill site of King 1984). The pollen and macrofossil record from the Delorme II site north of Lake Kaniapiscau, near the core of the final remnants of Laurentide Ice, showed that all the major species present in the region today were at the site from the beginning (Richard, Larouche, and Bouchard 1982). *Alnus crispa* and *Larix* formed the initial vegetation, followed at 5,500 yr BP by an expansion of *Picea mariana* to produce the maximum density of vegetation for the entire Holocene at this site. *Abies* appeared to expand slightly at this time, although then, as now, it was always very localized, probably near the northern limit of its range. At about 4,700 yr BP, the PAR values decreased, suggesting an opening of the vegetation cover with a decline in the amount of *Betula papyrifera*, essentially identical to the modern pattern of widespread shrubby tracts alternating with woody areas.

At the same latitude to the east of the Labrador Trough, the Gravel Ridge site of Lamb (1985) and the Ublik site of Short (1978) provide a representative portrayal of the sequence of vegetation from 10,000 to 6,000 yr BP, as the ice receded westward (Fig. 2.2), and thence to the present day, following complete glacier melting. It appears that the initial plant cover was sparse, probably discontinuous tundra with a predominance of grasses, sedges, herbs, and dwarf willows, not unlike the initial associations recorded elsewhere in eastern Canada immediately after deglaciation. Presumably the narrow strip of coastal landscape between the ice front and the sea provided an avenue for dispersal from southern stations. However, the question of source areas and routes of spread is reexamined in Chapter 8. Later, between 10,000 and 8,000 yr BP, dwarf birch and ericads increased, suggesting that a continuous shrub tundra developed, and at 6,700 yr BP, alder expanded to become locally common. *Picea* arrived at the site from the south at about 5,000 yr BP and increased, at first steadily, later rapidly to a maximum at about 4,000 yr BP. Later, changes in the vegetation at this coastal site comprise decreases in spruce and increases in birch, alder, willow, and sedges, indicating a decrease in tree cover and an expansion of tundra. The Gravel Ridge site in the forest–tundra zone (Fig. 5.5) shows an initial vegetation (6,400 to 5,750 yr BP) that Lamb (1985) suggests was a mosaic of species-rich communities, with scattered dwarf birch and alder, and locally continuous herb tundra with abundant grasses. By 5,700 yr BP, the shrub cover had increased and invasion by *Picea* had begun, involving both *P. glauca* and *P. mariana*, followed at 5,000 by *Abies* and *Larix*. A forest–tundra with *Picea mariana*, *Abies*, and *Larix* prevailed by 4,900 yr BP, with tree communities more widespread than today. The only subsequent change was a decrease in trees and a concomitant increase in shrub tundra cover, beginning at about 3,000 yr BP.

The northern portion of Quebec–Labrador remains incompletely known in its vegetational history, particularly on the west side towards Hudson Bay. However, the recent investigations reviewed above provide the basis for a preliminary attempt to reconstruct the past communities. What follows owes much to the detailed, knowledgeable discussion of all the sites from this area by Richard (1981a). The northernmost site (Diana 375) near Diana Bay, northwest Ungava Bay (Fig. 5.5, after Richard 1981a), was deglaciated at 7,000 yr BP, and the initial vegetation was a herb tundra with abundant grasses and sedges associated with *Oxyria*, Ranunculaceae, and Rosaceae, very similar to the herb tundra that occurs at the site today. This was followed at 6,000 yr BP, lasting until 3,500, by a dense forest–tundra with mesic sites occupied by continuous dwarf shrub communities with willow and dwarf birch and local alder stands, while upland sites continued to bear the herb tundra vegetation. Later (3,000 yr BP), the shrub tundra became more open and herb communities expanded. Farther south, at Rivière aux Feuilles (RAF II site, Richard 1981a, Fig. 5.5), in the modern forest–tundra zone, the initial vegetation occurred at 5,500 yr BP. A herb tundra persisted for a relatively short interval, being replaced by shrub tundra on mesic habitats, with dwarf birch and ericads, and local stands of *Alnus crispa* that progressively expanded in the landscape. At this time (4,800 yr BP), *Picea mariana* arrived, presumably from the southeast, and by 3,000 yr BP, the maximum development of arboreal vegetation had occurred, involving the spread and increase in density of black spruce and *Larix*. After 2,500 yr BP, the forest–tundra communities became less dense and the spruce and larch populations became restricted to their modern position on middle slopes, forming what Richard (1981a, p. 33) refers to as "une fôret-galerie ouverte" along drainage ways, at lake margins, and on the deepest soils.

Finally, our knowledge of the vegetational history of the western half of northern Quebec is based on a group of sites in the area of the Kaniapiscau River (Fig. 5.1 No. 65), and of these the *Bereziuk* site (Richard 1981a) reveals that, following ice retreat and the recession of the Tyrrell Sea about 7,000 yr BP, a herb tundra vegetation colonized the landscape followed shortly after by an immigration of *Populus tremuloides* and *Juniperus*, with abundant *Salix* and herbs. About 6,000 yr BP, *Picea* arrived and quickly expanded to establish an open taiga type of boreal forest, with local dominance of *Alnus crispa* and *Betula glandulosa*. At about 2,700 yr BP, declining pollen concentrations of *Picea* indicate a tendency for treeless scrub to expand at the expense of forest.

Northern Ontario and adjacent western Quebec

The Crates Lake site of Liu (1982), despite the problems with the ^{14}C ages at certain levels, illustrates the rather simple history of the modern boreal forest on the extensive till plains and rolling uplands of northern Ontario. A short-lived phase, when trees had not yet colonized the region, at approximately 8,000 yr BP, consisted of a predominantly herbaceous community, except for *Salix*. At about 7,500 yr BP, *Populus*, *Picea* (both species), *Pinus* (*banksiana/resinosa*), and *Betula papyrifera* occupied upland sites, and *Fraxinus nigra*, *Larix*, and scattered *Ulmus* colonized hydric habitats. At roughly 7,000 yr BP, *Thuja* and *Pinus strobus* became members of the forest vegetation, although restricted to local habitats. *Abies*, which had been present at low population densities from the beginning of forest development, expanded at 4,000 yr BP, as did *Betula* and both species of *Picea*, and these taxa remained as the dominants until the present day. Sporadic occurrences of *Fagus*, *Tsuga*, and *Acer* pollen at very low frequencies can probably be ascribed to transport from the south.

The Yelle site (Richard 1980a) in the southern region of the boreal forest, at the Quebec–Ontario boundary immediately to the south of Lake Abitibi, has yielded a vegetation sequence similar to that at Nina Lake. The lowest inorganic sediments below the dated 8,900-yr-BP-level contain high amounts (50%) of *Pinus banksiana*, with a mixture of *Ulmus*, *Quercus*, and *Picea*, and are probably not representative of the first vegetation as glacial Lake Ojibway was extensive in the region and Lac Yelle probably occupied a small island on the lake. Organic sedimentation began at 8,900 yr BP, and the first vegetation was a boreal forest assemblage with abundant *Populus* and a mixed community of *Pinus banksiana*, *Picea mariana*, and *Betula papyrifera*. At roughly 7,000 yr BP, *Pinus strobus* expanded into the area at low population densities, followed by *Thuja* (or *Juniperus*, if Richard's interpretation is followed). However, by 5,600 yr BP, their frequencies decreased and by 3,500 yr BP, the modern configuration of boreal vegetation was regionally stable, although no doubt undergoing local perturbations in response to natural fires. It is likely that very small populations of *Fagus*, *Tsuga*, and *Acer saccharum* were present with white pine in the mid-Holocene (6,000 to 5,000 yr BP) as their pollen frequencies, while very low (0.5%), were continuous in samples from that period. Macrofossil evidence of the northern extension of white pine, beyond its present range in Ontario, was presented earlier by Terasmae and Anderson (1970), from a site 75 km north-northeast of the Yelle site.

Arctic

High arctic As only a single detailed pollen record is available from the high arctic, sites from northwest Greenland are included.

Rock Basin Lake, near Baird Inlet, on East Ellesmere Island (Fig. 5.1, No. 66) was investigated by Hyvärinen (1985). The lake is small (3.7 ha) and deep (14 m), situated in a steep-sided bedrock basin with closed drainage. It is surrounded today by valley glaciers. The modern vegetation is tundra, with locally relatively productive heath communities on south-facing, sheltered slopes where *Empetrum*, *Cassiope tetragona*, and *Vaccinium uliginosum* are common. *Dryas* is frequent on upland sites, associated with various arctic herbs, grasses, and sedges. Organic, polliniferous sediment began to accumulate in the basin shortly after 9,000 yr BP and a total thickness of 51 cm was cored, spanning the entire nine millennia of Holocene time. Pollen concentration varied between 100 and 1,500 grains $\cdot$ cm^{-3}, but Hyvärinen (1985, p. 27) cautions that the close correlation between pollen deposition rate and sediment type implies that the in-washed, allochthonous pollen component has been both large and variable. The percentage diagram (Fig. 5.5) provides a sensitive record of vegetation change. The first pollen taxa from local sources were Gramineae, Cyperaceae, *Oxyria* (up to 70%) and *Papaver*, and Hyvärinen (1985, pp. 28–9) proposes that a pioneer vegetation dominated by herbs, grasses and sedge occupied the deglaciated local surfaces and persisted for about one thousand years. Then *Salix*, probably *Salix arctica*, expanded rapidly, followed shortly after by Ericaceae, interpreted by Hyvärinen as a partial replacement of the herb tundra by a heath tundra cover that persisted for about 3,000 yr, representing the "maximum density and extent of plant cover in the site history" (1985, p. 20). Several herb taxa (Caryophyllaceae, *Dryas*, *Papaver*, *Ranunculus*, and *Saxifraga oppositifolia*) increase slightly towards the top of the section, and Hyvärinen suggests an increase in open, herb tundra at the expense of heath tundra.

Palaeoecological interpretation of the record is best conveyed directly by Hyvärinen: "In terms of climatic interpretations, the early vegetational succession at Baird Inlet is uninformative; the *Salix* expansion is mainly a regional migration event, and the heath expansion may be simply explained as a local edaphic event. The only safe conclusion is that around 8,000 yr BP at the latest the climatic conditions were favourable for the immigration of *Salix* and Ericales" (Hyvärinen 1985, p. 29). He goes on to suggest that the slight increases in "pioneer herbs" following 4,000 yr BP are "retrogressive" and "climatically significant," indicating a cooling trend. An investigation of the algal flora of the same sediments, by Smol (1983), corroborates these conclusions; productivity decreased markedly at 3,500 yr BP and floristic changes in the diatoms are accounted for by increases in the persistence of lake ice in summer.

Mid-arctic A 1.9-m core from *Patricia Bay Lake*, a small (4 ha) lake in the Clyde Foreland of Baffin Island (Fig. 5.1, No. 67) yielded an informative record (Mode 1980; Short, Mode, and Davis 1985), although the 8,810 ± 205 yr BP date at 85 to 90 cm is rejected because it overlies a younger date, on a moss bed, of 6,320 ± 130. The percentage diagram is probably more reliable than the concentration values, for the reasons referred to by Hyvärinen in his discussion of the previous site. The lower levels, prior to the marine inundation reflected in the unit of marine sediment at 65 to 75 cm, have relatively high *Betula*, Gramineae, *Oxyria*, Cyperaceae, and *Lycopodium* (*clavatum*) values. *Salix*, *Oxyria*, and Cyperaceae increase towards the surface, whereas *Betula*, Ericaceae, and Gramineae decline variably. The modern vegetation is typically mid-arctic with herb tundra on uplands and with a significant dwarf willow component (*Salix arctica*, *S. herbacea*). The *Picea* and *Alnus* curves are interesting as they reach a maximum (4,000 yr BP) when spruce and alder were advancing northwards in Labrador and first reached the south end of Ungava Bay (Richard 1981a). The northern limit of dwarf birch on Baffin Island is 450 km north of Patricia Bay Lake, so this record indicates an early to mid-Holocene northward extension, presumably in response to a warmer climate. The change at or after 4,000 yr BP corresponds to the record from Baird Inlet (above) and Short, Mode, and Davis (1985) offer a similar explanation in terms of a cooler climate. The lowest spectra, with *Betula* (>20%), Gramineae (>20%), Ericales (5%) and *Salix*, (5%) are quite similar to modern spectra from the low- to mid-arctic transition except for the high values for *Lycopodium clavatum*.

The vegetation change indicated by this diagram is from shrub tundra in more sheltered sites, dominated by ericads and dwarf birch from 6,500 to 4,200 yr BP, to a herb tundra with abundant *Salix* from 3,000 yr BP to the present. Short, Mode, and Davis (1985) ascribe these changes to climate, with cooler conditions, in particular with colder summers and more extensive persistent snow cover beginning at 3,000 yr BP.

Low arctic On the east side of Cumberland Sound, Baffin Island, *Iglutalik Lake* (Fig. 5.1, No. 68), was investigated by Davis (1980) and reported on in summary form by Davis, Nichols, and Andrews (1980) and Short, Mode, and Davis (1985). A 2.9-m core yielded a detailed pollen diagram with five radiocarbon estimates. The lower two-thirds of the diagram has high values for Cyperaceae, Ericales, and Gramineae, with a large unexplained peak of Filicales (40%) at 240 to 250 cm (8,000 yr BP). The erratic concentration curves in the upper levels (Short, Mode, and Davis 1985, Fig. 22.6b) relate directly to major but short-lived changes in the sediments. At about 4,000 yr BP, *Salix* increases steadily to maximum values (20%). It is difficult to suggest an interpretation in terms of local vegetation change from the available record, but it would be interesting to recalculate the percentages excluding exotic taxa (*Picea*, *Alnus*, *Pinus*, and thermophilous trees).

Greenland It is useful to examine briefly the main findings from the detailed analyses of many sites in West Greenland conveniently summarized by Fredskild (1983). Three summary pollen diagrams are shown to illustrate the broad features of this rich record, but they are described briefly as the complete details are available in the original publications (Fredskild 1969, 1973, 1983, 1985; Fig. 5.5). The northernmost site is at *Klaresø* (Fig. 5.1, No. 69), where a roughly 5,000-year record was recovered from a small kettle that emerged from marine inundation immediately prior to the beginning of organic sedimentation at 4,950 ± 140. The modern vegetation is a discontinuous herb tundra with extensive areas of bare ground. Fredskild (1973) mapped sample areas round the lake and found the following percentage cover values in the only 2.8 percent of the total land area with any plant cover: Cyperaceae (38%); Gramineae (28%); *Dryas* (6.8%); *Salix*, the only shrub (6.1%); *Oxyria* (2.5%); and *Saxifraga oppositifolia* (0.4%). The dominant sedges and grasses, which presumably contribute most of the modern pollen in the uppermost samples, are *Eriophorum triste*, *Phippsia*, *Alopecurus*, *Colpodium*, and *Pleuropogon*. The Klaresø diagram (Fig. 5.5) shows an initial period of Gramineae–Cyperaceae–*Oxyria* dominance with relatively high exotic values (*Betula*, *Alnus*, *Pinus*, and others) followed at about 4,500 yr BP by an increase in *Salix*. The percentage diagram shows little change to the present day, but Fredskild (1973) notes a period from 4,000 to 2,100 yr BP with a maximum pollen influx and a maximum growth of *Calliergon trifarium* and other mosses in the lake that he ascribes to a period of more humid climate than before or at present. The *Johannes Iversen Sø* site (Fig. 5.1, No. 70) was originally analyzed by Iversen (1952–3), published in more detail by Fredskild (1973), and then recored for macrofossil, supplementary pollen, and radiocarbon analysis (nine levels). The site is of relevance to the present discussion because it is in the low arctic, with a continuous cover of dwarf-shrub heaths on upland sites, dominated by *Betula*, *Ledum*, *Vaccinium*, *Empetrum*, *Phyllodoce*, and *Loiseleuria*; *Juniperus communis* occurs on dry, rocky sites; and *Alnus crispa* and *Artemisia borealis* occur rarely, here at their eastern range limits. The diagram (Fig. 5.5) shows an initial herb–grass–sedge assemblage between 9,000 and roughly 7,500 yr BP, followed by successive increases of *Salix*, *Juniperus*, *Betula*, and finally *Alnus* at about 3,500 yr BP. Alder declines steadily to the present day after its early peak at 3,000 yr BP, and *Betula* and heaths increase. The macrofossil records confirm the presence and stratigraphic distribution of many taxa, including *Juniperus* and *Betula*. The *Klœftsø* site (Fig. 5.1, No. 71) is a long narrow pond (200 × 20 m) in a rocky landscape covered by a dwarf-shrub heath dominated by *Empetrum* and *Betula glandulosa*. It is outside the range of *Alnus*. The diagram (Fig. 5.5) shows the basic pattern – initial dominance of *Oxyria* after deglaciation at about 9,400 yr BP; immigration and spread of heath plants, followed later by *Salix* (at 7,500 yr BP), *Juniperus* (also recorded as macrofossils), and finally (at 4,000 yr BP) an immigration and expansion of *Betula glandulosa*, also confirmed by macrofossils.

The palaeoclimatic and phytogeographic implications of these records from both the Canadian and Greenland sectors of the arctic are examined in Chapter 8.

Treeline changes in Nouveau Québec Although the regional boundary between arctic tundra and the northern limit of living trees of arboreal habit, i.e., the treeline as defined by Payette (1983), is apparently determined by such macroclimatic characteristics as summer thermal conditions and winter snowfall (Hare and Ritchie 1972), the local pattern of treeline vegetation is a response to one of several possible influences. They can be illustrated by reviewing a number of detailed case studies completed by Payette and his colleagues along the treeline of northern Quebec that show that "the structure of the modern forest–tundra zone is the result of long-term tree regeneration dynamically related to Holocene climatic changes and to fire history." (Payette and Gagnon 1985, p. 2). The treeline of northern Quebec is one of the most recent to be established. Spruces, the chief treeline species, arrived from the southeast, and possibly from the southwest, along the east coast of Hudson Bay, as late as 4,000 yr BP (Richard 1981a) after the final residuum of Laurentide Ice melted from the central uplands south of Ungava Bay.

It has also been noted that pollen analysis, particularly of lake sediments, is not an effective tool for detecting small-scale changes. Experience has shown that while major advances and recessions of treeline in response to early Holocene summer warming and subsequent cooling have been demonstrated using detailed pollen accumulation data (Spear 1983), the small climatic changes that have occurred in the northern hemisphere at higher latitudes during the late Holocene are detectable in the pollen record only with difficulty (Lamb 1985) or by considerable acts of faith (Nichols 1976). Payette and his co-workers have used tree-ring analyses, fire history chronologies, and population dynamics to decipher the nature and controls of late-Holocene treeline changes in northern Quebec.

One of the central ideas being tested by these investigations is that "major periods of fire are climate-controlled, and depending on their occurrence during warm or cold climatic intervals, act selectively on tree or forest regeneration" (Payette 1980, p. 127). An analysis of radiocarbon-dated charcoal from 116 sites in the forest tundra of northern Quebec provided one test of the idea (Payette and Gagnon 1985). The results showed that a process of fire-induced deforestation began roughly 3,000 years ago, when *Picea mariana* and *Larix laricina* were eliminated from large areas by their failure to regenerate after fire. "From 3,000 BP to the present the expansion of treeless vegetation has primarily been the consequence of krummholz and forest burns . . . with the former krummholz being possibly the outcome of ancient lichen woodland that failed to regenerate as trees because of the onset of cooling" (Payette and Gagnon 1985, p. 2). The analysis also showed that the onset of treelessness was metachronous and that the process of increasing the area of tundra is not related to distance from the current treeline. The rate of expansion of tundra was maximal at 3,000 to 2,100, 1,800 to 1,500, 800 to 650, and 450 to about 100 (all yr BP).

The well-documented changes in mean annual temperature at high latitudes in eastern North America (Kelly et al. 1982) provide an opportunity to study vegetation response at sensitive locations. Payette and Filion (1985) describe a detailed analysis of a stand of *Picea glauca* at its northern limit in the

Figure 5.6. The age structure of a large, uneven-aged *Picea glauca* stand near the forest limit at 56°25′N in northeast Quebec, after Payette and Filion (1985). This stand illustrates a more general pattern of increased regeneration from 1880 to 1910 followed by a maximum regeneration between 1920 and 1960.

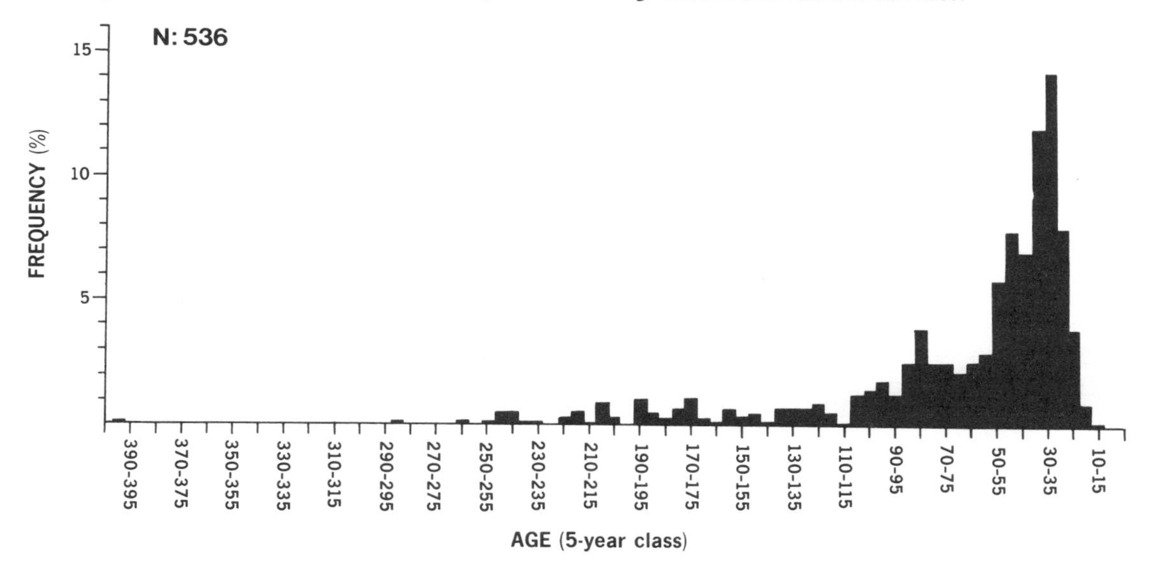

Richmond Gulf area of northeastern Quebec. They examined in detail the age structure of populations at the treeline and at the forest limit. A protected, forest-limit stand of uneven-aged trees provided 536 individuals of which 10 were too rotted for accuracy and 130 were too small to core. The age structure (Fig. 5.6) shows that a surge of recruitment occurred during the past 100 years, particularly during the period 1890 to 1910 AD. A second analysis of trees between the forest limit and treeline site produced 590 cored trees of which the vast majority were established after 1880. The main response to the twentieth-century climate warming, culminating in 1930 with cooling after 1960, was an increase of tree density, but only a small change in the position of treeline.

Another investigation, of the living and dead trunks of a rare old stand of *Picea mariana* lichen woodland near the present treeline in northern Quebec (at Bush Lake) that had escaped fire for about 900 years, produced a remarkable 1398 to 1922 AD tree-ring chronology (Payette et al. 1985). Living trees at the site are erect, about 5 m tall, whereas dead trunks were shorter, often oblique "with multiple stem leaders at a former snow–air interface level" (Payette et al. 1985, p. 135). Reproduction by layering occurred continuously over the past 600 years, but individuals established by seed (16% of the total) were established mainly between 1890 and 1910. An episode of mass mortality occurred between 1880 and 1890 and 70 percent of the trees to die then were older than 200 years and 30 percent were older than 300 years. Tree ring analysis shows a low-growth period during the Little Ice Age (1550 to 1850) and maximum values in the 1930 to 1940 period. The authors conclude that the stand responded to the Little Ice Age cooling by krummholz formation and repressed or episodic seed regeneration. The shift from krummholz to arboreal form has been a widespread response in treeline black spruce in northern Quebec to the twentieth-century warming, but very old trees (200 years toward the end of the Little Ice Age) were unable to respond to the climatic change, but, perhaps because of "limited photosynthetic mass, were already engaged in a lengthy process of dying." (Payette et al. 1985, p. 138). The authors suggest that this remarkable analysis "illustrates the importance of vegetation inertia, in a fire-free ambience, for persistence of communities in cold environments. . . . "

6 The Western Interior

The southern portion of the region included here (Fig. 6.1) is differentiated naturally from the adjacent Pacific–Cordilleran region to the west by the major, abrupt topographic break delimited by the lower Rocky Mountain foothills. To the east, the boundary is the western limit of the eastern temperate zone, effectively at the southern extremity of the Manitoba–Ontario border, where such wide-ranging temperate conifers typical of the eastern mixed forests as *Pinus strobus*, *P. resinosa*, and *Thuja occidentalis* have their western limits. Farther north, the eastern boundary of the Western Interior is an arbitrarily chosen line that coincides with Hudson Bay, thus bisecting the transcontinental boreal forest into eastern and western segments. However, as noted above, there is actually a gradual transition westward from Quebec to northern Alberta along gradients of decreasing annual precipitation and increasing continentality, accompanied by some floristic transitions such as the decreasing abundance and final disappearance of *Abies balsamea* (Fig. 1.3). The northern boundary is the limit of forest, thus leaving as a separate region the arctic and its southern treeline.

This chapter is short because the fossil pollen record and its interpretation have been reviewed recently (Ritchie 1976, 1983, 1985a; Liu 1980; Ashworth and Cvancara 1983; Webb, Cushing, and Wright 1983). A few new sites have been reported quite recently, so emphasis is placed on their results and implications. The locations of all sites considered here, either in some detail, including a summary diagram, or in passing, are shown in Figure 6.1. They fall readily into three main patterns that coincide roughly with their latitudinal positions – those near the modern forest–grassland transition, those near the modern forest–tundra transition, and those placed centrally in the boreal forest.

Sites near the forest–grassland transition

The *Pickerel Lake* site is in northeast South Dakota (Fig. 6.1, No. 1) on a glacial drift upland, the Coteau des Prairies. While the site is in the grassland zone, lying roughly 120 km west of the grassland–forest ecotone, the steep slopes and gullies surrounding the lake support deciduous woodlands, as reported by Watts and Bright (1968) who investigated the palaeoecology of the site from analysis of a 10-m core of sediment. The regional vegetation is a prairie, greatly altered by agriculture, but dominated in the few undisturbed residual patches by *Agro-*

Figure 6.1. A map of the Western Interior of Canada and adjacent United States showing the locations of the fossil sites referred to in the chapter.

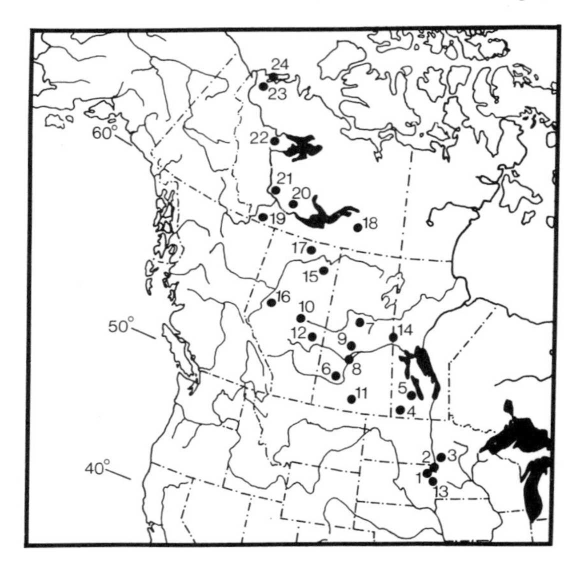

Figure 6.2. Summary percentage pollen diagrams for the sites indicated. In these, as in all subsequent composite pollen diagrams, the pollen percentages are plotted against a radiocarbon age scale. In preparing the diagrams from originals drawn on a depth scale, uniform sedimentation rate was assumed between dated levels and, where necessary, pollen curves were stretched or compressed to fit a constant time scale. The pollen sum in all diagrams includes all terrestrial taxa. Note occasional percentage scale changes used to improve visibility of changes in taxa with low percentage values (e.g., *Abies*, *Acer*). The details of the original investigations are given in the text, including the literature citations; locations are on Figure 6.1.

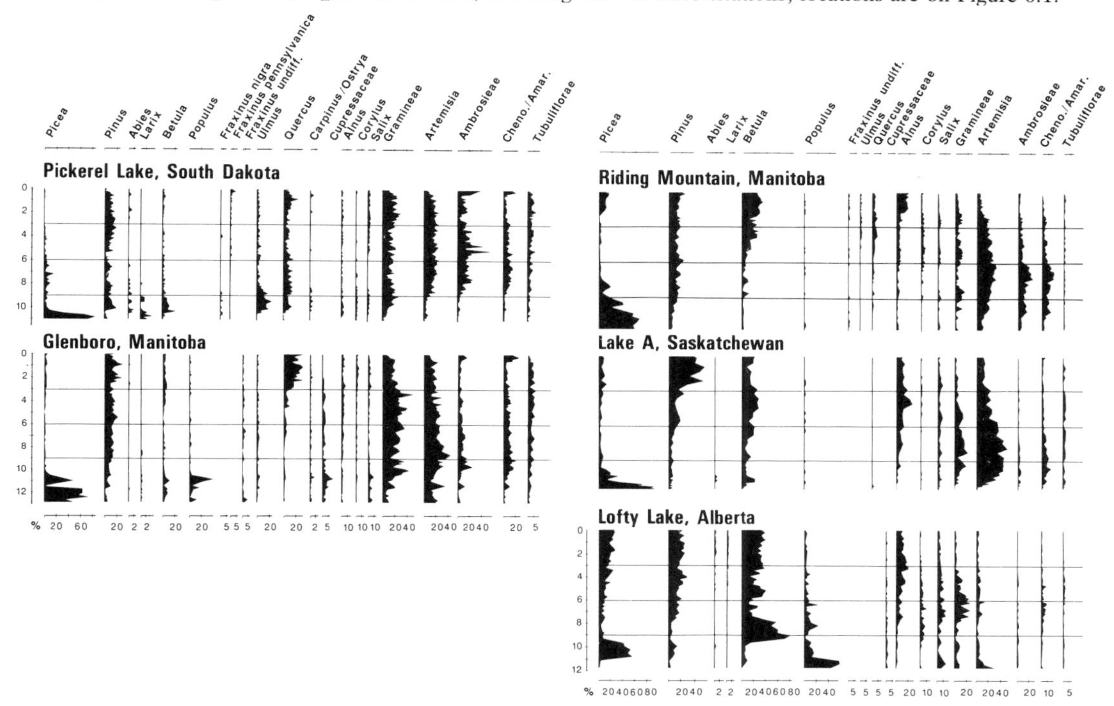

pyron, *Koeleria*, and *Stipa* – the tall bluegrass prairie zone. The local woodlands in sheltered habitats consist of *Quercus macrocarpa*, *Fraxinus pennsylvanica*, and *Tilia americana*. *Ostrya virginiana*, *Ulmus americana*, and *Acer negundo* occur in small local stands.

The pollen diagram (Fig. 6.2) has been described in detail by the original investigators and in summary form by Wright (1970). The lowest levels show a dominance of *Picea* associated with *Larix*, *Fraxinus nigra*, and small amounts of *Artemisia*. Macrofossil analysis of the lowest levels yielded *Picea* and *Larix* needles as well as fruits of the boreal taxa *Cornus canadensis* and *Rubus pubescens*. The spruce percentages decline steeply to less than 10 percent at 10,670 ± 140 yr BP and increases in *Betula*, *Pinus*, *Ulmus*, *Abies*, and *Quercus* follow immediately. Macrofossils of *Betula papyrifera* at this level (9.4 m) corroborate the pollen evidence that the tree was present, but the pine component is almost certainly derived by long-distance transport as its percentages in immediately presettlement spectra are as high as in the early Holocene. At about 9,400 yr BP, all the tree pollen percentages decrease to negligible values, except pine and oak. *Artemisia*, Ambrosieae, Gramineae, and Chenopodiineae increase, and a few taxa indicative of prairie appear – *Petalostemum*, *Amorpha*, and *Iva*. The macrofossil record from these levels (9,400 to about 3,000 yr BP) includes several typical prairie species. Toward the top of the section, at an estimated 3,000 yr BP, the modern presettlement spectra are established by the increases to low but continuous levels of pollen of *Fraxinus* (cf. *pennsylvanica*), *Ostrya*, and *Tilia*, with corresponding decreases in NAP. The uppermost levels register a rise in Ambrosieae pollen, typical of all pollen diagrams from the Western Interior, signalling the arrival of European immigrant farmers. A recent, so far unpublished thesis by Radle (1981) describes the pollen sequence (percentage and PAR) and sulphate concentration in sediments from *Medicine Lake* (Fig. 6.1, No. 2), roughly 35 km from Pickerel Lake. The sedimentological analyses permit some elaboration of the palaeoenvironmental record, showing that the early Holocene period (9,260 to roughly 3,500 yr BP) was characterized by a rapid expansion of prairie, increased salinity, and lower lake levels.

Thompson Pond, Minnesota (Fig. 6.1, No. 3), roughly 200 km north northeast of Pickerel Lake, lies on the prairie side of the grassland–forest boundary zone, and a complete pollen diagram was prepared from its sediments by McAndrews (1966). Although it lacks any radiocarbon dates, correlation with adjacent dated sites provides a rough chronology (Ashworth and Cvancara 1983). The record indicates that an early spruce assemblage, with significant larch, poplar, birch, and elm components, first occupied the site at about 11,000 yr BP and was replaced by a grassland assemblage after a short-lived, somewhat enigmatic, *Pinus–Pteridium* assemblage. The grassland assemblage prevails to the present day, with the addition of *Ostrya–Carpinus* and *Ulmus* in the late Holocene, attributed mainly to long-distance transport (Ashworth and Cvancara 1983, p. 61).

The *Glenboro* site in southern Manitoba remains the only detailed record from the eastern Canadian prairies (Fig. 6.1, No. 4). A 10-m core of sediment with five radiocarbon dates yielded a reasonably detailed record, confirmed by records from two adjacent sites in the same landscape region, the Tiger Hills upland of south Manitoba (Ritchie and Lichti-Federovich 1968). The area is an upland of irregular, morainic topography surrounded by lower, flat, glacial lake deposits. The vegetation of the Tiger Hills is a deciduous woodland with abundant *Populus tremuloides*, *Quercus macrocarpa*, and *Betula papyrifera* on uplands, and *Ulmus americana*, *Fraxinus pennsylvanica*, and *Acer negundo* on lower slopes and bottomlands. The pollen diagram (Fig. 6.2) shows an initial *Picea*-dominated assemblage very similar to that from Pickerel Lake. The radiocarbon chronology is imprecise, but it is likely that the spruce-dominated assemblage persisted from about 12,000 to 10,000 yr BP, as it did elsewhere in southern Saskatchewan, southern Manitoba, and the adjacent United States (see an excellent earlier review by Shay 1967; and recent reviews by Ashworth and Cvancara 1983, and Ritchie 1983). The pollen spectra from the Glenboro site include significant quantities of *Betula*, *Populus*, *Ulmus*, *Fraxinus*, *Carpinus–Ostrya*, *Salix*, *Juniperus*, Gramineae, and *Artemisia*, similar to the Pickerel Lake assemblage to the south and the Riding Mountain assemblage to the immediate north, but differing from late-glacial *Picea*-dominated spectra from sites farther west in that they have greatly reduced, or no representation of, deciduous tree taxa such as *Ulmus*, *Quercus*, and *Carpinus–Ostrya*.

Toward the end of the *Picea*-dominated period and immediately prior to a long early to mid-Holocene time of NAP domination, *Populus*, *Ulmus*, and *Carpinus–Ostrya* were common, with significant amounts of *Salix*, *Juniperus*, and *Shepherdia argentea* (the last named is not shown on Fig. 6.2, but can be seen on the original diagrams: Ritchie and Lichti-Federovich 1968, Figs. 2 and 3). From about 9,500 to 3,000 yr BP, nonarboreal taxa make up 60 to 70 percent of the spectra, dominated by *Artemisia*, Gramineae, Ambrosieae, and Chenopodiineae, with significant percentages (about 2%, equal to their modern values) of *Fraxinus*, *Betula*, and *Ulmus*. The only important change in the spectra, apart from the postsettlement increases in Ambrosieae and Chenopodiineae, is an increase in *Quercus* at about 3,000 yr BP from less than 5 percent to its modern values of 20 to 25 percent.

A detailed, well-dated record from a site in a topographically similar upland, roughly 100 km to the north, provides an interesting, informative comparison with both the Pickerel Lake and Glenboro sites. The *Riding Mountain* site (Fig. 6.1, No. 5) is in a Tertiary upland, capped by morainic and outwash glacial deposits and surrounded by lowlands of predominantly Glacial Lake Agassiz sediments (Teller et al. 1983). It is higher than the Tiger Hills area (720 vs 420 m) and a little farther north, so the slightly cooler and moister macroclimate supports a mixed conifer-deciduous forest, and Riding Mountain is classified as an outlier of the southern boreal forest (Rowe 1972). Upland mesic till soils support a mixed *Picea glauca–Populus tremuloides–Betula papyrifera* forest. *Abies balsamea* is infrequent, on local alluvial soils, and *Pinus banksiana* is rare, confined to outwash sands. Poorly drained, peaty soils are occupied by *Picea mariana–Larix* communities, and steep, south-facing gully slopes have local *Quercus macrocarpa* savannas. The pollen record (Fig. 6.2) has been published and described rather often, in the continued absence of a new record, so it will be summarized briefly. The early levels show a *Picea*-dominated assemblage, with macrofossils of *Picea glauca* (Ritchie 1964, Figs. 9 and 10), but it differs from the Glenboro record in that the deciduous tree element (*Ulmus*, *Ostrya*, *Fraxinus*) is less abundant though present (Ritchie 1969, PAR diagram). With no transitional stage of deciduous tree pollen abundance, the *Picea* assemblage is replaced at about 9,000 yr BP by an NAP assemblage that contains, in addition to the obviously dominant taxa, a few grassland elements in low frequencies – *Shepherdia argentea*, *Phlox hoodii*, *Collomia linearis*, and *Symphoricarpos*. In the mid-Holocene, at about 5,000 yr BP, *Quercus* and *Betula* increase slowly until, by 3,000 yr BP, they reach their modern values of 5 and 25 percent, respectively. At 2,500 yr BP, *Picea* and *Alnus* increase from negligible amounts to their modern proportions (20 and 12%, respectively).

A south to north transect of sites in central

Saskatchewan investigated by Mott (1973) provides a detailed record of the prairie–forest zone. One, *Clearwater Lake* (Fig. 6.1, No. 6), is in the heart of the grassland 200 km south of the prairie–woodland border, and it shows (Mott 1973, Fig. 2; and Ritchie 1976, a summary diagram in Fig. 1b) that the modern prairie has existed without major changes from at least 7,580 to the present. At the northern extremity of the transect, the *Cycloid Lake* diagram, from a site in the boreal forest 200 km north of the prairie–woodland transition (Fig. 6.1, No. 7), demonstrates that grassland has never extended as far north as that site during the Holocene (Mott 1973, Fig. 5; and Ritchie 1976, a summary version in Fig. 2a). *Lake A* (Fig. 6.1, No. 8) is within, but at the southern boundary of, the boreal forest in a position comparable to the Riding Mountain site. A summary version of the original diagram is included here (Fig. 6.2), although only a single, basal radiocarbon date is available. The modern vegetation consists of "forests of white spruce (*Picea glauca*), trembling aspen, balsam poplar (*Populus balsamifera*), white birch (*Betula papyrifera*), black spruce (*Picea mariana*), and some tamarack (*Larix laricina*) and balsam fir (*Abies balsamea*)" (Mott 1973, p. 2). The diagram is similar to that from Riding Mountain with a few interesting exceptions – it has no significant registration of pollen of *Quercus*, *Ulmus*, *Fraxinus*, or *Carpinus–Ostrya*, all trees whose western limit lies to the east and south of Lake A, but all of which occur in Riding Mountain. The early spruce assemblage is followed immediately by a grassland assemblage. Later, in the mid-Holocene, estimated here at about 6,000 yr BP, assuming a uniform rate of sedimentation, *Betula*, *Alnus*, and *Salix* increase significantly. At 2,500 yr BP, *Pinus* and *Picea* increase to establish the modern pollen assemblage. Both *Abies* and *Populus* are vastly underrepresented in the modern spectra at this site as elsewhere in the Western Interior.

Mott's (1973) northernmost site in this group is *Lake B* (Fig. 6.1, No. 9), about 80 km north of Lake A, well inside the mixed boreal forest, and its mid-Holocene record is intermediate between those of Lake A and Cycloid in that, following the initial spruce zone, herb pollen totals rarely exceed 50 percent, whereas *Betula* (10 to 20%), *Pinus* (20%), and *Alnus* (10%) make up most of the other half of the spectrum. In the late Holocene, the herbs decrease to modern values (total 15%), while *Picea* and *Pinus* increase to 20 and 40 percent, respectively.

The final site along the grassland–forest margin is the familiar *Lofty Lake* site in central Alberta, investigated by Lichti-Federovich (1970). The site is a kettle lake on glacial morainic uplands between the Athabasca River and the northern Saskatchewan River (Fig. 6.1, No. 10). It lies roughly 50 km north of the boundary between the mixed boreal forest and the aspen parkland zone. The modern vegetation is mixed forest with abundant *Populus tremuloides*, *Betula papyrifera*, and *Picea glauca* on upland tills. *Pinus banksiana* grows on sandy soils and *Picea mariana* and *Larix laricina* on fens and bogs. *Abies balsamea* occurs in the general area, particularly in old stands with *Picea glauca*, and it is noteworthy that it is near its western limit here. A 5.5-m section of sediment was dated at five levels, including the Mazama ash layer, and was pollen analyzed at close intervals.

The pollen record (Fig. 6.2 here and Fig. 2 in Lichti-Federovich 1970) is similar to Lake B above in that, between the initial *Picea–Populus*-dominated late-glacial assemblage (11,400 to 9,180 yr BP) and the late-Holocene (3,440 yr BP) establishment of the modern mixed boreal forest, there is a *Betula*-dominated assemblage then one with high NAP totaling 50 percent associated with birch, alder, and willow. An interesting feature of the record, not found in many others, is an episode of high *Populus* percentages (>40%) in the three oldest levels, presumably representing the first terrestrial vegetation following deglaciation. Also of interest is the *Abies* curve in the late Holocene.

Before considering the possible vegetational history of this southern segment of the interior, ancillary data from other sites are examined. The record, while often incomplete, supplements the basic outline that emerges from the five primary sites reviewed above.

A buried unit of laminated gyttja in the Missouri Coteau discovered, excavated, and partially analyzed (plant macrofossils) by de Vries, yielded a highly informative record of late-glacial vegetation (Ritchie and de Vries 1964; de Vries and Bird 1965). The *Hafichuk* site (Fig. 6.1, No. 11) is in the central grassland zone, and the original discovery of buried organic mud occurred during well drilling in a kettle depression. A subsequent pit excavation revealed over 1 m of continuously laminated lake mud with abundant macrofossils. The laminae dip obliquely, suggesting en bloc displacement after deposition, and, although it was not realized at the time of the investigation, the diatom analysis and interpretation of a similar deposit in Minnesota by Florin and Wright (1969) provide the likely explanation that the sediments were deposited in a supraglacial lake during the late-glacial stagnation of the Laurentide Ice, and subsequent melting resulted in the collapse and burial of the sediments. The unit spanned 11,650 to 10,000 yr BP and has a typical late-glacial *Picea*-dominated pollen assemblage with rare grains of *Abies* and frequent *Populus*, *Fraxinus*, and *Shepherdia canadensis*. However, the bryophyte and vascular plant records provide the details that secure the vegetation reconstruction. Sixty-one taxa of vascular plants were recovered, including abundant *Picea glauca*, *P. mariana*, *Populus tremuloides*, *P. balsamifera*, and *P. grandidentata*. Sixteen moss taxa were recovered (de Vries and Bird 1965) and they, and all the vascular plant macrofossils, are plants typical of the modern, closed boreal forest.

Elsewhere, truncated records from sites in south-central Saskatchewan and Manitoba reported by Kupsch (1960), Ritchie (1976), Mott and Christiansen (1981), Last and Schweyen (1985), and Terasme (unpublished) provide supplementary material, and the *Alpen Siding* site (Fig. 6.1, No. 12) reported by Lichti-Federovich (1972) confirms the Lofty Lake sequences in Alberta.

The inferred vegetation history of the Pickerel Lake site, following Watts and Bright's (1968, p. 871) tabular summary, is that the initial plant cover of the uplands was a coniferous, predominantly spruce forest with both *Picea glauca* and *P. mariana* forming closed-crown stands, associated locally with *Fraxinus nigra* and *Larix laricina* on wetter sites. This complex of essentially boreal vegetation persisted until about 10,500 yr BP when it was replaced, apparently quite rapidly. The interpretation suggested is that "*Betula*, *Alnus* and *Abies* immediately invaded the area to be joined by a full complement of mixed deciduous forest trees" (Watts and Bright 1968, p. 864). They point out that as *Abies* is usually underrepresented in pollen spectra, it was probably present in the landscape. Similar fir percentages were recorded at the *Kirchner Marsh* site, 350 km east of Pickerel Lake (Fig. 6.1, No. 13), and the macrofossils of *Abies balsamea* confirmed the pollen evidence conclusively (Watts and Winter 1966). No other sites in the grassland–forest region of the Western Interior have provided evidence to suggest a replacement of the early Holocene spruce forest by a rich "mesophytic" mixed forest including fir and resembling the modern forests. I return to this interesting question in Chapter 8 in a general examination of the postglacial history of several tree taxa in the Western Interior. The high percentages of *Pteridium*, Gramineae, *Artemisia*, and Ambrosieae pollen at the same time as the *Abies–Betula–Quercus–Ulmus* assemblage pose some interpretive difficulties, and it is possible that fire was more significant than was recognized by the original investigators.

The later decreases in arboreal pollen and increases of all NAP taxa are interpreted as a prairie complex on all upland soils with very few trees in the landscape. Some independent evidence for redeposition of pollen in the prairie zone of the diagram is used to explain the anomalous occurrence of scattered grains of *Picea*, *Larix*, and *Abies* in association with the typical prairie assemblage.

The most likely vegetation reconstruction for the Tiger Hills upland is a mixed forest from deglaciation until 9,500 yr BP, with abundant *Picea* on uplands, and local communities of *Larix*, *Ulmus*, *Populus*, *Ostrya*, *Betula*, *Salix*, and *Juniperus* on surfaces with particular edaphic conditions of moisture and/or texture. Xeric sites – knolls with gravelly soils – were probably treeless with *Juniperus* and *Artemisia* locally abundant. *Picea* declined and probably disappeared by 9,500 yr BP and upland xeric and mesic sites had a grassland complex, while lowland sites with moister soils supported shrub (*Salix*) and very local tree communities (*Ulmus*, *Fraxinus*, *Betula*). At about 3,500 yr BP, *Quercus* (*macrocarpa*) either expanded from existing scattered individuals or first arrived, probably from the southeast, forming oak savannas that persist to the present.

The possible reconstructions derived from the Riding Mountain records are as follows: The early Holocene, late-glacial vegetation consisted of a closed spruce (chiefly *Picea glauca*) forest on upland till soils; a *Larix–Fraxinus nigra* woodland in swales; an open *Juniperus–Artemisia* community on xeric, sandy soils; *Populus tremuloides* stands on gully slopes; and *Ulmus* and *Ostrya* localized in valley bottoms. By 9,000 yr BP, upland surfaces were in a rich grassland with *Symphoricarpos*, *Salix*, and *Corylus*. *Populus* was confined to steep north-facing slopes with local *Quercus* and *Betula papyrifera*. In the late Holocene, *Quercus* increased on upland surfaces to form a deciduous woodland with *Populus* and *Betula*. *Ulmus* and *Fraxinus* were common on lowland alluvial soils. At about 2,500 yr BP, *Picea*, *Pinus*, and *Alnus* expanded to form the modern mixed boreal forest, *Pinus* being restricted to the localized areas of outwash sands.

The vegetation reconstruction offered by Mott (1973, p. 11) from his Lake A site is similar to the Riding Mountain sequence except that some deciduous trees are absent or rare, particularly *Fraxinus*, *Quercus*, and *Ulmus*. At both Lake A and B, a late-Holocene spread southward of spruce and pine is evident, suggesting a late development of the modern southern boreal forest along its southern periphery. The Lofty Lake interpretation (Lichti-Federovich 1970, pp. 942–3) is broadly similar to Lake A, but the initial poplar phase is distinctive and might equally be a function of some peculiarity of the conditions for pollen preservation than of some distinctive, local pattern of vegetation history. The subsequent spruce forest phase (10,500 to 9,500 yr BP) is short-lived though spruce might have persisted there in local habitats throughout the Holocene as its pollen percentages never drop below 5 percent and exceed 10 percent in several spectra scattered throughout the mid-Holocene. The regional vegetation on uplands from 9,500 until about 3,500 yr BP was a deciduous woodland on mesic to hydric sites with abundant *Betula* and, by assumption, *Populus*, and extensive grasslands on xeric sites. The forest expanded slowly and *Picea*, and locally *Pinus*, became more important elements in the late Holocene.

The pollen and macrofossil records reviewed above suggest that three regional-scale assemblages and thus inferred regional vegetation, can be identified. The first is the late-glacial *Picea*-dominated forests, variably associated either contemporaneously or sequentially with deciduous forest elements. The extent of the development of the latter declines gradually from east to west until at Lofty Lake there is no evidence for the existence of *Fraxinus*, *Ulmus*, *Ostrya*, or *Quercus*. The floristic composition of this early spruce forest, revealed particularly by the detailed macrofossil analyses at Pickerel Lake and Hafichuk, indicates clearly a closed forest similar to the modern southern boreal forest of central Canada except for the absence of *Pinus*. The second is the grassland assemblage, whose identity as a temperate, southern prairie rather than an arctic "cold steppe" is established clearly by the presence at many sites of taxa confined today to the Great Plains grasslands. And the third is a forest or woodland assemblage made up of *Betula*, *Fraxinus*, *Tilia*, *Ostrya–Carpinus*, and *Quercus*, and the fossil versions of this assemblage display an interestingly similar early Holocene gradient of diversity, decreasing from Pickerel Lake to Lofty Lake, to the modern pattern of distribution in which *Tilia*, *Ostrya*, *Quercus*, and *Fraxinus* have western limits staggered across the eastern plains. The mixed boreal forest assemblage is represented at only those sites that occur within that zone today, and it is clear and well known that this particular community in the Western Interior was not assembled in its modern form until 3,000 years ago.

Sites within the modern boreal forest

These sites are treated separately because they reveal a generally similar postglacial development that differs from that to the south nearer the forest–grassland transition and to the north near the forest–tundra border.

The *Flin Flon*, Manitoba, site is on the southern periphery of the Shield in the central boreal forest (Fig. 6.1, No. 14) represented by fragmented stands of conifers and deciduous hardwoods forming a typical boreal forest mosaic that expresses both the heterogeneity of the landforms and soils and a complex fire history. Lacustrine soils deposited by glacial Lake Agassiz support mixed stands of *Picea glauca*, *Populus balsamifera*, and locally *Abies balsamea*. Upland, bedrock-controlled topography with variable thicknesses of drift has a pattern of closed *Picea mariana* forests alternating with recently disturbed sites occupied by young stands of *Betula papyrifera*, *Populus*, and *Salix*. Bogs and fens have *Larix* and *Picea mariana* stands or, in the wettest areas, are treeless.

Unfortunately, the first organic sediments are undated, but it may be surmized from the map and account of the geological history by Teller et al. (1983) that the area emerged from Glacial Lake Agassiz about 10,000 yr BP when the ice margin was very close to the site. The earliest pollen assemblages show a short-lived NAP phase followed by a typical *Picea*-dominated episode (Fig. 6.3). Nonarboreal elements remain abundant until about 8,000 yr BP when *Betula* and *Alnus* increase and then, at about 7,000 yr BP, *Pinus* increases abruptly to modern values (40%) and the entire pollen assemblage remains constant to the present day.

The nearby Cycloid Lake site, reported by Mott (1973), provides an almost identical record from an area with a similar geological history. The oldest organic sediments, 10 cm below a radiocarbon date of 8,520 ± 170, have spectra with high NAP values (*Artemisia*, 15%) and *Picea* (15 to 40%), but levels at and immediately above the dated sediment have higher spruce values (60%) and lower NAP. The rest of the diagram resembles the Flin Flon sequence in that *Betula* and *Alnus* increase, followed by *Pinus* at about 6,000 yr BP. From then to the present day, the pollen spectra remain constant. At both sites, the deciduous tree taxa *Quercus*, *Ulmus*, *Fraxinus*, and *Ostrya–Carpinus* occur as isolated, individual grains, probably of long-distance origin, and *Abies* has a discontinuous registration never exceeding 0.2 percent at any level.

Northwest Saskatchewan still lacks any continuous, complete pollen records from late- and postglacial sediments. However, Vance (1984, 1986) has made a significant contribution by providing a detailed percentage and PAR sequence from a site in northeast Alberta (the *Eaglenest Lake* site, Fig. 6.1, No. 15), thereby filling in what had long been a large gap in the record from western boreal Canada. The record shows an initial assemblage with low PAR values and dominance of *Populus*, *Salix*, and NAP, with a mixture of boreal and grassland indicator taxa (*Shepherdia canadensis* and *Rubus chamaemorus* on the one hand and *Sarcobatus* and *Ruppia* on the other). Abrupt increases in *Picea* and *Betula* (arboreal) occur at about 10,000 yr BP, followed by *Alnus* and, finally, *Pinus*, which increases to more than 20 percent at about 7,500 yr BP. A recent investigation by MacDonald (1984) has both filled in a major gap in the record from central Alberta, northeast British Columbia, and the Northwest

Figure 6.3. Summary percentage pollen diagrams for the sites indicated. Details as in the legend to Figure 6.2.

Territories and established the chronology of events securely. The *Lone Fox Lake* site (Fig. 6.1, No. 16) in the Clear Hills of west-central Alberta, investigated by MacDonald (1984), is actually in a sector of the boreal forest that is transitional to the Cordilleran ecoprovince in that *Pinus contorta*, a western species, more or less completely replaces *P. banksiana*. Otherwise, the vegetation is typical of the central boreal forest, i.e., closed, heterogeneous stands of *Picea glauca* and *Betula papyrifera* with pine on uplands, and *Picea mariana–Larix* woods on peat bogs. The 5-ha lake is at roughly 1000 m in a gently rolling glacial drift landscape. A 3.5-m core yielded six radiocarbon dates and PAR and percentage pollen records. A summary diagram (Fig. 6.3) displays a straightforward record similar in broad outlines to the Flin Flon and Cycloid sequences. The earliest spectra, dated at 10,700 ± 140, have high NAP values dominated by *Artemisia*, Gramineae, and Cyperaceae. *Picea glauca* and *P. mariana* increase from negligible frequencies to a total spruce of about 60 percent at about 9,500 yr BP, and arboreal birch increases significantly about one millennium later. *Pinus* begins to increase gradually from less than 5 percent at 8,000 to more than 40 percent by 7,000 yr BP. The record changes very little after about 6,500 yr BP, implying that the modern regional pollen spectra were established at that mid-Holocene stage. MacDonald (1984, p. 81) draws attention to the increase in PAR of *Sphagnum* spores and *Picea mariana* at the same time in the mid-Holocene. Two additional sites in the region were investigated by MacDonald (1984) and both their radiocarbon chronologies and pollen records are very similar to the Lone Fox Lake results.

A site in the Caribou Mountain upland that straddles the northern provincial boundary of Alberta provides the first complete, adequately dated pollen record from this phytogeographically interesting locale (MacDonald 1984). The *Wild Spear Lake* site (Fig. 6.1, No. 17) is near the east end of a Tertiary bedrock upland at 880 m, capped by rolling, ground moraine deposits. Extensive, poorly drained areas have peat soils with *Picea mariana–Sphagnum* communities; *Picea glauca* and *Betula papyrifera* are common on well-drained ridge and escarpment features. *Pinus contorta* is rare in the area, here at its eastern range limit. The 16-ha lake yielded a 3.2-m core and four levels were radiocarbon dated. The pollen record (Fig. 6.3) shows an initial assemblage (11,000 to 9,600 yr BP) dominated by NAP of which *Artemisia*, Gramineae, and Cyperaceae dominate with consistent representation of Cyperaceae, Chenopodiineae, and *Salix*. The PAR is relatively low in these lowest levels (<500). *Betula* (cf. *glandulosa*) increases at about 9,000 yr BP, followed by a short-lived peak of *Salix*. At an undated level between 10,000 and 8,000 yr BP (undated because the sediments from those levels were marly), *Picea glauca* increases from very low frequencies to about 30 percent. No further changes occur.

Similar records were obtained from a site 400 km west of the Caribou Mountains, on the Estho escarpment in northeast British Columbia (MacDonald 1984), and from two sites on the Alberta Plateau near the Peace River, British Columbia (White, Mathewes, and Mathews 1985). These recent findings by Vance, White, and MacDonald largely supercede the older literature, reviewed recently (Ritchie 1985a), which was based on large numbers of undated peat bog samples subjected to superficial, largely uninformative pollen analysis.

Three sites near latitude 60°N provide consistent records for this northwestern sector of the boreal forest, near the limit of *Pinus banksiana*. One, the *Porter Lake* site (Fig. 6.1, No. 18) lies east of Great Slave Lake and was not ice-free until about 8,000 yr BP (Ritchie 1980). The *John Klondike Bog* site (Fig. 6.1, No. 19) is in the western area and it yielded a detailed pollen and macrofossil record, but is truncated at the bottom (Matthews 1980). The central site, on the Horn Plateau, provides a record that is probably representative of upland sites that were deglaciated by roughly 11,000 years ago. The Horn Plateau is a bedrock upland capped by glacial drift and the site (Fig. 6.1, No. 20) is a small (5 ha) lake, referred to as *Lac Demain*, at 745 m, surrounded by closed forests of *Picea glauca*, *Betula papyrifera*, and occasional *Pinus banksiana* on upland, mesic sites, with extensive tracts of bog supporting *Picea mariana–Sphagnum* communities. A 2.2-m section of sediment provided percentage and PAR estimates and five radiocarbon dates indicate the chronology adequately. The relative pollen diagram (Fig. 6.3) is very similar to that from the Caribou Mountain. A preponderance of NAP at low PAR values in the oldest levels gives way to an increase in *Betula glandulosa* followed by *Picea* at roughly 9,000 yr BP and increases in *Alnus* and *Pinus* at 7,500 and 6,000 yr BP, respectively. The John Klondike sequence is very similar (Matthews 1980, Fig. 2) although it lacks the early record.

The broad outlines of the regional vegetational history of the central boreal forest (Western Interior) can be discerned from the four primary sites reviewed above and the ancillary records from the area. As uplands emerged from the receding ice and dwindling proglacial lakes, a short-lived treeless vegetation prevailed, but the available pollen documentation does not contain any indicator taxa that might enable us to offer an exact reconstruction of the plant cover. Abundant pollen of *Artemisia*, Gramineae, and Cyperaceae has been recorded all over the northern hemisphere in late-glacial sediments. Even the PAR data calculated by MacDonald (1984) and Vance (1986) are inconclusive in that they are higher than modern tundra values but lower than those from grasslands. Further research, particularly using macrofossil analysis, is required to improve on current interpretation as "a sparse herbaceous assemblage" (MacDonald 1984, p. 153). By about 10,500 to 9,500 yr BP, a *Picea*-dominated forest was widespread across the entire Western Interior between the Cordillera and the receding Laurentide Ice. Poplar-dominated phases occupied certain regions prior to or contemporaneously with the spruce forests. However, we should note Vance's (1986, p. 14) comment that his PAR data show that several of the taxa (*Populus*, *Salix*) dominating the percentage diagram "were no more abundant in the late Pleistocene than they are today." In all cases discussed above, as the respective authors emphasize, the PAR values derived from samples in the late-Pleistocene clayey and silty, usually undated, sediment should be regarded with caution. It appears that the particular topography and certain early Holocene seasonal wind patterns have contributed to the remarkable speed with which some forest taxa have colonized the Western Interior (Ritchie and MacDonald 1986; Vance 1986), but this topic is examined more fully in Chapter 8. Subsequent changes in the vegetation were controlled by the spread of tree birch and jackpine from the southeast and lodgepole pine from the Cordillera to establish the contemporary regional patterns. At the local scale, it can be assumed, in the absence of critical fossil data, that the effect of landforms and surface materials and later, as fuel accumulated, the influence of wildfire were important in determining the local patterns of vegetation. The major landforms and associated soils – bedrock ridges with little or no glacial drift, outwash sands, proglacial lake beds and former beaches, morainic ridges and till-capped Tertiary bedrock uplands – govern the distribution of boreal species at a particular locality within the broad, basically climatically controlled regional patterns.

Sites near the modern forest–tundra boundary

A concentrated effort to investigate late-Pleistocene vegetational history and related topics has been made in the far northwest, and a recent synthesis brings together most of the pertinent material (Ritchie 1984a). The following treatment, therefore, is concise and, while it uses new material, unavailable when the 1984 review was written, the basic sequence remains unchanged. It should be noted here that, although several sites that lie to the immediate west of this northern sector of the Western Interior are in mountainous regions (Selwyn, Mackenzie, and Richardson Mountains) and are therefore treated separately in Chapter 7 in the northern Cordilleran section, they are very similar to the sites dealt with in this section with respect to the dominant taxa of both the fossil record and the modern vegetation.

Eildun Lake (Fig. 6.1, No. 21) is near the Mackenzie River, north of the confluence of the Liard River and south of the Great Bear River, which flows from the west end of Great Bear Lake into the Mackenzie. It is a moderately sized lake (50 ha) at 302 m in a rolling to flat glaciated plateau, and it was investigated in detail by Slater (1985). It lies in a narrow tongue of boreal forest that extends along the Mackenzie River Valley and terminates about 50 km to the north of the site. The surrounding uplands have open parkland with tundra at high elevations. The local vegetation consists of extensive *Picea mariana–Sphagnum* communities on peats and mixed stands of *Picea glauca*, *Betula papyrifera*, *Populus tremuloides*, and *Pinus banksiana*, here close to its northern limit. Pollen spectra were tabulated at close intervals from a 240-cm section, of which the lowest 60 cm was a marl that, predictably, gave radiocarbon age estimates that Slater (1985, p. 666) correctly rejects as spuriously old. The pollen diagram is summarized in Figure 6.3, and I have adjusted the chronology of the lower part so that the *Picea* rise is assigned a date (9,200 yr BP) that is the average value of neighbouring sites (Lac Demain, 9,000; Lac Mélèze, 9,200; and John Klondike, where the rise is absent, but spruce was present by at least 9,500 yr BP). The pollen record shows an initial herb assemblage with abundant *Artemisia*, followed by peaks in *Betula glandulosa*, *Populus*, *Picea*, *Alnus*, and *Pinus*.

Farther north in the Mackenzie Valley, MacDonald (1984) reported six radiocarbon dates and pollen data from the *Lac Mélèze* site (Fig. 6.1, No. 22) near Norman Wells, a small (4 ha) lake at 650 m in glacial drift on the western flanks of the Franklin uplands. The vegetation is similar to Eildun Lake, except that *Pinus* is absent and *Alnus crispa* and *Betula glandulosa* are abundant around the lake. The pollen record (Fig. 6.3) is very similar to Eildun Lake except the *Pinus* values are very much less, and the reader can make a direct comparison of the summary diagrams.

At the edge of the boreal forest, in the Lower Mackenzie River area, a relatively dense network of sites has been investigated and a detailed recent review (Ritchie 1984a) attempts to present the central themes in the long postglacial history of the plant cover. By contrast with the sites described above from the so-called "ice-free corridor" (Rutter 1980), areas in the vicinity of the Mackenzie Delta were ice-free very early in the sequence of Laurentide Ice recession. The oldest reliable dates of lake sediment in the northern "corridor" area are about 11,000 yr BP (White, Mathewes, and Mathews 1979; MacDonald 1984, Lac Mélèze), but the first organic sediments in areas that had been covered by Late Wisconsinan Laurentide Ice (mapped in Prest 1984, Rampton 1982, and schematically in this book in Fig. 2.2) were as old as 15,000 yr BP.

The pollen record is familiar and I offer here one new record in summary form (*Kate's Pond* site, Fig. 6.3) along with a summary diagram of Spear's (1983) preliminary account of his meticulous investigations of the *Sleet Lake* site (Fig. 6.3).

Kate's Pond is a small (3 ha), shallow pond in slightly drumlinized till uplands east of the Mackenzie Delta (Fig. 6.1, No. 23), surrounded by open spruce woodlands (*Picea glauca*, *P. mariana*) with occasional *Betula papyrifera*. Parts of the landscape have been burned within the past few decades and the slow regeneration of trees results in extensive shrub areas dominated by *Alnus crispa*, *Salix*, *Betula glandulosa*, and ericoids. The summary pollen diagram is self-explanatory. Sleet Lake is small (1 ha), 8 m deep, in the forest–tundra zone (Fig. 6.1, No. 24). The surrounding vegetation is a low shrub heath of *Betula glandulosa* and ericoids and occasional spruce krummholz. The percentage pollen diagram, with supplementary macrofossil records, is also self-explanatory.

A possible vegetation reconstruction for this northern sector has been the subject of several recent publications (Spear 1983; Ritchie 1984a, 1984b, 1985b; Ritchie and MacDonald 1986), so the following will be as concise as possible. Before roughly 10,000 yr BP, upland surfaces were occupied by an initial herb vegetation, and at northern sites, where detailed pollen identification has been attempted (Ritchie 1977, 1985b), the pollen spectra are reasonably similar to modern ones from the mid- and high arctic of western Canada, indicating that a predominantly herb tundra occupied mesic sites in the late glacial. At about 12,000 yr BP, in the Lower Mackenzie River area, when the Lac Mélèze and Eildun Lake sites were still under Laurentide Ice, the vegetation became denser and more continuous by the expansion of dwarf birch (*Betula glandulosa*). A similar transition from herbaceous to shrubby vegetation was recorded at the southern sites (Lac Mélèze, Eildun), but the chronology is different, presumably because deglaciation occurred later at these localities.

A rapid transformation of the landscapes of this entire northern boreal sector occurred at roughly 10,000 yr BP when initially poplar (*Populus balsamifera* appears to have been more common, although *P. tremuloides* has been recorded) formed woodlands on upland sites, restricting the extent of the shrub tundra. About 9,500 to 9,000 yr BP, *Picea glauca* and *P. mariana* arrived, presumably from the south, their seeds propelled perhaps by a period of seasonally localized (fall–winter) surface winds from the south, across the relatively flat terrain of northern Alberta and the Northwest Territories (Ritchie and MacDonald 1986). Whether or not this speculative notion about the rapid spread of white, and probably black, spruce from the south survives the more rigorous analyses of the future, it is inescapable that a relatively abrupt change in vegetation occurred from treeless to boreal forest communities. Later, while the pollen spectra changed substantially as relatively high pollen producers expanded onto the landscape (tree birch, alder), the vegetational changes might well have been a gradual trend towards increasing involvement of the deciduous hardwoods, birch and poplar, driven partly by natural fires. At the northern boundary of the boreal zone, Spear (1983) and others have assembled fairly convincing evidence that the shrub tundra and forest–tundra zones have expanded towards the south. These themes are taken up again in Chapter 8 when the discussion involves palaeoenvironmental reconstruction.

7 Pacific–Cordilleran region

Although there has been considerable activity for several decades in the investigation of the vegetational history of this large, highly diversified region, the number of sites with both a detailed record of pollen and/or macrofossils and an ecologically informative and convincing interpretation remains small. The inevitable lack of radiocarbon control, certain problems of pollen identification, and a preoccupation in early investigations with bogs rather than with lakes render the voluminous older literature somewhat obsolete.

The Cordilleran Ice Sheet extended south into Washington during the Fraser glaciation, and at its maximum thickness (2,000 m) "it probably stood higher than the confining mountain ranges" (Prest 1984, p. 24). A period of extensive ice recession occurred, following the Vashon maximum at 14,500 yr BP, when lowlands became ice-free (Porter, Pierce, and Hamilton 1983; Waitt and Thorson 1983). Actually, it has been shown that the Vashon maximum represented the greatest extension of the southern edge of the Cordilleran Ice Sheet in the Pacific area (Fig. 7.1), when it stood at the north end of the Olympic Peninsula and along the Puget Lowlands to latitude 47°N (Hicock, Hebda, and Armstrong 1982). These authors discuss the roughly 3,000-yr lag in the Cordilleran maximum behind the 18,000 yr BP Laurentide maximum to the east, and suggest that it was due to a moisture deficit caused by eustatic sea level lowering. The recession of ice lobes to the east and west of the Cascade Mountains was not exactly synchronous, but there was a general retreat "from terminal positions to the international boundary during the interval 14,000 to 11,000 yr BP" (Waitt and Thorson 1983, p. 67). The pattern of deglaciation in southern British Columbia remains incompletely known.

I have subdivided the whole region into southern and northern segments and each into coastal (Pacific) zones and interior (Cordilleran) zones, following exactly the treatment of bioclimates described above in Chapter 2.

It should be noted that a more appropriate subdivision would undoubtedly be Krajina's (1969) biogeoclimatic classification of British Columbia into four formations – Alpine, Microthermal Coniferous Forest, Semiarid Cold Steppe, and Mesothermal. However, as no sites have been reported for two of these formations, it seems unnecessary to use the system here. On the other hand, at appropriate points in this chapter, I draw heavily on Krajina's authoritative descriptions of both the vegetation and environment of British Columbia and the autecology of the tree species.

Southern Pacific zone

This coastal rain forest region, equivalent to Krajina's (1969) coastal western hemlock zone, is the most productive of any coniferous forests in Canada. The primary tree species are *Tsuga heterophylla*, *Thuja plicata*, *Pseudotsuga menziesii* associated with *Abies grandis*, *A. amabilis*, *Picea sitchensis*, *Pinus monticola*, and *P. contorta*. *Alnus rubra* and *Acer macrophyllum* are locally common. The climate is summarized in Chapter 2, and the Victoria Airport and Alert Bay diagrams provide a convenient synopsis of the thermal and moisture means and ranges (Fig. 2.15).

Baker (1983), Heusser (1983, 1985), and Barnosky (1984) have provided reviews of the Holocene and late-Pleistocene pollen stratigraphy and vegetation reconstructions for north Washington and adjacent areas in the United States, so I shall avoid excessive repetition here. A few more recent sites are now available. I noted in Chapter 3 that uncertainties about the modern ecology and pollen repre-

Figure 7.1. A map of western Canada and adjacent United States showing the locations of the fossil sites referred to in the chapter.

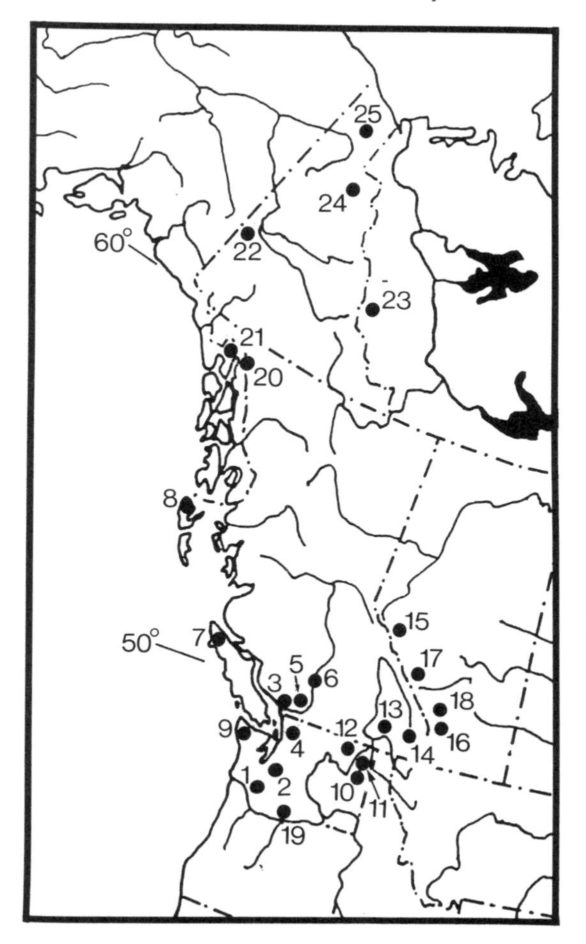

sentation (in lakes) of many taxa are aggravated by problems with the identification of some pollen types. I have chosen the following sites to illustrate what is known of the vegetational history of this complex region: *Surprise* and *Marion Lake* (Mathewes 1973; Heusser 1985), the Yale site on the Fraser River (Mathewes and Rouse 1975), *Mineral Lake* (Tsukada and Sugita 1982), *Davis Lake* (Barnosky 1981), *Hoh Bog* (Heusser 1974), and *Kirk Lake* (Cwynar in press). In addition, a full-glacial record from the *Port Moody/Mary Hill* sites in the Fraser Lowland near Vancouver (based on Hicock, Hebda, and Armstrong 1982), and a recent analysis of macrofossils and pollen from the Queen Charlotte Islands provide interesting additional documentation of late-glacial and Holocene floras. The *Battle Ground Lake* site farther south has already been referred to in Chapter 4 because its detailed pollen, macrofossil, and bryophyte record and closely controlled radiocarbon chronology provide an important record of the full-glacial to Holocene vegetation of an area well south of the ice margin (Barnosky 1985a; Janssens and Barnosky 1985). The locations of these sites are shown in Figure 7.1. All sites are at low elevations.

Summary pollen diagrams for those sites with complete records present the essentials of the relative pollen frequency data for the main taxa (Fig. 7.2). Macrofossil records, in one case in great detail (Cwynar in press), greatly strengthen the vegetation reconstructions.

The full-glacial records at Davis Lake (Fig. 7.1, No. 1) and Mineral Lake (Fig. 7.1, No. 2) are very similar, and they show that, between 20,000 and 16,000 yr BP, very low PAR values were registered for Gramineae, Cyperaceae, *Artemisia*, and a few herbs found today in the alpine zone – *Polygonum bistortoides*, *Epilobium alpinum*, *Valeriana*, and *Dryas*. *Picea*, probably *P. engelmannii* on the basis of macrofossil records at Davis Lake and the Port Moody/Mary Hill sites (Fig. 7.1, No. 3), *Abies* (*A. lasiocarpa* by macrofossils), and *Tsuga mertensiana* have low but continuous registration at Davis Lake and Mineral Lake, associated with high nonarboreal pollen frequencies. The Port Moody/Mary Hill record is based on units of organic silt and peat deposited during a brief episode about 18,000 yr BP. The pollen spectra show high percentages of *Picea*, *Abies*, and, at one site, nonarboreal taxa, while the macrofossil record yielded abundant needles and wood of *Abies lasiocarpa*, *Picea engelmannii*, and *Taxus brevifolia*. Between 16,000 and 10,000 yr BP, there are continuous increases in both the percentage and PAR values of *Pinus*, *Picea*, and *Tsuga*. Whereas the percentage of nonarboreal pollen decreases, the PAR values remain high. PAR curves from Mineral Lake, Davis Lake, and Kirk Lake (Fig. 7.1, No. 4) show decreases in *Picea*, *Tsuga mertensiana*, and *Pinus* immediately prior to the increases in *Pseudotsuga*, *Alnus*, and *Abies*.

All pollen records show a clear division into pre- and post-10,000-yr-BP segments and a change in the Holocene from an early dominance of *Pseudotsuga* to a later (roughly post-6,000 yr BP) dominance of Cupressaceae and *Tsuga heterophylla*. The usual assumption, at some sites corroborated by macrofossil records, is that the Cupressaceae represents *Thuja plicata*. A further consistent feature of all diagrams is the high frequencies of Diploxylon *Pinus* pollen between roughly 12,000 and 10,000 yr BP and the usual assumption, again substantiated at some sites by macrofossil records, is that the pine pollen represents chiefly *P. contorta*. In this same interval, 12,000 to 10,000 yr BP, *Picea* and *Tsuga mertensiana* show moderately high percentages (25 and 10%, respectively), but their PAR values are low at Davis Lake (Barnosky 1981) and Mineral Lake (Tsukada and Sugita 1982). Unspecified quantities of needles of *Picea engelmannii*, *Pinus contorta*, and *Abies lasiocarpa* were recorded in this interval at Davis Lake (Barnosky 1981, p. 253), but the same unit at Kirk Lake produced large numbers of macrofossils of *P. sitchensis* (5 to 10 per 100 cm^3 of sediment) along with smaller numbers of macrofossils of *Pinus contorta*, *Abies lasiocarpa*, and undifferentiated *Abies* and *Picea*.

All sites show "dramatic increases in percentages of *Pseudotsuga*, *Alnus*, *Abies* and trilete spores (cf. *Pteridium*)" at 10,000 yr BP (Barnosky 1981, p. 233). Tree taxa dominate the pollen proportions (>80%), but the percentages of *Pinus* and *Picea* at some sites drop to more than 2 percent, while *Abies*, *Pseudotsuga*, and *Alnus* (cf. *rubra*) dominate. Various Holocene zones are distinguished by Barnosky (1981), Tsukada and Sugita (1982), and others, indicated in Figure 7.2, following the original zonation scheme of Heusser (1977). These pollen zones are arbitrary subdivisions, useful for description, but rather uninformative about vegetational history. The consistent general pattern is that, following the decrease in *Pinus*, *Abies*, and *Picea* at 10,000 yr BP, *Pseudotsuga* (*menziesii*) increases rapidly and then decreases to minimal values at about 5,000 yr BP and later increases slightly to presettlement values of 10 percent. *Tsuga heterophylla*, by contrast, has low pollen values at 10,000 yr BP and increases very slowly to maximum percentages (30%) in the mid- to late Holocene (4,000 to 2,000 yr BP), decreasing to 15 to 20 percent in presettlement spectra (Hebda and Mathewes 1984). A minor peak in the mountain hemlock curve has been noted at the Marion and Surprise Lake sites (Fig. 7.1, No. 5) at 10,500 to 10,000 yr BP and has been discussed in relation to various palaeoclimatic reconstructions (Mathewes 1987). The question is examined in Chapter 8. Cupressaceae pollen (assumed to be represented by *Thuja plicata*) is absent or discontinuous until about 9,000 yr BP when it increases slowly at first, then rapidly, starting between 7,000 and 6,000 yr BP, to maximum values (40 to 60%) in the mid- to late Holocene (4,000) and finally decreasing to presettlement values of 30 to 40 percent. *Pinus monticola* type pollen is recognized by Mathewes (1973), and it increases gradually from its initial registration at 7,000 to a maximum of 5 percent at 2,000 yr BP, and then decreases to presettlement values of 2 percent.

The only detailed macrofossil diagram from this region, reported by Cwynar (in press) – and in fact it is one of the first investigations from western North America to combine detailed pollen (including PAR values) and quantitative macrofossil analyses – is the Kirk Lake site (Fig. 7.2 and 7.3). If it is assumed that it is representative of the Puget Lowland region, it amplifies significantly the previous records. The late-glacial presence of *Pinus contorta* along with *Picea sitchensis* and *Abies lasiocarpa* is established. The abundant macrofossils of *Pseudotsuga menziesii* and the absence of *Larix* macrofossils in the early Holocene suggest that the pollen records represent Douglas-fir, indistinguishable palynologically from *Larix*. Similarly, the high, continuous values of *Thuja plicata* macrofossils from the Mazama ash horizon to the present day confirm the usual assumption that the undifferentiated Cupressaceae pollen curve represents *Thuja*. The *Pinus monticola* and *Picea sitchensis* macrofossil records in the upper part of the diagram are of great interest because neither is represented in the pollen diagrams (percentage and PAR) from this site by correlative increases except for a change in the Haploxylon/ Diploxylon pine ratio.

The investigations reported by Mathewes and Rouse (1975) from the Yale area are of interest because both the modern setting and the fossil record are intermediate between the Pacific (coastal western hemlock) and Cordilleran (interior Douglas-fir) zones of Krajina (1965). The site (*Pinecrest Lake*, one of two cored sites; Fig. 7.1, No. 6) is at 320 m on the east slope of the Coast Mountains, in the Fraser River Canyon, only 50 km from the boundary of the interior Douglas-fir zone, where such Pacific taxa as *Acer macrophyllum* and *Tsuga heterophylla* have their range limits and where *Pinus ponderosa* first appears in the drier bioclimate of the interior zone (annual precipitation about 450 mm, compared to 1,600 mm at Hope, roughly 50 km south of the Yale sites). The pollen diagram (Fig. 7.2) has the same broad features as those from other Pacific sites, but a comparison with the nearest, at Marion Lake (Fig. 7.2), shows that *Tsuga mertensiana* was relatively infrequent at Pinecrest Lake but more common at Marion Lake; that *Pseudotsuga menziesii* was more frequent in the early Holocene pollen record at Pinecrest than at Marion Lake and maintained these higher levels (about 15 to 20%) to the present day; and that *Betula* pollen, relatively uncommon at Marion Lake, was frequent (20%) by the early Holocene at Pinecrest and has declined to only 10 percent in the late Holocene. Mathewes and Rouse (1975) draw attention to these points and conclude "that the presently transitional biogeoclimatic conditions in the Yale area have persisted for much of post-glacial time" (p. 754). Two cores of lake and bog sediment from a site near the northern tip of Vancouver Island (*Bear Cove*; Fig. 7.1, No. 7) produced an informative pollen record spanning almost 14,000 years to the present (Hebda 1983a). The investigation included a compilation of modern pollen spectra from surface moss samples. The initial as-

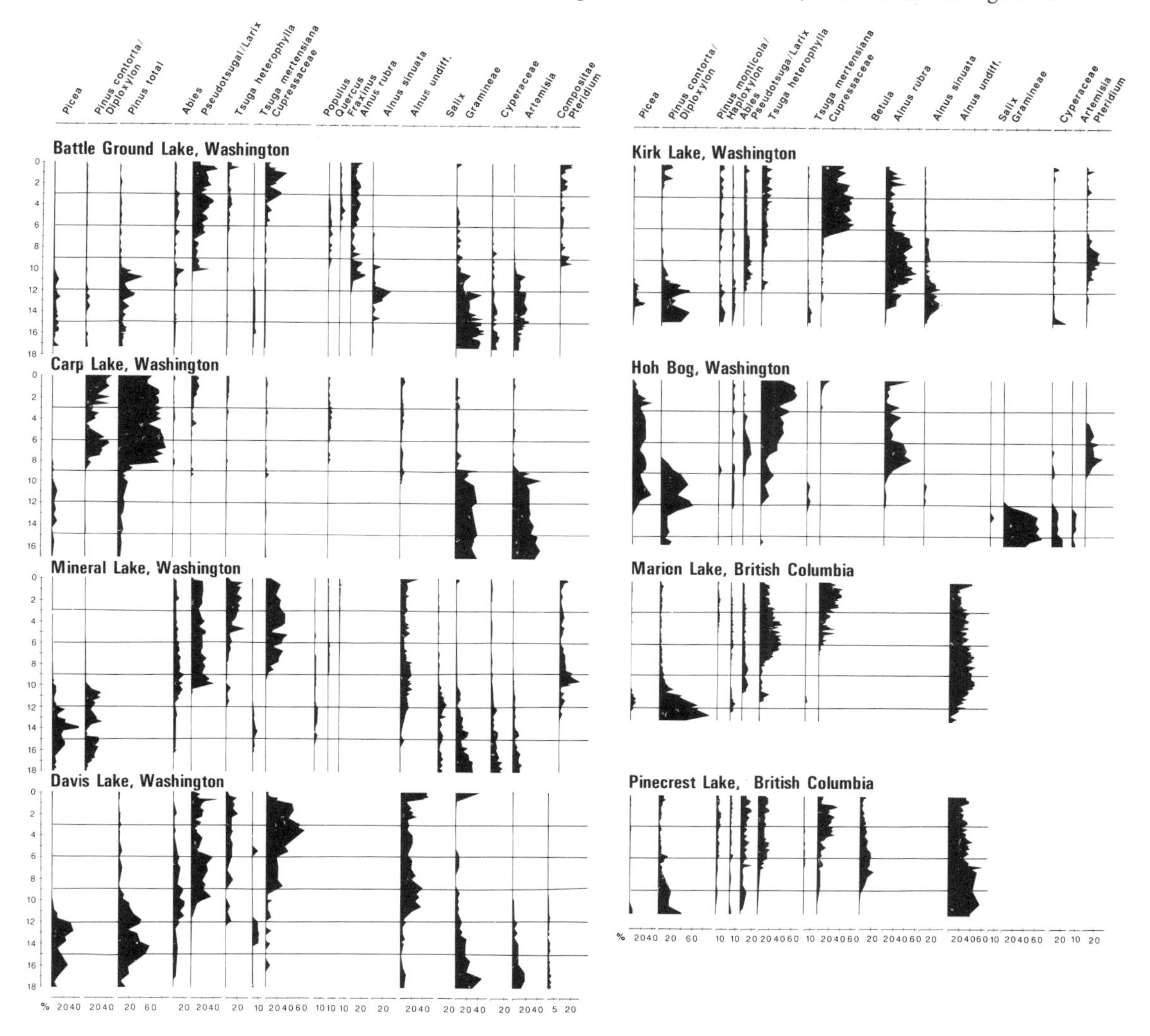

Figure 7.2. Summary percentage pollen diagrams for the sites indicated. In these, as in all subsequent composite pollen diagrams, the pollen percentages are plotted against a radiocarbon age scale. In preparing the diagrams from originals drawn on a depth scale, uniform sedimentation rate was assumed between dated levels and, where necessary, pollen curves were stretched or compressed to fit a constant time scale. The pollen sum in all diagrams includes all terrestrial taxa. Note occasional percentage scale changes used to improve visibility of changes in taxa with low percentage values (e.g., *Abies*, *Acer*). The details of the original investigations are given in the text, including the literature citations; locations are on Figure 7.1.

semblage was dominated by *Pinus contorta*, *Alnus*, and *Pteridium*, followed by *Picea sitchensis*, *Tsuga mertensiana*, and *T. heterophylla* at about 11,000 yr BP. *Pseudotsuga* is first registered prior to 8,120 yr BP, forming a maximum percentage at about 8,000 followed by increases in *Picea* and *Tsuga heterophylla*. Cupressaceae increases in the late Holocene.

Finally, Warner, Clague, and Mathewes (1984) have assembled a pollen and macrofossil record from the *Cape Ball* site (Fig. 7.1, No. 8) on the northeast coast of Graham Island, the large northern island of the Queen Charlotte group. A 3-m section of silts, peat, and marine deposits has yielded a pollen record spanning 15,400 yr to the present day as well as significant macrofossils from many levels. The lowest levels, from 16,000 to 11,500 yr BP, are dominated by grass, sedge, and willow pollen, with significant occurrences of both pollen and macrofossils of such herbs as *Polemonium caeruleum*, *P. pulcherrimum*, *Sagina* spp., *Hypericum scouleri*, and others. *Picea*, *Pinus*, *Tsuga heterophylla*, and *Alnus* quickly established themselves

Figure 7.3. The concentration of macrofossils of the tree taxa shown from the Kirk Lake, Washington, site referred to in the text. The pollen diagram from this site is shown in Figure 7.2. Both diagrams were made available by Dr. Les Cwynar in advance of the acceptance of his paper for publication.

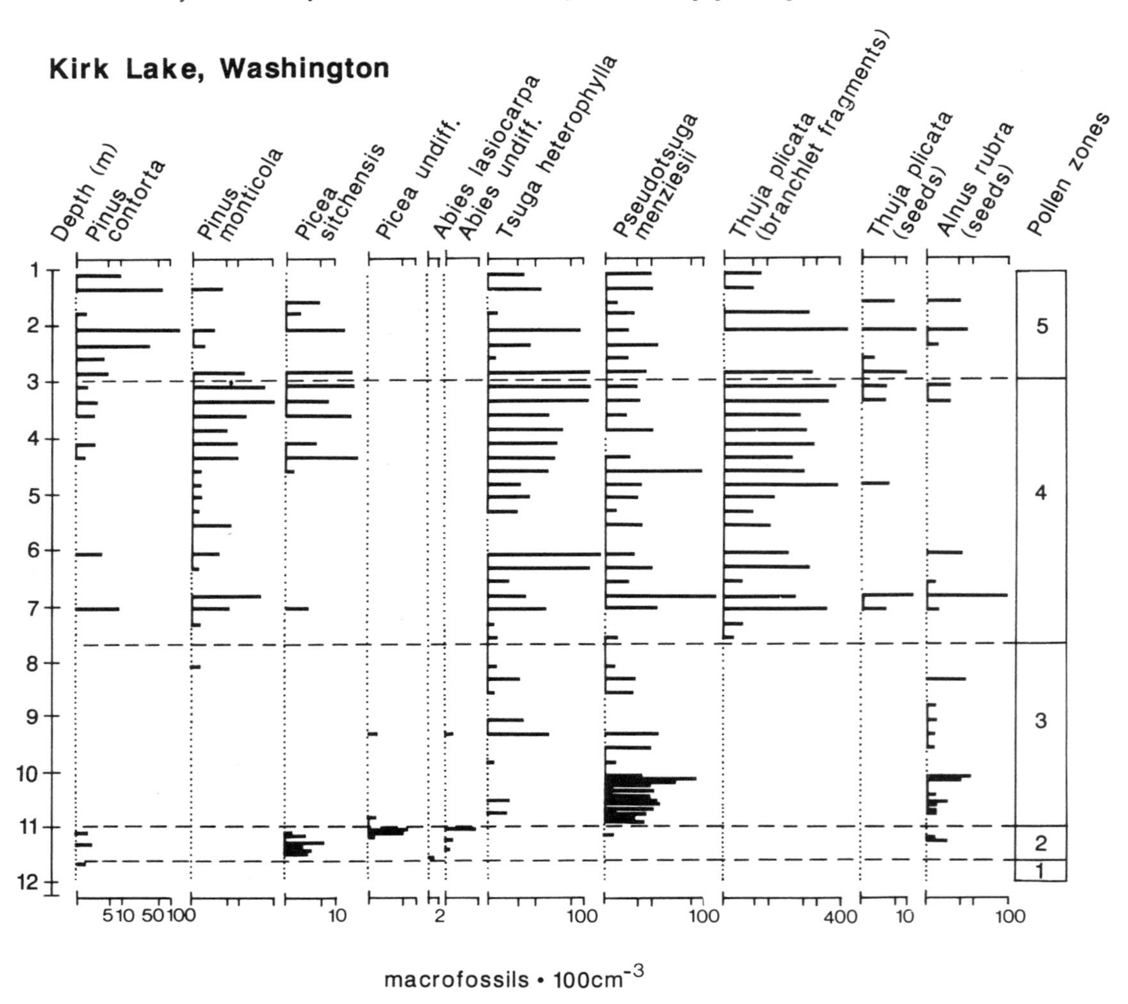

as pollen dominants after 9,000 yr BP, although the actual Cape Ball site was inundated by an eustatic rise in sea level. These findings show that there were ice-free areas on the Queen Charlotte Islands about 14,000 to 16,000 yr BP, but so far there is no evidence for a full-glacial refugium as some have speculated on the basis of the modern flora (reviewed in Mathewes in press).

The full-glacial assemblages recorded at sites in Washington slightly beyond the maximum extension of the Cordilleran Ice are interpreted by most investigators as representing a tundra–parkland, consisting of extensive treeless vegetation with scattered trees of *Picea engelmannii*, with considerable local latitudinal variations, dependent on altitude and soil. *Pinus contorta*, *Abies lasiocarpa*, and *Tsuga mertensiana* were present in small localized populations, and all three species increased steadily between 18,000 and 12,000 yr BP. By contrast, the Port Moody/Mary Hill record is interpreted as having been very similar to the modern Engelmann spruce–subalpine fir biogeoclimatic zone of Krajina (1969), a type found today at middle elevations in the interior of British Columbia, east of the crest of the Cascade Mountains, with a cold, continental climate. Hicock, Hebda, and Armstrong (1982, p. 2293) conclude: "Vegetation is interpreted as open, possibly a parkland mosaic of tree clumps and meadows typical of upper subalpine vegetation, thus indicating a treeline depression of 1,800 m." The younger portion of the late glacial (13,000 to 10,000 yr BP) is interpreted as a progressive increase of conifer in this open tundra–parkland, transforming it into a wooded landscape. *Picea engelmannii*, *Pinus contorta*, and *Abies lasiocarpa* were the dominant trees, but increases in *Tsuga mertensiana* were also registered indicating that it occurred locally "in more mesic sites and in areas of high snow accumulation" (Barnosky 1981, p. 233). In the modern Engelmann spruce–subalpine fir vegetation, *T. mertensiana* is "a relatively rare tree, dependent on such areas, which are very humid and where snow is deposited in large quantities before the ground freezes" (Krajina 1969, p. 125). The modern area of maximum abundance of *T. mertensiana* is in the coastal subalpine areas of the Pacific region.

The presence of abundant *Picea sitchensis* macrofossils at Kirk Lake (Cwynar in press) in the late glacial (13,000 to 11,000 yr BP), coexistent with *Pinus contorta* and *Abies lasiocarpa*, is not ecologically discordant with a reconstruction drawing on the modern Coastal Subalpine Forest region as an appropriate analogue. Krajina (1969, pp. 3–8) describes communities where these species coexist, although several are rare and confined to local habitats.

The early Holocene at all sites is characterized by the rapid increase of *Pseudotsuga*. For example, at Mineral Lake the PAR increase is from 11 at 11,000 to over 4,000 (grains $\cdot$ cm^{-2} $\cdot$ yr^{-1}) at 9,900 yr BP (Tsukada and Sugita 1982). The most widely accepted interpretation is that an open-crowned forest or woodland prevailed – an old idea (Hansen 1943) that has been rather uncritically accepted. The high *Pteridium* values are often cited as evidence of an open woodland. However, the macrofossil and PAR records at Kirk Lake are interpreted by Cwynar (in press) in terms of a closed forest of *Pseudotsuga* and *Tsuga heterophylla*, and he relates the *Pteridium* and *Alnus* curves to a fire-controlled gap phase or patch dynamics process. The charcoal record at Mineral Lake supports the suggestion of high fire frequency, although Tsukada and Sugita (1982, p. 407–8) also suggest an open woodland of *Pseudotsuga* with "only a few hundred trees per acre." In general, a reconstruction of a closed-crown type forest vegetation is, in the absence of clear evidence to the contrary, ecologically both a more parsimonious and a more rational proposition. Some inkling of such an interpretation is offered by Leopold et al. (1982) in a brief interpretation of a pollen diagram from Lake Washington, when they suggest: "The predominance of alder, Douglas-fir and grass pollen along with bracken fern spores 10,000 to 7,000 years ago suggests an open forest of Douglas-fir and alder or a forest mosaic . . . " (p. 1306).

Sites on the Olympic Peninsula provide evidence that *Pseudotsuga* might have been less common there than in the Puget Lowland and that *Picea sitchensis* and *Tsuga heterophylla* were the dominant trees. However, the Hoh Valley site (Fig. 7.1, No. 9) has ^{14}C dating problems at the late-glacial to Holocene boundary, so the record remains imprecisely established (Fig. 7.2).

All records on the Puget Lowland show forests becoming progressively richer in species by the mid-Holocene, and *Thuja plicata* in particular became an important codominant while, at some sites, *Pinus monticola* and *Picea sitchensis* became important in the late Holocene. The assemblage can be compared reasonably to the modern vegetation of the coastal western hemlock zone (Krajina 1969), although the modern dominant, *Abies grandis*, is absent from most fossil records. Hebda and Mathewes (1984) draw attention to the interesting correlation between an extensive prehistoric exploitation of red cedar wood for dugout canoes, monumental poles, and houses and the late-Holocene (5,000 to 2,500 yr BP) expansion of *Thuja plicata*.

Southern Cordilleran zone

Several sites or site clusters within this region have yielded variably informative, reasonably dated sequences. All sites lie within the maximum Cordilleran Ice limit, and none has a record older than 12,500 yr.

The *Waits Lake* site (Fig. 7.1, No. 10) has been described by Mack, Bryant, and Pell (1978) and as it is included by Baker (1983) and Mehringer (1985) in their recent reviews, it is dealt with only briefly here. The interpretation of the pollen record from this (Fig. 7.4) and the other sites in Washington has drawn on the surveys of modern (polster) pollen samples from this region by Mack and Bryant (1974) and Mack, Bryant, and Pell (1978) that I have used in Chapter 3, Figure 3.20. They show that some important trees (*Tsuga heterophylla*, *Picea engelmannii*, and *Abies*) are underrepresented in modern pollen spectra, whereas *Pinus* shows the opposite relationship.

Waits Lake is in the interior *Pseudotsuga* zone with annual precipitation of 500 mm and mean July and January temperatures of 18 to 20°C and –4 to –5°C, respectively. From 12,500 to 10,000 yr BP, low PAR values, a dominance of *Artemisia*, and moderate amounts of Gramineae and Cyperaceae are the main features of the record. Pollen identification of nonarboreal types was not attempted beyond a rather rudimentary level, so the interpretation of the late-glacial record as a treeless vegetation, though discussed at length, is necessarily inconclusive. The early Holocene pollen spectra (10,000 to 6,000 yr BP) are dominated by Diploxylon *Pinus* (>50%), *Alnus*, *Artemisia*, and Gramineae. *Abies* and *Picea* have continuous curves at low frequencies. *Artemisia* fluctuates around 20 percent until about 5,000 yr BP

Figure 7.4. Summary percentage pollen diagrams for the sites indicated. Details as in the legend to Figure 7.2.

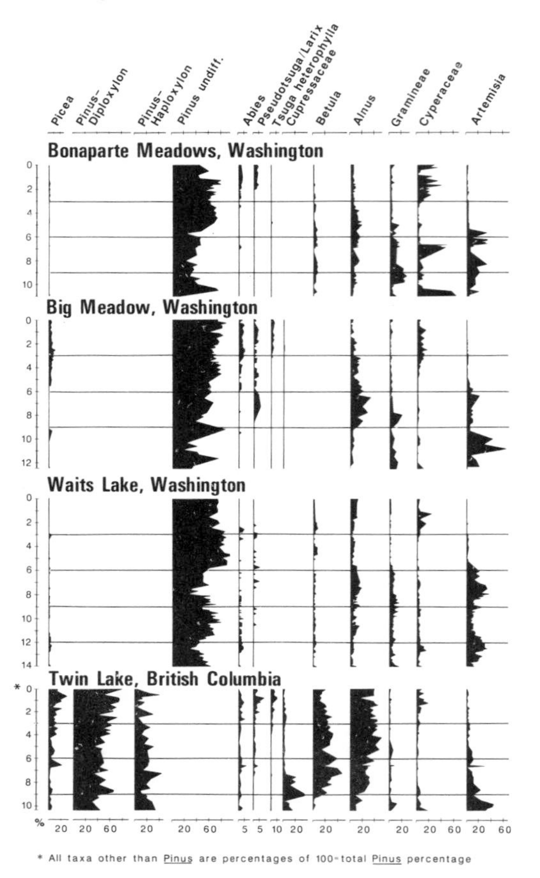

when it decreases to less than 3 percent and *Pinus* increases to more than 70 percent. Although Mack, Bryant, and Pell (1978) state that "*Abies*, *Picea*, *Alnus* and *Larix*/*Pseudotsuga*-type occur frequently from 230 cm to the surface" (p. 501), in fact the evidence presented in the pollen diagram does not sustain such an assertion. There are no significant changes in the percentage curves of these taxa from, in the case of *Picea* and *Abies*, the 9,880 ± 150 yr BP dated level and from the Mazama ash horizon in the case of *Pseudotsuga*. So the summary of the "vegetation dominants" offered is difficult to reconcile with the facts, as is implied by the inconclusive discussion of the *Pseudotsuga*/*Larix* record. The inferred palaeoclimates – cooler, moister from 11,000 to 9,000 yr BP, warmer, drier from 8,000 to 3,500 and cooler, moister to the present – rest on very tenuous evidence in the pollen data.

The *Big Meadow* site (Fig. 7.1, No. 11) is a mire north of the Waits Lake site, at an elevation of 1,000 m, in the interior of the *Tsuga heterophylla* zone, with an annual precipitation of 700 mm, and mean monthly July and January temperatures of 21°C and –5°C, respectively (Mack et al. 1978). *Abies grandis*, *Pseudotsuga menziesii*, *Picea engelmannii*, and *Pinus monticola* are common trees at the site.

In the lowest sediments, from the beginning of sedimentation (presumed to be 12,500) to 10,000 yr BP, at the Glacier Peak tephra layer, the pollen spectra are dominated by *Artemisia*, Gramineae, and *Pinus* (mainly Haploxylon) (Fig. 7.4). The authors reach the same conclusion that a treeless, tundra-like vegetation occupied the area as it did at Waits Lake. Diploxylon *Pinus* and *Pseudotsuga* appear at 9,000 yr BP, and *Artemisia* declines sharply. *Picea* and *Abies* show roughly synchronous increases to 10 percent and 5 percent, respectively, between 4,000 and 2,500 yr BP, and then *Tsuga heterophylla*, which shows sporadic pollen registration from about 5,000 yr BP, has a continuous curve of about 2 to 5 percent. The authors suggest also that a *Pinus ponderosa* and/or *P. contorta* vegetation occupied the area in the early Holocene followed quickly by the establishment of *Pseudotsuga* forests at about 7,000 yr BP. They suggest that the *Picea–Abies* pollen maxima between 4,000 and 2,500 yr BP represent the establishment of *Abies lasiocarpa* and *Picea engelmannii* forests, although they caution that identification of fir and spruce pollen to species is not possible. Finally, *Tsuga heterophylla* expands at 2,400 yr BP to become "a major forest member concomitant with the decline of *Picea* and *Abies*" (Mack et al. 1978, p. 963).

The *Mud Lake* and *Bonaparte Meadows* sites (Fig. 7.1, No. 12) are close together in the same drainage system at 650-m elevation in the Okanogan Valley, Washington (Mack, Rutter, and Valastro 1979). They lie in the interior *Pseudotsuga menziesii* zone, with *Tsuga heterophylla–Abies grandis–A. lasiocarpa* forests on adjacent mountains and *Pinus ponderosa–Artemisia tridentata* savannas. Precipitation and temperature means near the sites are 414 mm and –6°C (January) and 20°C (July), respectively. The Bonaparte Meadows percentage diagram is reproduced in summary form (Fig. 7.4). Fourteen radiocarbon dates, all in correct sequence, provide a detailed chronology and enabled the authors to prepare a PAR diagram (Mack, Rutter and Valastro 1979, Fig. 4).

Both sites register a basal pollen assemblage from greater than 11,000 to 10,000 yr BP dominated by Haploxylon *Pinus*, *Artemisia*, Cyperaceae, and *Shepherdia canadensis*, essentially similar to the late-glacial spectra recorded at the Waits Lake and Big Meadow sites (above). At about 10,000 yr BP, the *Pinus* component changes to a predominantly Diploxylon type and *Betula*, *Alnus*, *Artemisia*, and Gramineae all increase. Changes in the pollen record between 8,000 and 6,700 yr BP appear to be related primarily to a change in the surface of the fen, including deflation, with an accumulation of sedge peat and a resulting increase in sedge pollen and decreases in *Artemisia* and *Pinus*. It would have been more appropriate at this site to exclude Cyperaceae from the pollen sum as it is one of the main local pollen producers. A real change occurs at the 2- to 2.5-m level (3,000 to 2,000 yr BP) when both percentage and PAR values of *Pinus*, *Picea*, and *Pseudotsuga* increase. The authors identify the beginning of this afforestation trend at 4,800 yr BP when they note that "sediment containing evidence of *Artemisia*-dominated vegetation was replaced by a unit in which diploxylon pine and probably *Pseudotsuga menziesii* became prominent." In fact, the first significant increase in *Pseudotsuga* percentages is at 2.5 m (3,500 yr BP) and later in the PAR diagram, but the general trend of increasing arboreal PAR values begins after the Mazama ash layer and reaches maximum values at 2 m (2,500 yr BP). Although the authors make no reference to it, *Abies* and *Picea* (both underrepresented pollen taxa) increase markedly at about the 3,000-yr-BP-level in both the percentage and PAR diagrams and presumably represent late-Holocene expansions of *Picea engelmannii* and *Abies grandis* and/or *A. lasiocarpa*, all of which grow near the sites today.

The acute difficulties in interpretation posed by these data, due to the problems of pollen identification among the species of *Pinus* and *Artemisia*, leave us with a large degree of uncertainty about the nature of the past vegetation. The conventional wisdom would have us accept late-glacial "open tundra-like environments, perhaps with scattered *Pinus albicaulis*," followed by an early Holocene *Artemisia* steppe with *Pinus contorta* and/or *P. ponderosa*, and "a replacement of steppes and grasslands by forests" by about 5,000 yr BP (Baker 1983, pp. 112–13).

Mack, Rutter, and Valastro (1979) maintain "that those *Artemisia* species which contributed to the late-glacial pollen zone at both Mud Lake and Bonaparte Meadows were different from those species following ice recession," and they suggest *A. tridentata* and *A. tripartita* as the most likely taxa. Making this assumption, they then see a close resemblance of spectra to modern steppe samples, and proceed to make palaeoclimatic assumptions about increased warmth and dryness.

The *Kelowna Bog* site (Fig. 7.1, No. 13) (Alley 1976) lies in the Okanagan Valley about 150 km directly north of the Mud Lake–Bonaparte Meadows region. It is in the *Pinus ponderosa*–bunchgrass zone (Krajina 1969), which is the driest and warmest of the Pacific–Cordilleran bioclimates (Figs. 2.3a and 2.15). The annual precipitation is 190 to 360 mm, and the mean July and January temperatures are 18 to 22°C and –8 to –3°C, respectively. The bottomland vegetation of the Okanagan Valley is a *Pinus ponderosa–Artemisia tridentata* savanna or treed steppe, with local occurrences of *Pseudotsuga menziesii* and *Thuja plicata* along rivers. The flanking mountains show a zonation of forest communities, made up of the interior Douglas-fir (at 1,000 m) and Engelmann spruce–alpine fir (above 1,350 m) zones.

Alley suggests that the lowest part of the pollen record, from 8,900 yr BP, when a proglacial lake that occupied the valley floor drained, until roughly 7,500 yr BP, was dominated by *Pinus ponderosa* with sporadic *Abies* and *Pseudotsuga*. He separates a zone KB2, from 2.3 m to the Mazama ash layer at 6,600 yr BP, but appears to use assumptions about the climate to arrive at vegetational reconstructions: "This (KB2) pollen assemblage is interpreted as resulting from a warmer and drier period in which non arboreal vegetation predominated." A grass maximum at roughly 7,000 yr BP is interpreted as an expansion of grassland in the valley bottom with a retreat of pine and Douglas-fir to the surrounding uplands. The spectra from 6,000 to 0 yr BP are grouped in a separate pollen zone (KB3), which is "overall . . . dominated by *Pinus ponderosa*, with fluctuations in alder and birch, and significant increases in spruce and Douglas-fir." The author recognizes three intervals in KB3 that he describes as responses of local bog communities (*Alnus*, *Betula*, *Myrica*, *Corylus*) to a changing water table, which he then correlates with "three known stades of the Neoglaciation." A vegetational reconstruction for zone KB3 is never actually suggested, so it may be supposed the author assumes that the modern pattern was established following the Mazama ash deposition.

The original zoning of this diagram determined largely the author's subsequent interpretation and discussion of the pollen data. This is unfortunate as possible useful information is obscured. His zone KB2 is characterized by a decline in the frequency of arboreal pollen due, as he points out, to a major sedge increase and later a grass peak, although he notes that the sedge curve was likely a reflection of local conditions at the site and was unrelated to regional vegetational change. Mack, Rutter, and Valastro (1979) have questioned the reconstruction of the lowest assemblage mainly because they are unconvinced that *Pinus ponderosa* and *P. contorta* can be separated. It might be further suggested that the slight decreases in tree pollen (*Pinus*, *Picea*) from 2.3 to 1.75 m are artefacts of a pollen sum that includes Cyperaceae, and that some sedimentological change would be expected in the upper half of the section if the birch, alder, and hazel fluctuations reflect local water-table changes. It is clear that additional work at this site would be helpful. A further curious feature of the record is the low values of *Artemisia*, particularly in the upper levels when, presumably, *A. tridentata* was locally abundant as it is today.

It is unfortunate that one of the most effective investigations of Southern Cordilleran vegetation history has not yet been published. Hazell (1979) reported the Holocene pollen and macrofossil record of two small lakes in the Columbia–Kootenay Valley (Fig. 7.1, No. 14), along with a transect of surface pollen samples. The sites lie in the Rocky Mountain Trench, in *Dunbar Valley*, a tributary valley of the Columbia River. They are at 1,100 m, in the bioclimatic zone represented by Cranbrook (Fig. 2.15), with a mean annual precipitation of 440 mm and mean July and January temperatures of 20° and –10°C, respectively. The great topographic and elevational contrasts result in complex vegetational patterns within the general interior Douglas-fir biogeoclimatic zone (Krajina 1969). Upper montane surfaces surrounding the sites are covered by alpine tundra, beginning at 2,200 m, and a subalpine forest of *Picea engelmannii* and *Abies lasiocarpa* with *Larix lyallii* from 1,250 to 2,200 m. The sites are in the *Pseudotsuga menziesii–Pinus albicaulis* zone with abundant *Pinus contorta* and *Picea glauca*. Dunbar Valley is in a transitional vegetational region – it lies south of the interior western hemlock zone, whereas the northern limit of *Pinus ponderosa* lies to the south.

Organic sedimentation began at the two sites about 10,000 yr BP when the Columbia–Kootenay Valley was finally free of glaciers. The *Twin Lake* percentage diagram (Fig. 7.4) illustrates the results. The lowest levels have high *Artemisia* (40%) with Cupressaceae, Gramineae, Cyperaceae, and *Shepherdia canadensis*. *Artemisia* decreases while Cupressaceae reaches its maximum (35%) along with increases in *Betula*, *Picea*, and *Alnus*. *Pseudotsuga* appears first in the record just below the Mazama ash horizon (6,600 yr BP), increasing steadily to a presettlement maximum of 20 percent. *Pinus* values are high throughout, with Diploxylon values increasing steadily from 35 percent at the base to 50 percent at the top while the converse trend is seen in the Haploxylon percentage. *Abies* values vary only slightly throughout, between 3 and 10 percent. *Tsuga heterophylla* appears about 4,000 yr BP and increases to its maximum values (10%) at presettlement (Fig. 7.4).

Hazell (1979) points out that ice remained in the valleys later than on the uplands and, supported by his discovery of *Pinus contorta* macrofossils in the local sediments, he suggests that the initial vegetation in the region was a park tundra with *P. contorta*, *Juniperus*, *Artemisia*, *Shepherdia*, *Selaginella densa*, and Gramineae. He speculates that "whitebark pine, subalpine fir, and spruce may have been associated with lodgepole pine" (p. 63). The early Holocene (10,000 to 8,000 yr BP) pollen record, confirmed by macrofossils, is interpreted by Hazell (1979) as a trend toward closing of the vegetation, with abundant shrubs (*Juniperus*, *Shepherdia*, *Salix*, *Betula*, *Alnus*), and *Pinus contorta*, and *P. albicaulis* on more sheltered sites.

Closed-canopy forests of *Pseudotsuga*, *Pinus contorta*, *Picea*, and *Betula* spread at about 7,500 yr BP with a decrease in open vegetation. The increases in *Picea* and *Abies* (slight) at 4,700 yr BP signal the establishment of the modern, presettlement vegetation, with "a gradual lowering of the elevational limits and a retreat further to the south in the Columbia–Kootenay Valley of the sage-and-grass-dominated communities" (Hazel 1979, p. 68). *Tsuga heterophylla*, which is absent or rare today in the Dunbar Valley, became common at least on the western slopes of the adjacent (to the west) Purcell Mountains after its appearance at 4,000 yr BP. Hebda (1983b) refers to several of his own unpublished records from the interior of British Columbia in the course of reviewing the history of grassland vegetation in the region. The primary data and diagrams were not available at the time of writing.

A long-term investigation by Luckman and his associates in the *Jasper National Park* area of the Rocky Mountains in the Southern Cordilleran region (Fig. 7.1, No. 15) uses pollen, macrofossil, and tree-ring analyses to elucidate the record of vegetational history, treeline change, and past environments (Luckman and Kearney 1986). Several of these studies have produced interesting results.

Kearney and Luckman (1983a) report the pollen and macrofossils in Holocene peat deposits in the Tonquin Pass, British Columbia. The site is at 1,935 m in the *Picea engelmannii–Abies lasiocarpa* zone on the British Columbia–Alberta border at 52°40′N. It consists of vertical exposures of peat, 75 to 116 cm deep, containing both the Mazama and St. Helen's tephra layers. A basal peat sample yielded a maximum radiocarbon age of 9,660 ± 280 yr BP. The lowest pollen zone (only two samples) poses some difficulties of interpretation for the authors in that it consists of Diploxylon *Pinus* (assumed to be *P. contorta*) at 75 to 80 percent along with Abies (5 to 6%). Later, Haploxylon *Pinus*, assumed to be *P. albicaulis* on the basis of a few macrofossils, replaces the Diploxylon pine, and the authors suggest a *Pinus albicaulis–Abies lasiocarpa–Picea engelmannii* vegetation. They recognize a marked vegetation change at about 8,000 yr BP "distinguished by an abrupt rise in *Picea* percentages." These changes are described as "evidence for Hypsithermal conditions between 8,040 and 4,300 yr BP." The macrofossil record (Kearney and Luckman 1983a, p. 780, Table 2) seems inadequate to sustain such conclusions as only five levels produced remains of only five taxa. The percentage pollen curves (Kearney and Luckman 1983a, Fig. 5) appear to me to be largely devoid of significant change and most of the variations could probably be ascribed to sample noise. For example, the discontinuous *Pseudotsuga/Larix* record throughout the section can have no stratigraphic meaning, yet zone 3a is described as having a significantly higher frequency of this taxon. Similarly, the *Abies* frequencies appear, from the diagram, to show only the slightest trend in the section, but one that would be difficult to interpret precisely.

Investigations by the same group at other sites within the National Park used pollen, fossil logs, and macrofossils to document treeline changes, and oxygen isotope analyses of fossil wood were used to derive palaeoclimatic reconstructions (Kearney and Luckman 1983a and 1983b; Luckman and Kearney 1986). Pollen diagrams from analyses of two peats at the *Watchtower Basin* and *Excelsior Basin* sites show only very gradual trends of small magnitude in the percentage curves of the conifer taxa. The authors use the ratio of *Abies* to *Pinus* to measure treeline change, making certain rather tenuous assumptions about which pine species is represented by Haploxylon-type pollen (*P. albicaulis*). Dated fossil logs and the presence of abundant needles of *Picea* and *Abies* in the peat sections at levels below the Mazama ash horizon and the absence of tree macrofossils above a level dated at 5,800 yr BP are interpreted as an early Holocene phase of generally higher treelines.

Several sites along the eastern margin of the Cordilleran area have been investigated recently and they provide important information on the "ice-free corridor" concept. Recent investigations by MacDonald (1984) and White, Mathewes, and Mathews (1986) show that so far, no conclusive evidence has come to hand for a full-glacial ice-free tract between the Laurentide and Cordilleran Ice Sheets, extending from southern Alberta to the northwest. However, on the questionable assumption that the 18,000-yr-BP radiocarbon age is accurate, the *Chalmers Bog* site (Fig. 7.1, No. 16) in the Rocky Mountain Foothills provides pollen evidence for an early ice-free landscape (Mott and Jackson 1982). Chalmers Bog is in the Engelmann spruce–white spruce transitional zone (Rowe 1972) between the strictly Cordilleran vegetation to the immediate west and the boreal forest to the east. *Pinus contorta* is a common successional dominant of these forests in areas where wildfire has temporarily removed the *Picea engelmannii–P. glauca* mature forests. The pollen record shows a treeless assemblage dominant

Figure 7.5. Summary percentage pollen diagrams for the sites indicated. Details as in the legend to Figure 7.2.

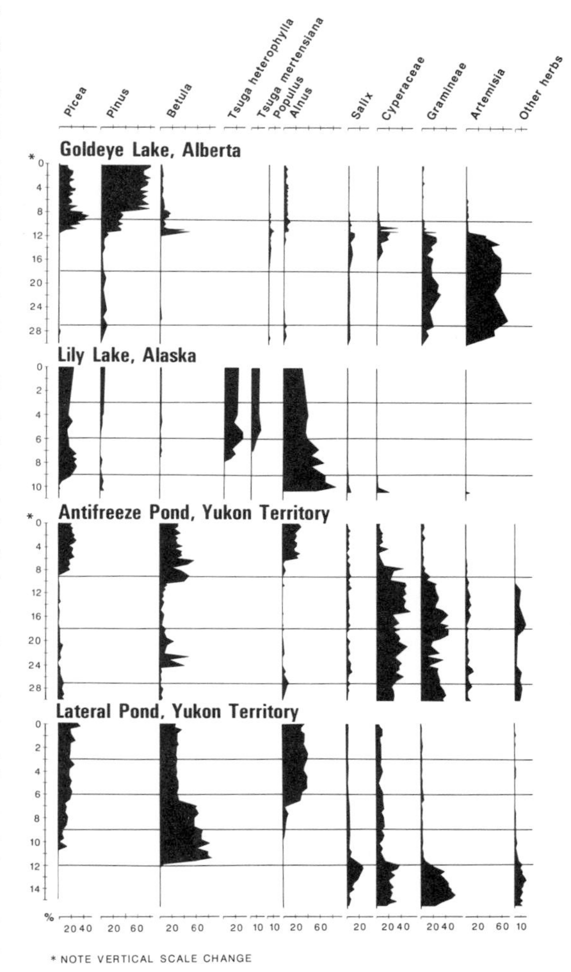

from 18,000 yr BP until an undated level earlier than 8,220 ± 80 when *Pinus*, *Picea*, and *Betula* increase, preceded by a *Salix* peak. A similar record has been recorded by Schweger (1986, in press) from *Goldeye Lake* (Fig. 7.1, No. 17) where a full- and late-glacial, treeless assemblage (*Artemisia*, Gramineae, Cyperaceae) is followed by willow, poplar, and birch peaks at 12,000 yr BP, a spruce zone at 11,000 to 10,000 and the modern pine-dominated assemblage begins at 7,500 (Fig. 7.5). The vegetation reconstruction proposed by both sets of investigators is a tundra in full- and late-glacial times followed by a succession of forest and shrub phases until, at about 8,000 yr BP, lodgepole pine expanded and the modern transitional Cordilleran–boreal forest was established. A similar sequence, but lacking a long earlier treeless phase, was recorded at *Yamnuska Bog* (Fig. 7.1, No. 18) near Calgary, by MacDonald (1982).

Figure 7.6. A schematic summary of the full-, late-glacial, and modern vegetation across a transect of Pacific–Cordilleran southwest Washington, reproduced with permission from Barnosky (1984).

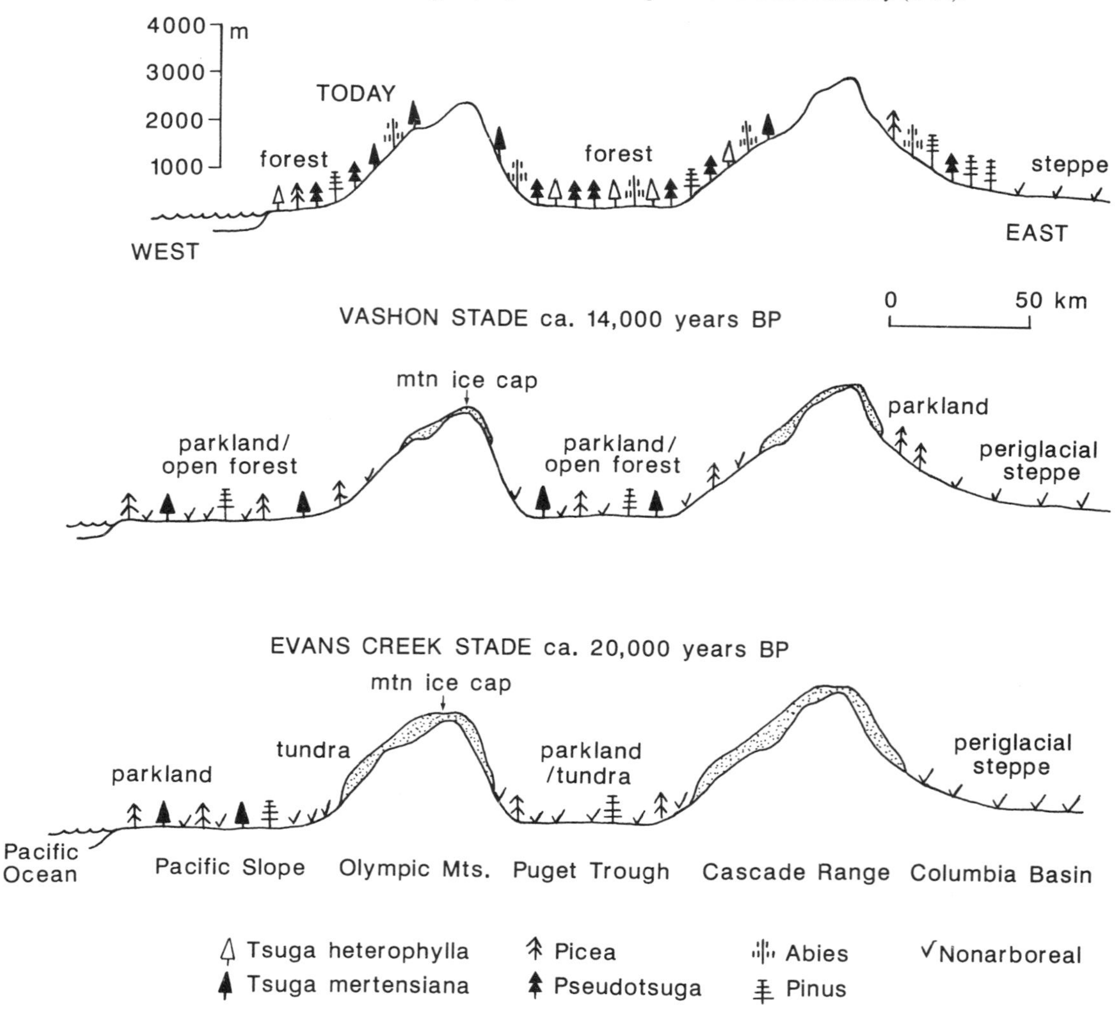

Attempts to sum up concisely the vegetational history of these extremely complex montane regions, even if a detailed record were available, are certain to oversimplify. In the case of the southern Cordilleran region only the broadest outlines of the vegetational history are perceptible, all based on records from low-elevation sites (up to about 1,100 m).

The full-glacial vegetation in the southwestern Columbia Basin has been reconstructed by Barnosky (1984), summarized here in Figure 7.6, chiefly from her *Carp Lake* site (Barnosky 1985b; Fig. 7.1, No. 19 and Fig. 7.2), as a "periglacial steppe," and she suggests that "the forest zone (of the eastern Cascade Range) was greatly compressed and that alpine vegetation may have merged with steppe in many places." Sites near the modern Canadian border were not ice-free until after 12,000 yr BP, and the first vegetation at low elevations was a treeless complex dominated by *Artemisia*, grasses, and some Haploxylon *Pinus*, probably *P. albicaulis* in northeastern Washington, whereas, at higher elevations in the Columbia–Kootenay valley, *Pinus contorta* might have been more common forming a park–tundra vegetation.

At roughly 10,000 yr BP, *Pinus* became more common and the preponderant Diploxylon type is interpreted by most investigators as belonging to *P. contorta* and *P. ponderosa*. It is suggested that "more mesic sites" supported pine forests by about 7,000 yr BP in the valleys of northwest Washington while, farther north in the Columbia–Kootenay valley, Douglas-fir increased to form closed forests with lodgepole pine, with a corresponding diminution of grassland. A similar change in north-central and northeastern Washington, to modern *Pseudotsuga* forests in areas previously occupied by open sage–grasslands, occurred at about 5,000 yr BP. Hebda (1983a) provides a very similar summary in his interesting review of grassland history, and he extends his survey farther south into Montana and Wyoming than I have considered appropriate in the present context. In addition, the interested reader can find a useful review of the pollen evidence from sites to the south and east, in Idaho and Utah, comprising a large portion of the Great Basin region, in the recent compilation by Mehringer (1985). He records a broad sequence of vegetational change that is similar to the summary abstracted above from the Barnosky (1984) review of the Columbia Basin.

The above inconclusive record from the Interior Cordilleran region is partly the result of the inherent difficulties of palynological investigations in montane terrain, partly explained by the acute problems of pollen identification among the important conifers, and partly due to the scarcity of small, closed-drainage lake basins. However, I suggest that the unsatisfactory record is also due to the tendency of some investigators to so intermingle vegetation reconstruction and palaeoclimatic inference that they lapse into circularity of reasoning. Such older notions, themselves based on flimsy evidence, as a Holocene hypsithermal period of dry conditions, have been accepted uncritically and incorporated into the vegetation reconstructions. It may prove, of course, that these palaeoclimatic inferences will survive a more rigorous analysis with new data independent of pollen. Nonetheless, a clearer separation of vegetation reconstruction and palaeoclimate would better serve the interests of sound methodology.

Northern Pacific zone

This zone is characterized by a coastal strip, mainly in Alaska, of many deeply incised coastal inlets with the lower elevations covered by northern coastal rain forest vegetation, and a narrow strip inland of mountainous terrain up to 3,000 m with a characteristic zonation of subalpine spruce forests giving way above 1,500 m to alpine tundra and local ice caps.

The early work in the fifties and sixties by Heusser, based on mire investigations, was summarized by him monographically (1960) and very briefly in recent reviews by Ager (1983) and Heusser (1985). Ager uses the upper *Montana Creek* site (Fig. 7.1, No. 20) to summarize the record and it shows (Ager 1983, Fig. 9.3, p. 131) that an early *Alnus–Pinus contorta*–Cyperaceae assemblage (10,000 yr BP) was replaced by a succession of *Picea sitchensis*, *Tsuga heterophylla*, and *T. mertensiana* to form the modern configuration of forest vegetation by mid- to late Holocene. Peteet (1986) provides a useful addition to this record of pollen and macrofossil analyses of radiocarbon-dated muskeg section in the *Malaspina Glacier* region. A recent investigation by Cwynar (personal communication) has brought chronological precision to the Holocene record. A well-dated sequence is reported from *Lily Lake*, a small lake on the Chilkat Peninsula of coastal Alaska and therefore in the north Pacific bioclimate (Fig. 7.1, No. 21). The record is based on 4 m of lake sediment with 14 radiocarbon dates providing a precise documentation of the spreading of particular tree species into the area (Cwynar personal communication). The oldest spectra (10,660 to 9,100 yr BP) were dominated by *Alnus* (>80%) with small amounts of *Salix* and *Pinus*. *Picea* increased rapidly from less than 5 percent at 9,300 to 30 percent at 8,300. *Tsuga heterophylla* appears first in the record at 7,800 yr BP and increases rapidly to almost 20 percent at 7,000. *Pinus* disappears from the record at about 7,800 yr BP and reappears later. *T. mertensiana* follows at 7,000 increasing to about 10 percent by 6,000, and finally *Pinus* reappears at about 4,000 (Fig. 7.5).

Northern Cordilleran zone

The *Antifreeze Pond* site reported by Rampton (1971) provides one of the longest records of continuous pollen data from Canada. The site is a small (<2 ha), shallow (2 m) pond at 720 m on moraine north of the Kluane Range Mountains in the southwest corner of the Yukon Territory (Fig. 7.1, No. 22). The climate is cool, continental, adequately summarized by the climate diagram from nearby Whitehorse (Fig. 2.3a). The vegetation is altitudinally zoned with a closed spruce forest at 600 to 1,000 m, dominated by *Picea mariana* and *P. glauca* and occasional *Betula papyrifera* and *Populus tremuloides*. The forest is open at higher elevations and treeline is at about 1,300 m. The tundra zones fall into a lower shrub unit to about 1,500 m and a herb unit at higher elevations.

A 5.2-m section of organic sediments yielded an interesting pollen record spanning 30,000 years. The earliest spectra are dominated by herb pollen with several characteristic tundra taxa associated with abundant grass and sedge pollen. The precise chronology of the older records remains obscure because of the inverted order of three dates, but the pollen record indicates that *Betula* cf. *glandulosa* became more common in the early NAP assemblage and these spectra persisted with little significant change until about 10,000 yr BP when *Betula* increased from 10 to about 50 percent and the herbs decreased proportionately. Then, at 8,700 yr BP *Picea* increased from low values to its modern level (25%) and later (5,600) *Alnus* increased from negligible quantities to its present-day proportion of about 25 percent. *Picea* and *Alnus* show increases in the oldest levels to 10 percent, associated with high *Lycopodium* values, and Rampton (1971) accounts for their presence in these samples in terms of redeposition from older sediments (Fig. 7.5).

The *Natla Bog* site reported by MacDonald (1983) is an exposure of Holocene peat near the Natla River in the Selwyn Mountains (Fig. 7.1, No. 23). The site is just above the limit of spruce forest. The chronology is based on six radiocarbon dates and the presence of the White River ash in the uppermost 20 cm of the section. The lowest levels, from 8,630 to 7,700 yr BP are dominated by *Betula*, probably *B. glandulosa*, associated with *Salix*, *Artemisia*, Gramineae, and Ericales. The upper part of the record, from about 7,000 yr BP, has high percentages and PAR of *Picea* and resembles closely the modern spectra.

The northern Yukon sector of the Cordilleran region is represented by the Mackenzie, Ogilvie, Richardson, and British Mountains. The earliest investigation in the area was of two sites in the Ogilvie Mountains by Terasmae and Hughes (1966) and, while relatively few samples were dated, this analysis established that the basic pollen stratigraphic sequence recorded earlier for late- and postglacial sediments in Alaska by Livingstone (1955) was also applicable in the adjacent north Yukon. The sequence, subsequently confirmed with only minor local variants at many sites below treeline in eastern Alaska and northwest Canada (Ager 1982; Ritchie 1984a), is a full- and late-glacial herb assemblage followed by a herb–birch late-glacial phase, an early Holocene *Picea–Populus*, and finally a mid-Holocene to present-day *Picea–Betula–Alnus* assemblage.

The *Lateral Pond* site in the Richardson Mountains serves as a representative example of this regional pattern (Ritchie 1982). A small (3.25 ha), shallow (3 m) pond at 600 m in a glaciated valley on the west flanks of the Richardson Mountains (Fig. 7.1, No. 24) produced a 15,500-yr-old record, analyzed in detail palynologically (Ritchie 1982, Fig. 9). The uplands were unglaciated in the late Pleistocene and only the valley bottoms were covered by tongues of ice representing the distal extensions of the Laurentide Ice in the northwest of Canada (mapped in Hughes 1972 and Hughes et al. 1981). The vegetation in the vicinity of Lateral Pond is relatively complex because of varied topography and bedrock, including calcareous and noncalcareous types. Upper slopes and ridges have rich herb and shrub tundra with floristically distinctive communities associated with particular bedrock types (described in detail in Ritchie 1982, pp. 577 ff.). Treeline is at 700 m, formed by *Picea glauca* on most surfaces, with a local *Larix laricina* treeline on slopes of north aspect. The pollen record in summary form (Fig. 7.5): from 15,500 to 12,000 yr BP, a treeless assemblage changing gradually from a predominantly herb composition with low PAR values to a herb–*Salix* and then, at 12,000 yr BP, a *Betula* (cf. *glandulosa*)–herb assemblage. *Picea* increases from negligible occurrences to 15 percent at about 9,500 yr BP, joined later by *Betula papyrifera* and *Alnus* (at 6,500 yr BP).

In spite of the great topographic and edaphic diversity of the North Yukon and adjacent Alaska, sites from within the boreal forest and forest–tundra zones show a remarkable constancy in the pattern of late- and postglacial pollen records, as the recent reviews by Ager (1982, 1983), Ager and Brubaker (1985), and Ritchie (1984a) demonstrate. In the Yukon, the Holocene pollen sequence from mires in the *Old Crow* and *Bluefish* proglacial basins compiled by Ovenden (1982, 1985) confirm the general outlines of the Lateral Pond sequence, as does the late- and postglacial records from the *Bluefish Cave* site in the same region (Ritchie, Cinq-Mars, and Cwynar 1982). The *Birch Lake* and *Lake George* sites in the Tanana River Valley in adjacent eastern interior Alaska, reported by Ager (1975), reveal a similar record. Recently investigated sites farther west in Alaska, in both the Brooks Range and the Interior (Brubaker, Garfinkel, and Edwards 1983; Anderson 1985; and Edwards et al. 1985), provide sequences that are essentially similar with the interesting difference that the increase of *Picea* occurs at progressively younger times in the Holocene towards western Alaska, indicating that it (both species) spread from the Mackenzie–North Yukon region of Canada after about 9,000 yr BP.

The northernmost site of the Cordilleran axis is just beyond the northern extremity of the Richardson Mountains, less than 100 km from the Beaufort Sea. It is the *Hanging Lake* site, an isolated lake in a bedrock plateau on the unglaciated plains between the Richardson and British Mountains (Fig. 7.1, No. 25). The 60-ha lake, 9.5 m deep, yielded a 4-m section of sediment that was investigated in detail by Cwynar (1982). Numerous radiocarbon dates, large pollen sums, and exhaustive pollen identification produced a very thorough and detailed analysis that serves as the yardstick for such studies in the northwest.

The oldest levels, from about 30,000 to 18,500 yr BP, have a pollen assemblage with low PAR values (5 to 100) and a predominance of herb taxa, chiefly arctic–alpine genera. An interesting incidence of elevated *Picea*, *Alnus*, and *Betula* pollen during this NAP stage, at about 21,500 to 18,500 yr BP, is similar to results from Rampton's (1971) Antifreeze Pond site in the southern Yukon (Fig. 7.5) and Anderson's (1985) *Squirrel Lake* site in the northern Brooks Range of Alaska. The possible phytogeographic implications of these data are examined in Chapter 8. *Salix* and Cyperaceae pollen increase markedly between 18,500 and 14,600 yr BP and then, from 14,600 to 11,000, *Betula* (cf. *glandulosa*) increases rapidly to maximum values of 60 to 80 percent, but as Cwynar (1982, p. 11) stresses, the maximum increases in dwarf birch PAR values occur at 11,600. Ericoid taxa increased significantly in both percentage and PAR between 11,000 and 9,000 yr BP and *Picea* and later *Alnus* increased and the Holocene spectra remained unchanged.

The pollen record from the northwest Cordilleran region reveals a remarkable degree of regional constancy on the Canadian side of the international border. The full- and late-glacial records suggest a sparse herb tundra on mesic uplands. Recent investigations in Alaska (Edwards and Brubaker 1984; Edwards and McDowell 1984; and Anderson 1985) have confirmed the conclusion of Cwynar and Ritchie (1980) that, whatever nomenclature one might select to label it, the full-glacial upland vegetation of the eastern Beringian region was a sparse herb tundra very similar both physiognomically and floristically to the modern mid- to high-arctic vegetation of Banks and Victoria Islands (Ritchie, Hadden, and Gajewski in press). The transition from the late glacial to the Holocene was marked by increases in plant cover, probably without very major floristic changes. Dwarf birch increased and a closed-tundra cover probably developed on mesic sites, with dense willow thickets and sedge-grass marshes in lowlands. A major change occurred at lower latitudes and elevations at about 9,000 yr BP when *Picea*, preceded at some localities by *Populus balsamifera*, expanded rapidly to transform mesic sites from tundra to forest or woodland. Subsequent changes on uplands involved minor adjustments in species composition as paludification increased, favouring *Picea mariana* over *P. glauca*, and *Alnus* increased in abundance, possibly favoured by increases in wildfires. Farther north, beyond the range of trees, the discontinuous full- and late-glacial herb tundras developed into closed shrub tundras on mesic sites. In the extensive lowlands formed after the draining of proglacial lakes, a sequence of mire development from marsh vegetation to bog heaths and *Picea mariana* woodlands occurred during the Holocene, influenced variably by permafrost aggradation in the deeper mires (Ovenden 1982, 1985).

8 Vegetation reconstruction and palaeoenvironments

Introduction

It is now appropriate to examine the broader questions of the origins and history of the major Canadian vegetation regions and to explore the environmental controls that might have influenced the patterns of vegetational or floristic change.

Two major classes of questions emerge from the preceding review of Canada's fossil record: (1) What reconstructions of past vegetation can be made and with what degrees of certainty? The closely related topic arises – can the routes and chronologies be discerned for the spreading from full-glacial refugia of the main taxa? (2) What palaeoenvironmental changes can be inferred from the record? And the corollary – can the record be used to test existing models of palaeoenvironmental, particularly palaeoclimatic, change?

The following four sections of this chapter deal respectively with the postglacial origins and history of the vegetation, the possible effects of environmental factors on the vegetational history, an examination of various outstanding problems and gaps in knowledge, and a brief concluding summary.

Origins and history

Eastern temperate forests

It is worth reiterating here that the eastern temperate forests make up a roughly triangular wedge centred on the southern Great Lakes, having ecotones to the north with the boreal forest, to the west with the prairie, and to the south with the subtropical forests (Bailey 1980). The temperate forests are divided into the northern mixed forest and southern deciduous forest zones. Only a small area of Canada, in the southern extremity of Ontario, has deciduous forest vegetation.

The pollen record of the mixed forest region is the fullest in Canada, although detailed vegetation reconstruction remains an elusive objective. Several recent compilations and reviews of parts or all of the pollen data, variously treated, are at hand (Davis 1983; Webb, Richard, and Mott 1983; Wright 1984; Anderson 1985; Davis and Jacobson 1985; Gaudreau and Webb 1985; Holloway and Bryant 1985; and Richard 1985, in press). Consequently, what follows is as concise as possible without losing coherence and, of course, I have drawn, in various ways, on some of the insights and conclusions of the above authors, wittingly or otherwise.

Reference was made in Chapter 4 to the major unsolved problem of the whereabouts of the full-glacial refugia of the main taxa of the modern eastern temperate forests. So far, only three sites have been found, in Tennessee, Alabama, and Florida, where some of the taxa have been recorded (see recent reviews by Watts 1983; Delcourt and Delcourt 1985). The evidence for most species is totally inadequate to determine accurate, or even approximate, locations of refugia. As a result, migration maps for these taxa (e.g., Davis 1981, 1983), while they may provide a roughly accurate portrayal of Holocene routes and chronology, are probably quite misleading for the full glacial as they are based on so few sites. Davis, in my view, errs in emphasizing negative evidence – e.g., "Hemlock did not grow in Georgia or Tennessee during the full glacial" (1981, p. 146) – when there are only five sites in these states with late- and full-glacial records. Tennessee and Georgia alone comprise 250,000 km^2 of varied, partially mountainous topography and are more than four times larger than Switzerland, where after five decades of pollen analysis resulting in 500 sites, the postglacial record of beech, for example, can still be extended back 1,000 years in time by new discoveries of macrofossils (Lang 1985). The potential refugial area in eastern North America is a topographically complex region composed of the southern Appalachian ridge of mountains with deep valleys and coves, the incised Piedmont Plateau, and the Coastal Plain. The entire area, comprising at a minimum most of Alabama, Georgia, Tennessee, and the Carolinas, occupies roughly 400,000 km^2 of ecologically highly diverse topography and microclimates. The use of negative evidence in pollen analysis is not recommended in general (Faegri 1985), and it seems very likely that, when the number of sites has been increased tenfold and more, it will be seen that the full-glacial ranges of these taxa were more widespread than the present sparse record indicates. Bennett (1985, in press) makes the same point in his discussion of the possible full-glacial refugium of *Fagus*: its sole full-glacial records in the Tunica Hills of Louisiana (Delcourt and Delcourt 1977) and at Anderson Pond, Tennessee (Delcourt 1979), should not lead to the conclusion that its full-glacial range was so limited. As I will discuss further, recent evidence supports Bennett's (1985) suggestion that several trees (e.g., *Tsuga*, *Fagus*) with roughly equal pollen/vegetation representation indices could have been present in the landscapes around sites though their pollen registration was in very low frequencies or undetectable in low to moderate pollen sums per sample. Others that are significantly underrepresented (*Acer saccharum*, *Tilia*, *Thuja*, *Abies*) will almost certainly remain undetected in normal pollen sums.

The salient feature of the pollen record is that all sites within the modern temperate mixed forest region have the same basic pollen stratigraphy, from the familiar Rogers Lake sequence in the south to the Lac Colin and Ramsay Lake sites (Fig. 5.2) in the north. The sequence of initial registrations and later increases is similar throughout, illustrated here by the Chase Lake sequence from Maine (Fig. 8.1a, from R. B. Davis and Kuhns, personal communication), but the pollen percentage values and the radiocarbon chronologies vary from site to site, and trends in the latter have been used as the basis for compiling postglacial, so called migration, maps for the main tree taxa (Davis 1976, 1983). The record has been interpreted in differing ways, giving rise to alternative but partially overlapping interpretations that are examined later. Most investigators agree that between 13,000 and 11,000 yr BP, when receding ice exposed southern Quebec as far north as the shore of the St. Lawrence River, all of New Brunswick, much of Nova Scotia, and the coastal regions of Newfoundland, a sequence occurred of herb tundra (using the term tundra here only in its botanical sense), shrub tundra and/or poplar woodlands, and finally spruce woodlands or forests. The evidence is still inadequate to ascertain whether these structural units were zonally arranged, according to elevation in the highlands and, elsewhere, according to latitude or distance from the ice fronts. Palaeovegetation maps by Davis and Jacobson (1985) and Richard (1985) show such a zonal pattern, but these reconstructions depend heavily on the interpretation of data from very few sites. It is of interest to note that the initial assemblage, *to the extent that the pollen and macrofossil records are representative of the vegetation cover*, was very similar qualitatively and quantitatively to the pioneer assemblages that occurred a few millennia later in the eastern Arctic of Canada and in West Greenland, discussed below in this chapter (pp. 131 to 133).

A different pattern is seen in the records from southern Ontario and the adjacent Great Lakes (United States) region in general. While the overall floristic composition of the late-glacial vegetation was similar, the initial vegetation in southern Ontario was probably more of a mosaic of spruce groves, poplar woodlands, and tundra than a zonal pattern, implying certain possible palaeoclimatic differences between the maritime and the Great Lakes areas, to be explored below.

By 10,000 yr BP, all of the region now occupied by the temperate mixed forests was ice-free, and, as it happens, the ice margin lay along the modern northern limit of *Pinus strobus*. By then *Picea* and associated boreal taxa (*Larix*, *Betula papyrifera*, *Abies*, *Pinus banksiana*, along with *Fraxinus nigra*, *Ostrya*, and *Ulmus*) spread north into Ontario, Quebec south of the St. Lawrence, New Brunswick, and, in the case of most taxa, Nova Scotia. At that time, many of the dominant trees of the temperate forests were in adjacent areas to the south, in upper New York, Vermont, and Maine, but the precise chronology of their spread northward depends on how the pollen data are interpreted, as will be discussed below.

The Laurentide Ice had receded to about 49°N in Quebec and Ontario (Fig. 2.2) and rapid changes occurred in the species composition of the forests. *Quercus*, *Acer saccharum*, and *Pinus strobus* spread from the south, and the process of transformation of the predominantly coniferous forests of the late glacial began, involving changes in relative abundance of the taxa through responses to environmental change and competitive interaction. Later *Tsuga*, *Betula alleghaniensis*, *Fagus*, *Thuja*, *Carya*, and *Castanea* arrived, and the ensuing changes in pollen frequencies reflected changes in forest populations in response to climatic change, fire, competitive interaction, and biotic factors (examined below). The boreal taxa, with the exception of *Pinus banksiana*

Figure 8.1a. A simplified percentage pollen diagram based on the detailed analysis of the Chase Lake, Maine, site of R. B. Davis and Kuhns (personal communication).

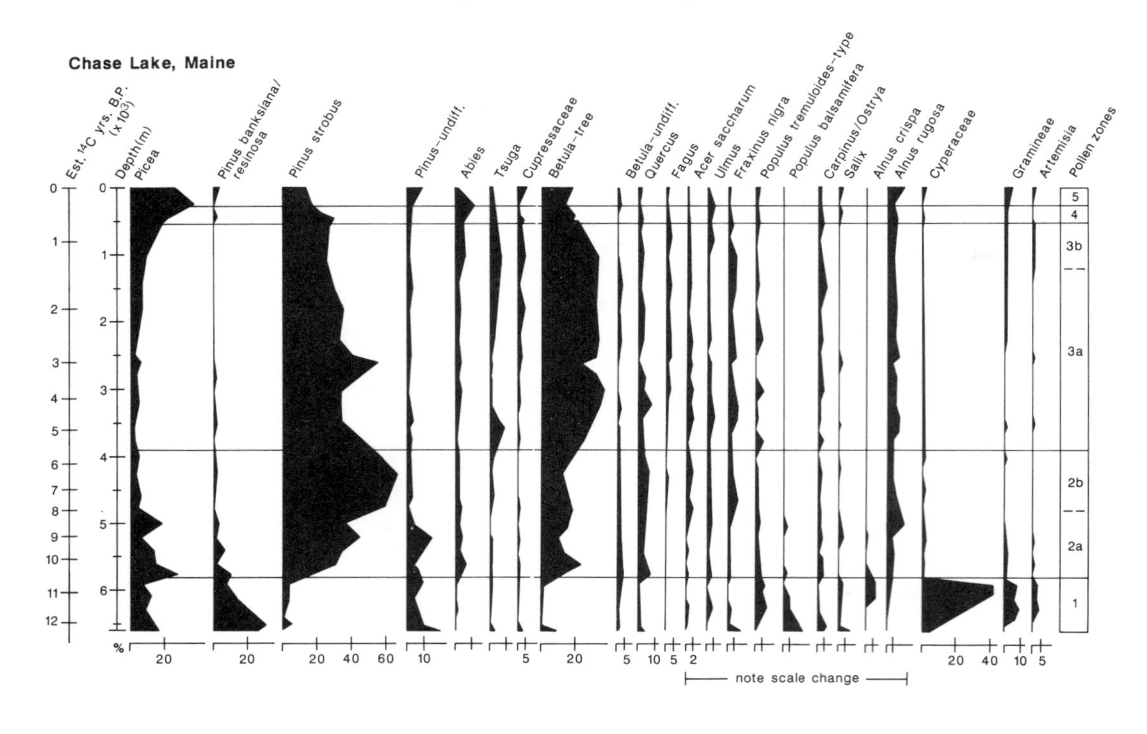

Figure 8.1b. A simplified macrofossil diagram based on the detailed analysis of the Chase Lake, Maine, site of R.B. Davis and Kuhns (personal communication).

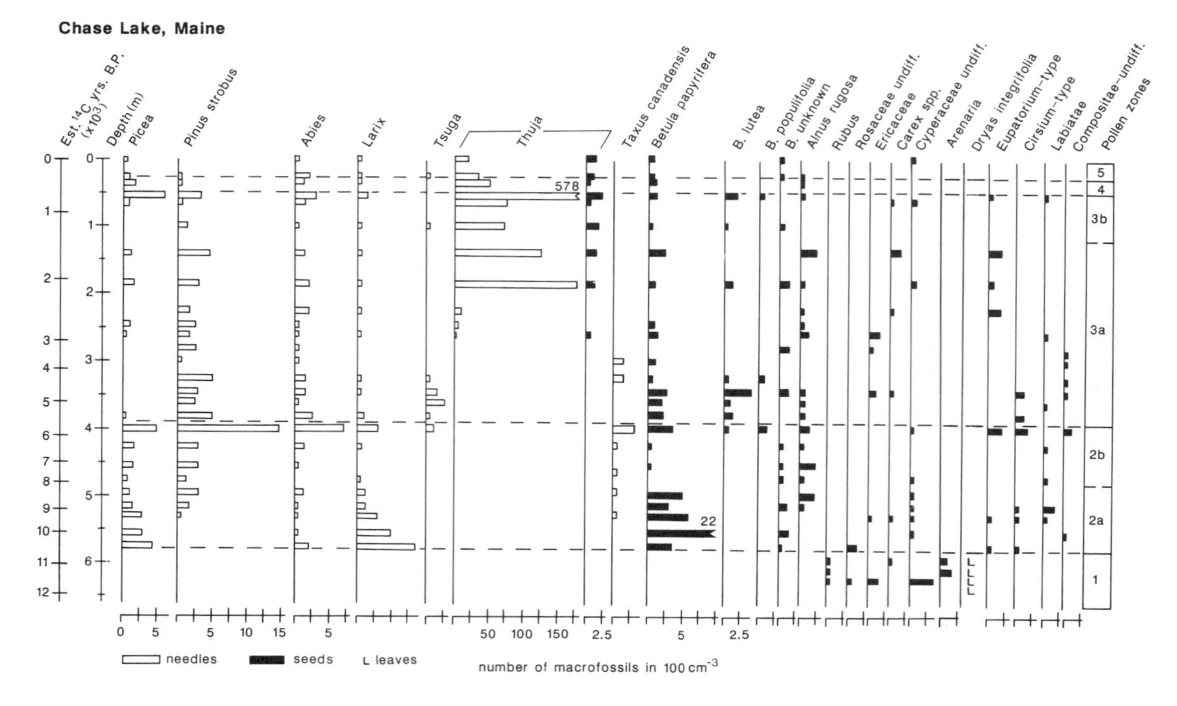

and *Juniperus*, remained in all of ice-free eastern Canada, and they persisted throughout the following 10,000 years in greatly reduced and fluctuating populations, and, in some cases (*Picea mariana*, *Ulmus*), in increasingly restricted habitat ranges.

The chronology and nature of this transition from a boreal to a temperate forest assemblage depends on the choice of criteria when a species is present at the site being investigated. In the absence of more precise PAR and macrofossil data from a much denser network of sites than at present, it will remain a question of subjective choice. The conservative option is advocated by Davis (1976, 1983), who used "the first sharp increase in pollen influx" when available, and/or the first occurrences of macrofossils, or, when only percentage data are available, "increases in pollen percentages." In a recent investigation of a small area of the Upper Great Lakes, she used 1 percent for *Tsuga* and 0.5 percent for *Fagus* as lower limits of pollen values that indicate the presence of the tree at the site, provided that there was "no other large population within 100 km that could have served as a source for pollen grains" (Davis et al. 1986b). The less precise threshold was used by Davis (1983) to compile maps showing the times of first arrival of the main tree taxa of eastern North America to the nearest thousand years. The spatial resolution is equally coarse, so the resulting pollen contour maps indicate only broad routes and migration chronologies. Bennett (1985, in press) and Green (in press) advocate a "more flexible" approach and suggest that low continuous records should be interpreted as indicating the presence of a taxon in the immediate area, except in cases where long-distance pollen transport is likely (e.g., *Pinus*). With this approach Bennett (1985, in press) calculates *Fagus* arrival times in southern Canada that are roughly 1,500 years earlier than those of Davis (1983). The modern and late-Holocene pollen record from eastern Canada confirms the point made by Bennett (1985) that, for several important taxa (*Fagus*, *Tsuga*, *Acer*, *Abies*), "very little pollen is dispersed beyond the species range limits." For example, Richard (1980a) investigated a site near Lake Abitibi (Yelle Lake) at the northern limit of the mixed forest zone. Very small stands of hemlock, maple, and black ash, somewhat isolated from the main populations to the south, occur in the catchment area of the lake. They are registered in the top samples with pollen frequencies of less than 0.2 percent. A continuous but very low percentage curve, preceding a significant rise, is a common feature of pollen curves for several taxa (*Fagus*, *Tsuga*, Cupressaceae, *Acer*). The "flexible approach" suggests that these first registrations be interpreted as indicating the presence of the initial scattered populations of a taxon and that the later expansions reflect a rapid population growth. It is of interest to note that a very recent investigation of a site in Cape Cod, Massachusetts (Winkler 1985b), shows that beech, hickory, and chestnut were present much earlier in the Holocene than the interpretation of the Rogers Lake results suggest (Davis 1983). Macrofossil analysis can be used to strengthen or weaken such an interpretation, but not of course to prove or disprove it because the chance of macrofossils reaching a lake-centre core position from a few scattered trees in the landscape are remote, unless they are shoreline or streamside taxa. Such investigations have been made in eastern North America (Davis, Spear, and Shane 1980; Liu 1982; Jackson 1983, 1986; and Davis et al. 1986). The absence of macrofossils from sediment coeval with a pollen curve that is continuous, but at very low values, provides no secure basis for any conclusions, in common with all negative fossil evidence, but presence is of value. The proper investigation of the meaning of these low pollen values preceding the initial rise by the macrofossil method is laborious but potentially rewarding, as Schneider and Tobolski (1985) have shown from their elegant and definitive combined investigation of the pollen and macrofossils in the sediments of Lago di Ganna in southern Switzerland. After analysis of fourteen complete profiles from marginal and central locations in the basin, it was possible to make detailed quantitative comparisons of the pollen and macrofossils from the same levels in all profiles; they show the presence of bud scales of *Fagus* in sediment of the low-frequency part of the percentage pollen curve in a near-shore profile, followed by a large increase in both pollen and macrofossils roughly 1,000 years later. They conclude that scattered trees of beech were present near one side of the lake and that the familiar beech expansion occurred later. The very detailed investigations of the postglacial vegetation of the Vosges Mountains in northern France by the Utrecht University group have shown a similar pattern for *Fagus*, and Edelman (1985), for example, suggests that the pollen curves, with a mid-Holocene expansion preceded by a long continuous curve at less than 1 percent, means that by 10,000 yr BP, "in the lower river valleys, e.g., the Vologne cleft, *Betula* forests are likely to be present with the thermophile trees like *Quercus*, *Ulmus* and *Tilia*. *Corylus*, *Fagus* and *Abies* are also present" (p. 129).

It is of relevance to note that pollen of *Tsuga*, one of the important taxa still unrecorded in any full-glacial site, is absent from or discontinuously registered in 0 to 2,000 yr BP sediments from both the northern and southern extremities of its range. I have already noted this fact, for the Yelle Lake site, and a similar situation prevails near the modern southern limit of hemlock. Delcourt (1979) notes that "communities of mixed hardwoods, hemlock, and hemlock–hardwoods typically are restricted to north- and northeast-facing slopes of steep sided gorges" in middle Tennessee. Yet the pollen record from her Mingo Pond site in that region shows no hemlock for the past 2,000 years, and at Anderson Pond it occurred discontinuously at very low frequencies over the same period. The mapped pollen frequencies and growing stock volumes (GSV) for hemlock, given the coarse scale of the maps, show a similar relationship (Delcourt, Delcourt, and Webb 1984). However, the *Tsuga* example suggests that the authors' use of forest inventory data of merchantable timber as the basis of estimates of the representation of a taxon in the landscape involves a significant underestimation. Their map for hemlock (Fig. 18) shows no GSV, and apparent absence of the tree, in Ohio, Indiana, and Kentucky. However, a detailed map of hemlock distribution (Olson, Stearns, and Nienstaedt 1959a) and references in the literature (e.g., Black and Mack 1976) indicate that local hemlock stands are found in fifteen counties of Ohio, ten in Indiana, and at least half the counties in Kentucky. Yet, hemlock pollen is absent from modern samples in all three states and even absent, or of sporadic occurrence, in Tennessee, where the tree occurs in roughly twenty counties, frequently in some. In other words, scattered hemlock stands over a wide area of diverse topography have gone unrecorded in the roughly thirty-five modern pollen stations recorded within the area by Delcourt, Delcourt, and Webb (1984, Fig. 3). There is no reason to assume that similar examples of scattered populations of low density would be registered differently in the pollen rain.

Nonetheless, firm conclusions about the correct interpretation of the continuous but very low pollen frequencies of tree taxa in early Holocene sediments from eastern Canada and adjacent United States are not possible until dense networks of sites with detailed pollen and macrofossil analyses of near-shore and central cores have been completed. A grid of closely spaced sites is necessary to distinguish between pollen sources, such as "a small, nearby colony, and a far-distant but very large population of trees" (Davis et al. 1986). It is obvious that the early Holocene history of the forests of eastern Canada and adjacent United States is very inadequately known, and investigations of sediments in small repositories should prove to be informative, particularly if pollen concentrations are determined at high enough levels to provide adequate registrations of such important but overlooked taxa as *Abies*, *Larix*, *Acer*, and *Thuja* (*Juniperus*). All are recorded in early Holocene eastern sites at higher values than in any modern samples, from forests where all are abundant, yet the prevailing view suggests that the early Holocene forests of eastern North America were spruce-dominated (Delcourt and Delcourt 1981; Delcourt, Delcourt, and Webb 1984). Only Richard (1978, 1981a) gives due recognition to the problems of over- and underrepresentation of taxa.

Boreal forest

The hypotheses that some boreal elements, particularly white spruce (Raup and Argus 1982) and balsam poplar (Hopkins, Smith, and Matthews 1981), might have survived the latest ice age in the northwest remain open to question in light of the continuing absence of tree macrofossils from full-glacial sediments in Beringia. However, I acknowledge the limitations of such negative evidence. MacDonald and Cwynar (1985) offer a more compelling case, based on both migrational and genetic data, against an older view that lodgepole pine survived in Beringia. On the other hand, the few detailed studies available (Cwynar 1982; Ritchie 1982) provide evidence that some of the important boreal chamaephytic elements (dwarf birch, ericads) and miscellaneous herbs persisted in Beringia and probably spread eastwards into the interior, as MacDonald (1984) and Edwards and Brubaker (1986) have suggested for *Betula glandulosa* in the Mackenzie River and Porcupine River valleys, respectively.

The current evidence indicates that most of the taxa, and probably all of the tree and tall shrub species, spread northward from a refugial area in the midcontinent in the same general pattern of spread, though not necessarily with the same chronology, as has been suggested for white spruce (Fig. 8.2 from Ritchie and MacDonald 1986). The review of full-glacial vegetation in Chapter 4 and the several recent accounts in the literature cited in that chapter document adequately that all of the important boreal elements were widespread in the ice-free continental region from Kansas to the Carolinas and, variably, as far south as Texas and Missouri. While the zonation, structure, and composition of the vegetation cover remain unknown because of the inadequacies of the record, the presence in some fossil records of plant and animal taxa with good indicator value permits some attempts (below) to reconstruct the environment. The records from the northernmost full- and late-glacial sites suggest that boreal vegetation (defined simply as predominantly coniferous and arboreal) did not extend to the ice margin and that a belt of treeless vegetation of unknown extent intervened. For example, the detailed, well-dated studies at Wolf Creek, Minnesota (Birks 1976), and Conklin Quarry, Iowa (Baker et al. 1986), could be interpreted as a north–south gradient of vegetation, at

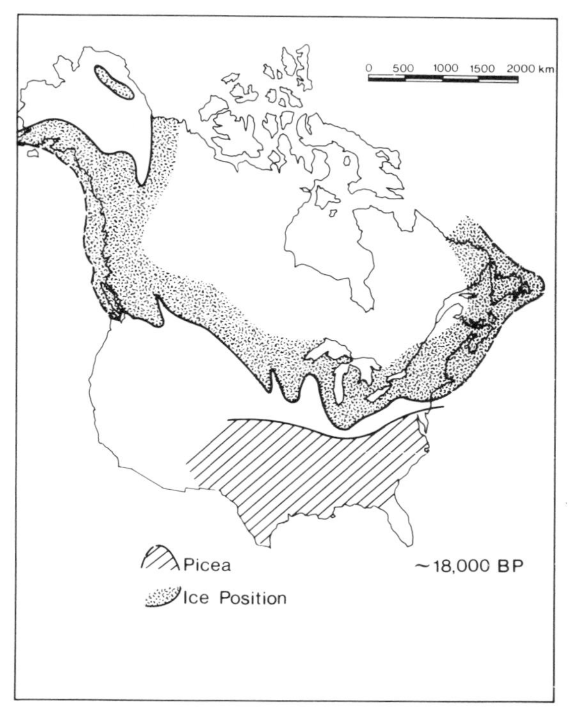

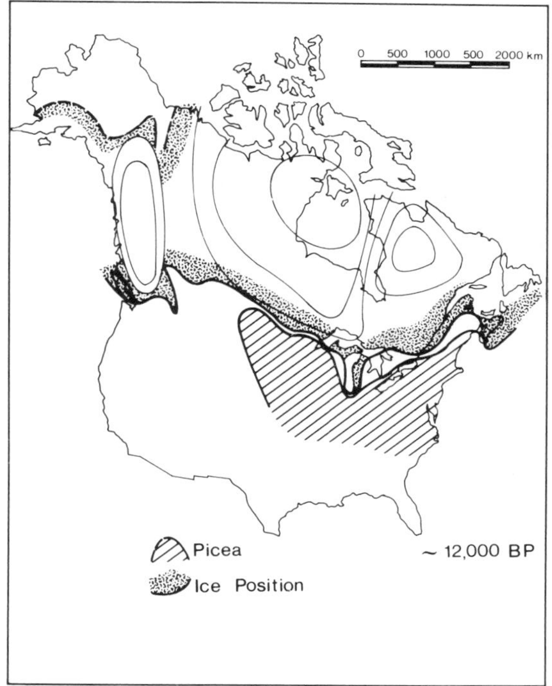

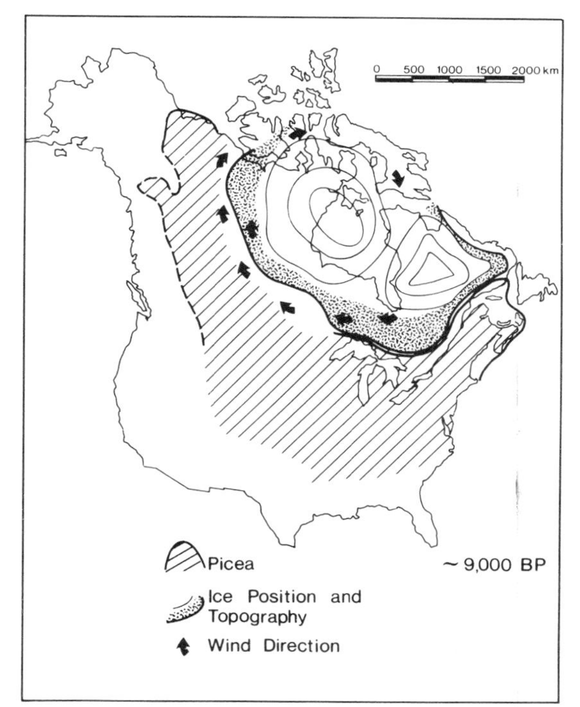

Figure 8.2. A simplified mapped summary of the history of *Picea* (chiefly *P. glauca*) in North America, showing 18,000, 12,000 and 9,000 yr BP approximate ice positions and the model-simulated fall and winter wind directions from Kutzbach and Guetter (1986). The maps were published originally in Ritchie and MacDonald (1986).

about 18,000 yr BP, from a discontinuous spruce–larch woodland or forest–tundra in Iowa, analogous to the modern forest–tundra, to a predominantly treeless landscape 500 km farther north at Wolf Creek. However, by 13,000 yr BP, when small land areas along the southern fringes of Canada were first exposed, a belt of spruce-dominated forest extended across the interior plains from the Dakotas, through the southern Great Lakes area to the Atlantic, and, while details of its structure and composition remain uncertain, it is likely that all of the main taxa (*Picea*, *Pinus*, *Larix*, *Abies*, *Betula*, *Alnus*, *Populus*, *Salix*) as well as *Fraxinus nigra*, *Ulmus*, *Ostrya* (and/or *Carpinus*) were present. By 12,000 yr BP, this essentially boreal assemblage, no doubt varied longitudinally, extended from southern Alberta to the Atlantic coast, but all of the land eventually to become occupied by the modern boreal forest, except in the far northwest, still lay to the north under ice. Most of the boreal elements persisted in the east after 12,000 yr BP, being joined by and locally, but not zonally or regionally, replaced by trees of the mixed temperate forests – oak, white pine, maples (sugar and red chiefly), yellow birch, beech, and hemlock (Davis and Jacobson 1985). It is important to stress that none of the boreal trees, except locally jackpine, became completely absent in southern Ontario, southern Quebec, New Brunswick, or the Maritimes during the Holocene as they might have had the climate changed to a greater extent. Their relative abundances declined sharply as they were replaced by more efficient forest competitors, particularly on mesic sites. But they persisted, as they do today, often in restricted habitats.

A thoroughly documented example of the persistence mechanisms of boreal trees in the temperate mixed forest region is the long-term permanent plot investigations set up in northern Lower Michigan in 1919 to follow the succession after earlier logging and burning. Sakai and Sulak (1985) show that the boreal taxa *Abies balsamea*, *Thuja occidentalis*, *Picea mariana*, *Fraxinus nigra*, and *Populus tremuloides* have all persisted for over four decades in lowland sites with poor drainage and peaty soils, and Sakai, Roberts, and Jolls (1985) have shown that, on mesic sites in the same area, early successional boreal species (aspen and white birch) became established initially by vigorous vegetative growth, but are now showing high mortality as shade-tolerant, longer-lived species (beech, sugar maple, and others) increase. The early Holocene spreading of temperate forest dominants into the northern Great Lakes area and adjacent eastern Canada displaced the boreal species from all sites except localized hydric habitats and early successional stages following disturbance. This provides an ecologically more acceptable concept of the late glacial–Holocene transition than the notion of whole biomes being replaced (Delcourt, Delcourt, and Webb 1984). Farther north, beyond the range of climatic tolerance of yellow birch, beech, hemlock, sugar maple, and white pine, the boreal taxa assumed and retained dominance as the landscape became ice-free. Several of these basic ideas about the ecological status of boreal species in the eastern temperate forest zone were enunciated, in a different context, over fifty years ago in a classic essay by Nichols (1935).

In Quebec–Labrador, the modern boreal forest reached its approximate present-day extent about 5,000 years ago. In contrast, in western Canada, the early spruce forests that occupied the landscape in southern Manitoba, Alberta, and Saskatchewan were replaced entirely about 9,000 yr BP when grassland and parkland expanded. As the ice receded in the Western Interior, the boreal elements (first poplar, then the spruces, tree birch, and later jackpine) spread onto the landscape. It has been suggested independently by Vance (1986) and by Ritchie and MacDonald (1986) that the very rapid spreading into the northwest of at least white spruce was controlled in part by increased southerly winds in fall and winter in the early Holocene as a result of adiabatic flow from the large remnant of Laurentide Ice that occupied the northern part of the Western Interior (Fig. 8.2, from Ritchie and MacDonald 1986, based on palaeoclimatic reconstructions by Kutzbach and Guetter 1986, and the original suggestion about Holocene wind directions by David 1981). Kutzbach and Wright (1986) provide a detailed discussion of the postulated early Holocene environment to explain the formation of parabolic dunes in the Lake Athabasca region. They show a reconstruction of the 9,500-yr-BP Laurentide Ice Sheet with anticyclonic southeasterly winds along the ice margin in western Canada, similar to the reconstruction reproduced here in Figure 8.5.

The history of boreal forest development varies across Canada as a function of the chronology of deglaciation and latitudinal position. The length of time available for vegetation development since deglaciation probably determined the sequences. For example, in the far northwest, surfaces now occupied by boreal forest (Lower Mackenzie region, northern Yukon) were treeless for several millennia before poplar and spruce arrived, and a sequence of herb tundra–shrub tundra–poplar–spruce is typical, roughly similar to the late-glacial sequence at some sites in eastern Canada and adjacent United States that were deglaciated by 12,000 yr BP (e.g., Basswood Road Lake, Fig. 5.2). By contrast, sites in regions where deglaciation occurred late (west-central Nouveau Québec) show that the boreal tree and shrub vegetation developed almost immediately (e.g., Richard, Larouche, and Bouchard 1982, Delorme II site, Fig 5.5).

The western boreal forest provides a clear example to illustrate a point made recently by Faegri (1985) when he used a theoretical boreal forest to note that the registration in pollen diagrams of a single environmental (climatic) change causing ecotone movement varies with the latitude of the sample site, such that investigators working at stations in the middle of the boreal forest or well beyond its boundary "will maintain that there has been no climatic event," whereas those working near but at varied distances from the ecotone "will agree that there has been a climatic event but disagree as to its duration and importance." Along the southern ecotone with the grasslands, the boreal forest has existed at two different times since deglaciation in its two forms – the early version from 11,000 to 9,500 yr BP (Fig. 6.2) was dominated by *Picea glauca* with small amounts of other boreal taxa and probably no pine. Following several millennia of absence, when grassland and parkland prevailed, it reappeared in its modern guise with a larger representation of birch, poplar, and jackpine. The expression of that change (presumably climatic, discussed below) is barely perceptible at sites in the interior of the boreal forest (Fig. 6.3). At its northern limit in both the western and eastern sectors, the boreal forest has receded since the mid-Holocene. On the other hand, most sites distant from the ecotones, within the boreal forest and across its entire width, show an early Holocene dominance of spruce, often white spruce, followed by a more evenly mixed assemblage of the main boreal taxa (birch, pine, alder, spruce) that persists to the present day. The Lake A site in Saskatchewan (Fig. 6.2) and the Flin Flon site in Manitoba (Fig. 6.3) illustrate the different histories of the boreal forest, and a graphical portrayal of the principal components analysis scores of the two sets of samples provides a useful alternative method of expressing the contrasting sequences (Fig. 8.3 from Ritchie and Yarranton 1978). The Lake A sequence proceeds from the early boreal to the modern boreal forest by way of a grassland phase, while the Flin Flon pattern is from the early boreal forest to modern by modifications in tree composition and abundance. Similarly, at the northern ecotone, the response of the early Holocene boreal forest to a mid-Holocene climatic cooling is expressed directly at the margin by a regression to tundra, while at sites farther south, within the forest, the changes are solely in tree composition and abundance. The primary pollen data show these variations in history (Figs. 6.2 and 6.3), and a numerical comparison provides an effective supplemental method of presenting the data (Ritchie 1977; MacDonald and Ritchie 1986).

The eastern part of the boreal forest reveals patterns of change that differ from those in the west. Sites at the southern limit have early and late-Holocene boreal forest stages, but, while in western Canada the intervening phase was one of prairie, in eastern Canada it was mixed (temperate) forest. Liu and Lam (1985) demonstrate this sequence effectively by applying discriminant analysis to their Jack Lake data. The analysis provides a numerically derived vegetation reconstruction based on modern samples (Fig. 8.4) and shows that the early boreal forest was transformed into a mixed (temperate) forest in the mid-Holocene and reverted to the modern boreal forest in the late Holocene.

The differences between the western and eastern sectors of the northern ecotone can be related primarily to the great differences in deglaciation history. As the pollen record from Labrador and Nouveau Québec showed (Fig. 5.5), the boreal elements arrived late in the region, probably spreading north along the Atlantic zone to roughly Ungava Bay by 6,800 yr BP and thence westward into the central uplands as they became exposed by the final ice disintegration at 5,000 to 4,000 yr BP (suggested routes are mapped in Richard 1981a, p. 119). He demonstrates that the maximum density of the forest–tundra to the southwest of Ungava Bay occurred at about 4,200 to 4,000 yr BP, a conclusion that agrees closely with the findings of Short and Nichols (1977), Short (1978), and Lamb (1985) from other sites in subarctic Labrador–Quebec. Since 3,000 yr BP, the northern limits of the boreal forest and forest tundra have retreated. Toward the end of last century and the early decades of the present one, there has been a minor northward expansion (Payette and Filion 1985; Payette and Gagnon 1985). Late-Holocene treeline changes have been adduced from tree stumps, charcoal layers, and fossil forest soils in modern tundra areas of Keewatin, and, while the fine details of the record are variously reconstructed by different authors, there is agreement that the treeline retreated significantly between 3,500 and 3,000 yr BP (Sorenson and Knox 1974). As Bradley (1985) points out, trees near the treeline can be "out of phase with contemporary climatic conditions." Reliable interpretation requires detailed analyses of several aspects of tree response to change, as the investigations by Payette's group, cited above, illustrate so convincingly. Also, as Gajewski (1983) demonstrates, climatic change can be detected within a vegetation region as well as at its ecotone if annually varved sediments are available for analysis.

In summary, the northern and southern boundaries of the boreal forest have changed position during the Holocene, and, while interior sites distant from the margins have remained forested throughout postglacial history, compositional

Figure 8.3. Ordinations of the Lake A and Flin Flon fossil pollen samples grouped in pollen zones, each with the mean weighting (solid dot) and standard deviations (bars), and of the modern pollen samples from the main boreal forest and the grasslands. The radiocarbon age range of each zone is shown in yr × 10^3. The plots illustrate the similarities of the initial and final stages and the differences between the intermediate stages of the two vegetational histories.

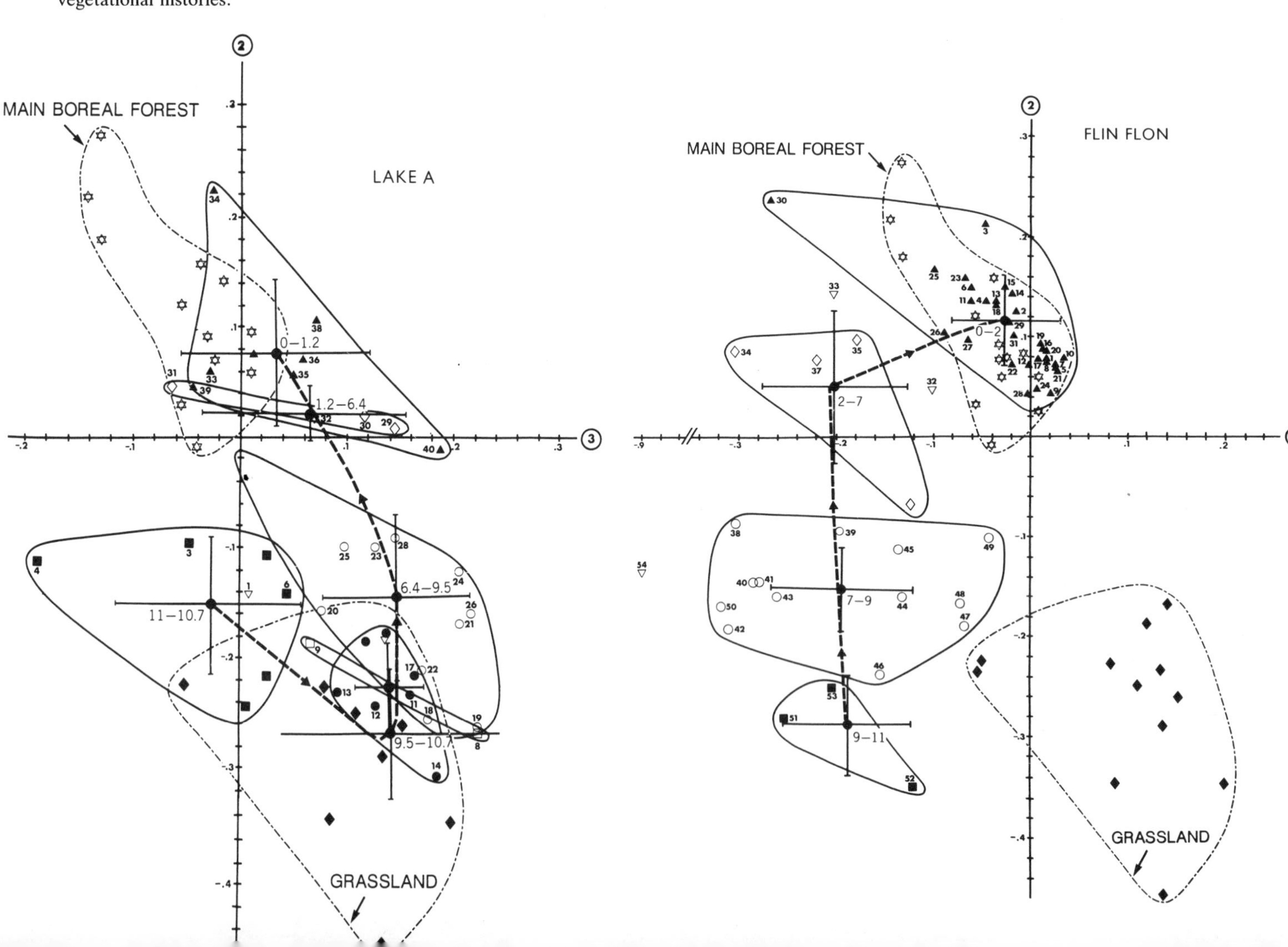

Figure 8.4. Sites near the southern edge of the present-day boreal forest of eastern Canada show a mid-Holocene northern extension of the mixed temperate forest. For example, the Jack Lake site in northern Ontario, investigated by Liu and Lam (1985), shows such a sequence, redrawn here from their paper. The vegetation zonal index is a statistical measure of the probability that a fossil spectrum belongs to a particular vegetation type, based on a large set of modern pollen samples. Full details are given in Liu and Lam (1985).

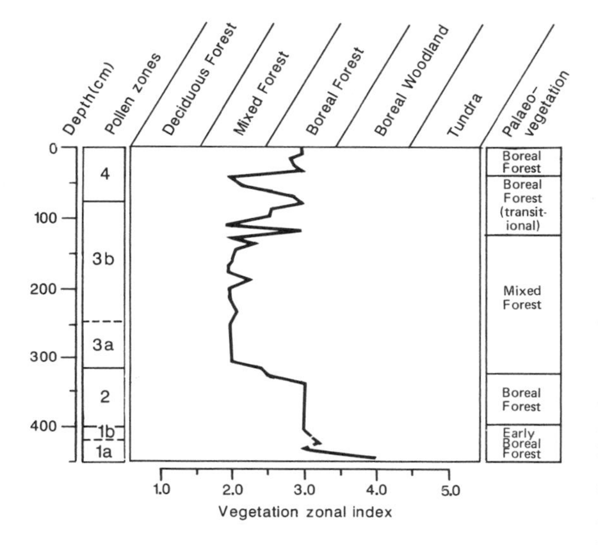

changes have occurred such that the modern regional assemblage reflected in pollen samples has been stable for only about five or six millennia.

Grasslands and parklands

The grasslands of British Columbia are considered in the section on the Pacific–Cordilleran region, although, as Hebda (1983b) has pointed out, it is probable that the history of the Great Plains grasslands was connected with that of the intermontane grasslands. However, knowledge of the origins of the modern Canadian prairies is fragmentary because of the great dearth of fossil sites between the eastern foothills of the Rocky Mountains and the Missouri River – the northern Great Plains of much of Wyoming, Montana, the Dakotas, and adjacent Nebraska. With the exception of a site in Wyoming investigated recently by Markgraf and Lennon (1986), discussed below, the sites in the western region are either at high elevations or lie beyond the Rockies on the northern edge of the Great Basin (reviewed by Baker 1983; Baker and Waln 1985; Mehringer 1985).

The postglacial history of the Canadian prairies and parklands is simple, as described in Chapter 6. The early version of the boreal forest that occupied a wide belt across the centre of the continent from 14,000 to 12,000 yr BP was replaced between 10,000 and 9,000 yr BP by a parkland or prairie vegetation on all sites except bottomlands and north-facing valley slopes, where groves of *Fraxinus*, *Ulmus*, and, in the eastern sector, *Quercus macrocarpa* prevailed as they do today. Outlying populations of *Picea glauca* and *Populus balsamifera* in the Black Hills of South Dakota might well be survivors of the early boreal assemblage. The records reviewed in Chapter 6, from sites in the Dakotas and the southern parts of the three Prairie Provinces of Canada, indicate that the aspen parkland and grassland developed from those residual elements of the late-glacial assemblage that could tolerate a drier, possibly warmer summer climate. Grimm (1983) offers the interesting interpretation of both the older record and his new site, *Cottonwood Lake* in central South Dakota, that the early Holocene dominance of Gramineae and *Artemisia* with minor but constant amounts of deciduous trees probably represented an early version of aspen parkland, expressed imperfectly in the pollen record because of poor poplar pollen preservation. He suggests that the later increase of Ambrosieae associated with such rare but ecologically sensitive indicator taxa as *Amorpha*, *Petalostemum*, and *Sphaeralcea* represented a replacement of the parkland by prairie. A similar pattern is evident in some of the Canadian pollen sequences (Fig. 6.2) and a similar vegetational history might also describe these sequences accurately. At least two sites at the southern fringe of the boreal forest in western Canada (Lake A and Riding Mountain, Fig. 6.2) provide clear evidence that the prairie zone extended farther north in the mid-Holocene than today, and that the modern boreal forest at these sites has a short history. In a similar way, the prairie zone expanded eastward in the mid-Holocene into the midwestern United States, shrinking back again from about 3,500 years ago (Webb, Cushing, and Wright 1983). The authors make the important point, derived from detailed investigations by Grimm (1983) of the prairie–forest border in south-central Minnesota that, while the overriding regional factor controlling these changes was climate, the local expression was influenced strongly by topographic factors that determined the extent of wildfires.

A recent investigation from the prairie–forest border in Wisconsin documents these conclusions in greater detail by the presentation of pollen, charcoal, and sedimentological analyses (Winkler, Swain, and Kutzbach 1986). Earlier descriptions of the aspen parkland in Canada (Bird 1961) included assertions that the southern boundary of the boreal forest had extended southward since European settlement as a result of decreases in wildfires set in earlier times by indigenous people. Zoltai (1975) has reexamined the areas mapped at the beginning of this century that formed the basis for these claims and found them to be inaccurate.

In the absence of secure data from much of the northern Great Plains, one can only surmise that, during late- and full-glacial times, the important dominants of the modern grasslands of southern Canada, now greatly reduced in area by agriculture, occupied smaller areas in the plains region near the Rocky Mountain foothills but to the north of the Great Basin. Recent collations (Spaulding, Leopold and Van Devender 1983; Spaulding and Graumlich 1986) have shown that full-glacial environments in the Great Basin region must have been cooler and moister than present, on the basis of plant macrofossil (packrat midden) and microfossil data; these results correlate well with palaeoclimatic reconstructions based on lake-level analyses (Smith and Street-Perrott 1983). On the other hand, the evidence from the Columbia Basin region suggests that the full-glacial vegetation was a steppe assemblage in a cold, dry climate. It is possible that some elements of the northern Great Plains vegetation were derived from this region. A recent investigation by Markgraf and Lennon (1986) of a site in central Wyoming provides pollen, ostracod, and macrofossil evidence that sagebrush–grassland vegetation has occupied the area unchanged since 13,000 yr BP. In addition to filling an important gap in our knowledge of the vegetational history of the northern plains, this study offers the intriguing possibility that the full- and late-glacial spruce-dominated assemblages recorded from the eastern plains did not extend westward beyond the Dakotas.

Pacific–Cordilleran complex

The northern and southern extremities of this linear, complex montane region have moderately dense networks of fossil sites, but the central zone is sparsely represented, except for bog sites along the coast. As it happens, the direction of ice recession was roughly from the extremities towards the centre, so the longest records are found in Washington and in the northern Yukon. The reconstructed vegetation of northwestern Washington and southern British Columbia during full- and late-glacial times was reviewed in Chapter 7, and it was concluded that only the most general outlines have been established, with no clues as to the occurrence of refugial populations of the temperate conifers that dominate the modern forests. Heusser (1985, p. 160) notes, however, that the record from older, interglacial sediments reveals great similarities with modern forests and he concludes that "Pacific coastal forest interglacial communities . . . have undergone little change as a result of physical upheavals during ice ages."

The vegetation of lower elevation sites changed steadily during the late-glacial period from a parkland with abundant lodgepole pine and mountain hemlock to a forest or woodland with Engelmann spruce and alpine fir as well as the pine and hemlock. But evidence of other taxa in the macrofossil record of at least one site (Kirk Lake, Fig. 7.3; Cwynar in press) makes it clear that much local variation in structure and composition, and in elevational differentiation probably occurred, so far largely undetected in the pollen record. Changes in the Holocene occurred rapidly at first as Douglas-fir, Sitka spruce, and western hemlock expanded to dominate lowland sites. The ensuing changes were slower as *Thuja plicata* expanded to assume local dominance. In the northern Pacific region, interesting sequences of migration and spread have been suggested (Heusser 1985; Cwynar personal communication) from an initial pine with alder at 10,000 yr BP followed by *Picea*, *Tsuga heterophylla*, *T. mertensiana*, and a reappearance of *Pinus*. Heusser (1985) notes that some of these migrations, or increases, occurred very late in the Holocene in the Gulf of Alaska, for example, at Prince William Sound, where Sitka spruce and mountain hemlock first appeared at 3,000 yr BP but reached Kodiak Island only within the past 1,000 years, documented by both occurrence of the pollen increase below but close to the Mount Katmai tephra of 1912 AD and by age structure studies of spruce populations.

Inland, in the northern Cordilleran, an easier interpretive task is presented by the long, relatively consistent records in the northern Yukon, indicating a sequence from full-glacial herb tundras through late-glacial dwarf shrub tundras to either boreal forest, forest–tundra, or tundra in the Holocene, depending on the altitude or latitude of the site.

A tidy summary of the vegetational history of the southern Cordilleran is not possible, partly of course because the region is extremely complex topographically, with a wide range of meso-climatic zones, but partly because of inherent difficulties in the pollen record.

It is likely that further investigations, some currently in process, will provide the data needed to address some of the problems touched upon above. Secure documentation on the timing and routes of spread of the major trees requires the addition of carefully chosen sites, well spaced both horizontally and vertically to detect the critical registrations of pollen and macrofossils. Cwynar (personal communication) has in hand data from several closely dated lake sites in northern British Columbia and south Yukon that should provide a detailed chronology of the spread of *Pinus contorta*, the *Tsuga* species, and *Picea*.

Tundra (arctic)

The origins and history of the arctic flora have intrigued plant geographers for many decades and a large body of highly discursive literature continues to grow as new investigators revisit old ideas. Interested readers can pick up the threads of this set of speculations and assertions in the contributions of Löve and Löve (1974), Murray (1981), and Yurtsev (1982). Little significant progress has been made in elaborating and testing the germinal propositions of Hultén (1937) about the glacial refugia and postglacial radiation of arctic and boreal plants except that some of the suggested centres of persistence in eastern Canada (Newfoundland, Labrador, Gaspé) have been shown to have been ice-covered in the latest full-glacial period (Ives 1974; Prest 1984), and Hultén's (1937) suggestion that conifers persisted in Beringia during the full-glacial period is not supported by the fossil record. Late Wisconsinan nunataks or local ice-free areas, potential refugia for plants, occurred in Gaspé (Mont Jacques-Cartier), coastal southwest Nova Scotia, and West Greenland. Brassard (1984) draws attention to the evidence from moss floristics in Newfoundland and adjacent regions that appears to support the suggestion that parts of coastal Newfoundland remained ice-free in the full glacial. He writes: "In addition to the possible refugia in the Long Range Mountains . . . other species with a coastal distribution in Newfoundland might have survived in refugia located along the south coast or on the Grand Banks, parts of which would have been ice-free and above sea level during the last glaciation."

What of the herb and shrub tundras that occupy a vast expanse of northern Canada and much smaller peripheral areas of Alaska and Greenland? Can the origins of these vegetation types since the maximum of the latest glaciation be reconstructed? The fullest records are in eastern Beringia, where the long and detailed pollen sequence from Hanging Lake and Antifreeze Pond (Fig. 7.4) of Cwynar (1982) and Rampton (1971), respectively, supplemented by shorter records from elsewhere in the northern Yukon and adjacent Alaska (see Chapter 7), provide convincing evidence that herb and shrub tundras were extensive during the full- and late-glacial intervals (Ager 1986). But, so far, no reliable records of pollen or macrofossils have been reported from full-glacial sediments in any of the arctic islands in western Canada that were ice-free (Fig. 1.1).

A growing number of pollen and macrofossil analyses of full-glacial sediments from sites south of the Laurentide Ice (Chapters 4 and 5) provide abundant evidence that many taxa with modern arctic ranges were common in this southern refugium. Assemblages of these taxa, often associated with nonarctic species (e.g., the Wolf Creek assemblage, Birks 1976; the Columbia Bridge assemblage, Miller and Thompson 1979; and the St. Eugene assemblage, Mott, Anderson and Matthews 1981, in Table 5.1) appeared to spread north and east into eastern Quebec and, presumably along the Labrador coast, on newly deglaciated surfaces. It is possible, though still undocumented, that the coastal plain, exposed during the full glacial, served as a refugium and migration route for arctic taxa that appeared in the late-glacial records of Nova Scotia and Newfoundland (Holland 1981). A nonarboreal pollen assemblage preceded the early *Picea*-dominated assemblage at most sites in eastern Canada, with the notable exception of northwestern Nouveau Québec, where trees and shrubs initiated the process of revegetation. However, it is unlikely that a direct connection between these partly arctic assemblages of southern origin and the Beringian radiants could have occurred until later than 10,000 yr BP when a large mass of confluent Laurentide Ice still occupied much of northeastern Canada (Fig. 2.2). The pollen and macrofossil record from the interior plains of Canada and the adjacent United States provides some indication that the full-glacial tundra assemblage recorded at Wolf Creek, Conklin Quarry, and elsewhere became less frequent in occurrence northward, as the Laurentide Ice diminished in area. Many pollen sites lack an initial NAP assemblage, as several investigators have noted (Terasmae 1967; Wright 1968; and Saarnisto 1974). It is relevant to note that the more sensitive fossil beetle record is interpreted by Ashworth and Schwert (1986) in terms of the regional extinction of certain taxa south of the Laurentide Ice, at about 15,000 yr BP and that "the unglaciated Alaska–Yukon region served as the principal refugium for the arctic–subarctic beetle fauna." It is possible that parts of the arctic–subarctic flora that occupied full-glacial ice marginal habitats in the continental interior became regionally extinct as the climate warmed and forests spread northward as rapidly as the ice sheet diminished. Some taxa, on the other hand, particularly in the Great Lakes area, appear to have spread northward to the arctic, leaving relict populations in localized habitats (as documented in detail by Given and Soper 1981). A hypothesis of regional extinction is testable with beetle data, but it would require a dense network of pollen *and* macrofossil sites to establish that regional absence of a plant taxon was due to extinction rather than simply a gap in the fossil site inventory or to truncated sedimentary sequences.

It should be noted that the depiction of ice sheets in Figure 2.2 is schematic and conforms to the "maximum portrayal" of Prest (1984). The "minimal portrayal" for the eastern arctic shows large ice-free areas on Devon, Ellesmere, and Axel Heiberg islands. In fact the earliest eastern arctic fossil (pollen) records of tundra vegetation cluster round 9,000 to 10,000 yr BP (Fredskild 1969, 1973, 1983, 1985; and Funder 1978, 1979, sites in Greenland; Hyvärinen 1985, the Baird Inlet, Ellesmere Island site). The implication of the above rough schema is that isolated genotypes of many wide-ranging arctic species would have become genotypically diffentiated in the Pleistocene. Areal continuity, with the resulting potential for gene exchange, would have been established only in the early Holocene. Evidence for such processes in *Oxyria digyna* was noted in Chapter 3, but other examples are rare. It is likely that the northwestern refugium, including eastern Beringia and the unglaciated parts of the Arctic Archipelago (on Banks, Victoria, Melville, Prince Patrick, Bathurst, Ellef Ringnes, Amund Ringnes, Cornwall, and Cornwallis islands), provided the main sources of the arctic vegetation, the disseminules being spread efficiently by wind over open, often frozen and snow-covered, landscapes (Savile 1972; Glaser 1981).

Despite the paucity of sites in the central and eastern arctic, some attempt to recognize patterns across the entire North American–Greenland arctic zone is of interest. Just as it was noted that the differences in the development of the boreal forest were related to relative deglaciation history between eastern and western sectors, so with the arctic. It has been shown from pollen and macrofossil analysis of roughly thirty sites in Greenland (reviewed recently by Kelly and Funder 1974; Funder 1978, 1979; Fredskild 1983, 1985) that, while ice-free areas might have persisted locally in eastern Greenland, the earliest recorded vegetation dates from 10,000 yr BP and that a broadly similar pattern of vegetation has been recorded, varying chiefly with distance from the coast. The pattern consists of an initial or pioneering assemblage of predominantly herbaceous plants dominated by grasses, sedges, *Oxyria*, *Saxifraga oppositifolia*, and others, with a few ericaceous taxa, in particular *Vaccinium uliginosum*. It is of interest to note that this assemblage, with only minor variations, is recorded not only throughout Greenland but at the Baird Inlet site on Ellesmere (Hyvärinen 1985) and in many of the late-glacial and early Holocene sediments from sites in eastern Canada (see, for example, Table 5.1 in Chapter 5; the late-glacial sites in Newfoundland; and the initial assemblages from sites near Ungava Bay, Quebec, described in Chapter 5, pp. 83–4). In Greenland and Ellesmere, this initial assemblage occurred from 10,000 to roughly 8,000 yr BP when the vegetation became shrubby with the arrival and spread of *Salix* (often *S. arctica*) and additional ericaceous shrubs. In low-arctic west Greenland and Baffin Island, *Betula glandulosa* increased in the mid-Holocene (about 5,000 yr BP), and, at sites between 61° and 63°N on west Greenland, *Juniperus* and then *Alnus crispa* increased significantly at about 4,500 yr BP. Later, declining frequencies of *Betula* and *Juniperus* on west Greenland and increases in fell-field taxa at high latitudes indicate a significant change. The general palaeoenvironmental reconstructions offered are that the initial assemblage reflects simply a group of tolerant, opportunistic, wide-ranging species; that the spread of willows and heaths is mainly a regional migrational event, accelerated by soil development, and that "around 8,000 yr BP at the latest the climatic conditions were favourable for the immigration of *Salix* and heaths"; and that the later (3,500 yr BP) retrogressive change "is climatically significant, particularly as the change is preceded by a lengthy period of vegetational stability" (both quotations from Hyvärinen 1985).

The vegetational history of the western arctic is based on very few sites, as reviewed in Chapter 6. The eastern arctic sites have only Holocene records, at even fewer sites, but a common pattern is found – an early herb assemblage followed at 7,000 to 8,000 yr BP by a *Salix arctica*-dominated assemblage; ericads increase at 7,000, and at 3,500 to 4,000 yr BP, "indicators of bare ground and fell-field increase." In southwest Greenland, Fredskild (1983) describes in great detail a sequence from a pioneer, predominantly herbaceous stage with *Oxyria*, *Saxifraga*, *Silene*, *Minuartia* and sedges at 9,400 to 8,000 yr BP followed by increases in ericads and then (8,000 to 6,300) a *Salix*–Cyperaceae stage. *Betula* and *Juniperus* increase at 6,000 to 3,500 yr BP followed by *Alnus*–*Betula* (3,500 to 1,800) and finally a *Betula*–*Empetrum* stage. A large number of these dominant taxa immigrated to west Greenland in the early to mid-Holocene, some (*Alnus crispa*, *Betula glandulosa*) from eastern North America, apparently by the dispersal mechanism suggested by Savile (1972) – wind transport over extensive frozen ocean and land surfaces. It is of interest to note that the initial plant assemblages recorded in northwest Europe following deglaciation include several of the taxa noted above for sites in eastern North America and Greenland, particularly *Oxyria*, *Dryas*, *Saxifraga*, and various members of the Caryophyllaceae (reviewed by Birks 1986). Also, subsequent stages, interpreted in terms of responses to soil development, were dominated by identical or equivalent taxa (*Empetrum, Salix*, *Betula nana*, and *Juniperus*) in the two regions. However, these similarities are primarily between pollen assemblages at regional scales, and local communities probably differed markedly, as shown for

Greenland by the detailed analyses of Fredskild (1973) and others.

Palaeoenvironmental controls

Climate

Before examining possible palaeoclimates in Canada during full-glacial, late-glacial, and Holocene times, a brief sketch of the now familiar Milankovitch model of climatic change might be useful.

The Milankovitch model

Although recent analyses suggest that the climate rhythms predicted by the Milankovitch model might have occurred in older geological records (Herbert and Fischer 1986, and others), the main focus of investigation has been on cycles of climatic change during the Quaternary. A recent two-volume conflation includes a convenient description of the model:

> that the major fluctuations in global climate associated with the ice age cycle are caused by variations in the pattern of incoming solar radiation – variations that are, in turn, caused by slow changes in the geometry of the earth's orbit that occur in response to predictable changes in the gravitational field experienced by the earth (Berger et al. 1984, p. ix).

Interested readers can grasp from these volumes the excitement and momentum of recent research in palaeoclimatology centred on this theory as the rapid expansion of the proxy database from marine and nonmarine sediments, combined with the increased capacity of high-speed computers in modeling experiments, have opened up new avenues of enquiry. A recent set of general circulation model experiments, spanning the past 18,000 years, provides a convenient set of climate predictions for the northern sector of the North American continent (Kutzbach and Guetter 1986). These simulations consisted of 3,000-year interval estimates, from 18,000 to 0 yr BP, of the global climate with a resolution of 4.45 × 7.5 degrees. I propose here to simply draw attention to the main results of these simulations and indicate the extent to which the observed vegetation changes and inferred palaeoclimates of Canada can be correlated. A complete examination of these comparisons, using all available proxy data on a global scale, are being assembled by the Cooperative Holocene Mapping Project (COHMAP), and will be published shortly.

Kutzbach and Guetter (1986) carefully stress the limitations of a model that uses glacial ice, sea ice, ocean surface temperature, and land albedo values adjusted at 3,000-yr intervals according to the geological record and then varies the solar radiation at each 3,000-yr interval according to the orbital (Milankovitch) parameters of tilt, eccentricity, and

Figure 8.5. A schematic representation of the course of solar radiation in summer (June, July, August) and winter (December, January, February, lower curve) for the northern hemisphere between 18,000 yr BP and the present, according to the Milankovitch theory of external forcing. The diagram is based on Figure 1 in Kutzbach and Guetter (1986).

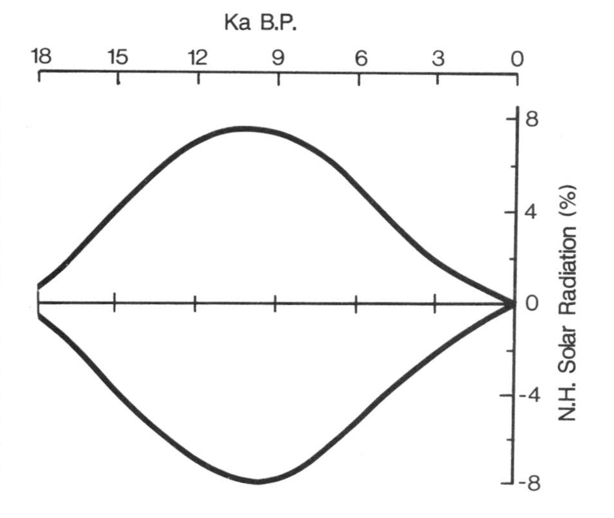

date of perihelion. Nevertheless, they conclude that "many features of the results were in agreement with geologic evidence." The important conclusions relevant to the present deliberations are as follows:

1. The seasonal cycle of solar radiation was amplified considerably in the northern hemisphere about 16,000 and 6,000 yr BP such that at 9,000 yr BP there was a maximum of summer (+8% over present), and a minimum of winter (−8%) insolation in higher latitudes of the northern hemisphere (Fig. 8.5 from Kutzbach and Guetter 1986).
2. Northern hemisphere land surface temperatures (July) were much lower than present at higher latitudes at 18,000 yr BP.
3. Northern hemisphere land surface temperatures showed large increases between 18,000 and 9,000 yr BP, with July temperatures 2.5°C greater than present.
4. Precipitation decreased at high latitudes for both January and July from 15,000 to 12,000 yr BP.
5. Ice-free land surfaces between 30° and 60°N were significantly colder in January (−5°C below present) at 18,000 and 15,000 yr BP.

In the following summary I will examine what is known or inferred about the past vegetation and climate of Canada, and the adjacent United States, from the height of the latest glacial to the present day.

Full-glacial conditions

Our knowledge of the nature of the vegetation and climate of North America at 18,000 yr BP is very incomplete, but certain patterns are emerging (Fig. 8.6). To the northwest of the ice cover, in Beringia, a treeless, predominantly herb tundra prevailed on uplands, in a cold, dry polar climate analogous to the modern mid-arctic continental climate of the western Canadian arctic. Reconstructions from interior Alaska differ slightly from those in northwest Canada, because of extensive loess and sand dunes in the former area, derived from major river basins. In general, it is suggested from both the physical and biological evidence that temperatures were lower and precipitation less than today (reviewed by Kutzbach and Wright 1986). South of the ice in the western Pacific region Barnosky (1984) suggests a parkland tundra on the Pacific Slope with *Tsuga mertensiana*, *Pinus contorta*, *Picea*, and nonarboreal elements, and a periglacial steppe dominated by *Artemisia* and grasses in the Interior Columbia Basin (Fig. 7.5). She suggests a cold continental climate, influenced in part by the lowered full-glacial sea-surface temperatures in the adjacent Pacific Ocean and in part by the Cordilleran Ice Sheets, causing intensified anticyclonic circulation, thus shifting the westerlies to the south. The full-glacial vegetation and climate of the northern plains region (comprising most of Montana and the ice-free fringe of adjacent Canada, eastern Wyoming, and the Dakotas) is almost completely unknown because of the absence of investigated sites, but, on the tenuous assumption that the radiocarbon age estimates are free of old carbon error, two sites provide evidence that *Artemisia*–grass tundra or steppe assemblage occupied some of southern Alberta (Mott and Jackson 1982; Schweger in press). Evidence of periglacial permafrost and extensive fossil dunes in the plains region of the northwest supports the suggestion that the mean annual temperature was about 10°C lower than modern values, with very strong winds from the northwest, and the simulation model results are in agreement with that reconstruction and the botanical evidence cited above for drier conditions than at present (Kutzbach and Wright 1986).

Figure 8.6. A highly schematic reconstruction of the palaeoclimates and associated pollen assemblages at 18,000 yr BP in North America, excluding the southwest. The approximate limits are shown for glaciers and pollen assemblages. The climate diagrams include the modern curves for temperature (dashed lines), but it should be noted carefully that the palaeoclimate curves are simply speculative suggestions for discussion purposes.

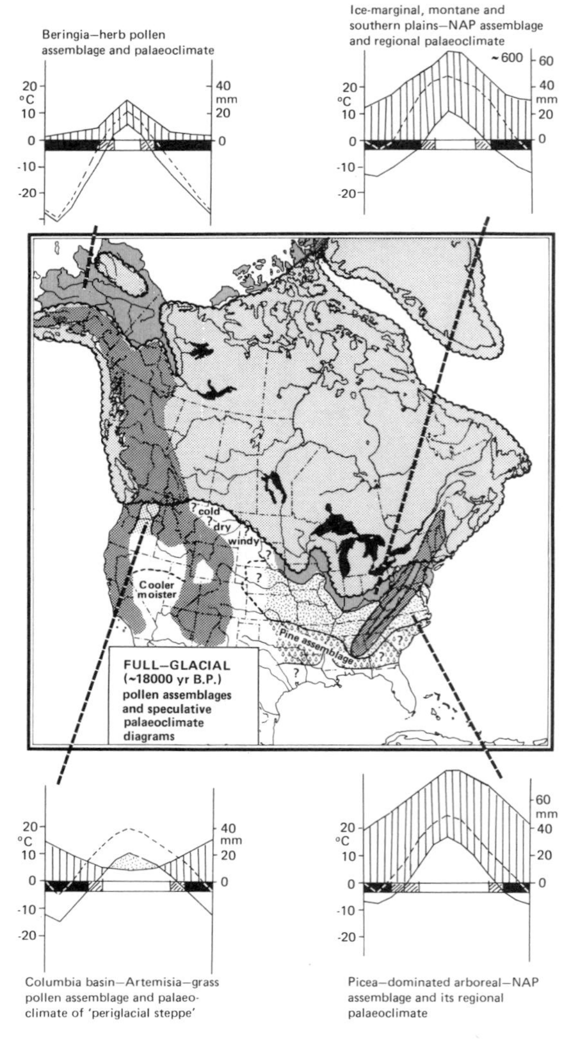

It is of great interest that a recent conflation of the pollen and macrofossil (packrat midden) evidence from the American southwest, comprising the Great Basin and southern desert areas of Oregon, Idaho, Nevada, Utah, California, and Arizona, shows that the full-glacial climate was cooler in both summer and winter than at present and that precipitation was greater, effected by winter rains (Spaulding, Leopold, and Van Devender 1983). These findings agree closely with the analysis of Smith and Street-Perrott (1983) of the pluvial lake levels in the same region; they conclude that a pluvial maximum occurred at 18,000 yr BP and that water-table levels in the southwest were probably as much as 10 m higher than at present. The contrast between the full-glacial reconstructed climates for the region immediately south of the Cordilleran Ice (northern Washington) and the Great Basin and areas to the south is of interest and shows remarkably close agreement with the simulation model output for 18,000 yr BP when the North American Ice Sheet caused the westerly flow to split with the southern branch developing over the southwestern United States, producing a band of increased precipitation (Kutzbach and Guetter 1986).

The eastern half of the southern ice-free region was occupied by forests up to within approximately 100 km of the ice margin, where a treeless tundra associated with periglacial features prevailed. *Picea* dominated a central swathe of the ice-free forested area from western Kansas through Tennessee and adjacent states to Virginia. Farther south, *Pinus* became more important in the pollen spectra. However, as I have suggested above, the record is still inadequate for precise reconstructions of the vegetation. What is evident is that almost nothing is known of the possible full-glacial refugia of temperate western and eastern forest dominants and of the northern grasslands.

The full-glacial records from the southeastern United States, in particular the region immediately south of Tennessee, are few in number, but the continued search for sites will eventually provide an adequate basis for reconstructing the origins of the temperate forests. It seems likely that the full-glacial forests in the southeast might have been very similar in species composition to those found today in the northern Great Lakes area, but the relative abundances of the taxa were reversed – that is, while today in the northern hardwoods the temperate trees predominate and the boreal taxa are less common, at 18,000 yr BP in the southeast the boreal taxa dominated but the temperate taxa were scattered throughout a large, topographically varied region centred in Tennessee, often in such low densities that their pollen was rare or absent in lake sediments. In addition to the temperate taxa with modern ranges in the northern Appalachians and Great Lakes regions, many of the trees today characteristic of the deciduous forest zone presumably occurred also in these putative refugia. The concept of "intermingled biotas" (Graham 1986) is a useful hypothesis to explain the full-glacial record, and an appropriate palaeoclimatic reconstruction is emerging from both the fossil record and simulation model results. The climate simulation suggests a regime that would sustain such a mixture of elements, with mean January temperatures about 5°C lower (to values of roughly 0°C) and mean July temperatures lower by the same amount (to values about 20°C) (Kutzbach and Guetter 1986). As Kutzbach and Wright (1986, p. 171) point out in their comparison of the geological evidence for full-glacial environments with the simulation output, "decreased seasonality of temperature, i.e., great equability of climate, is cited as a possible explanation for the 'disharmonious' vertebrate faunas that characterized much of the plains area during the last glaciation (Graham and Lundelius 1984)." On the other hand, both Beringia and the ice-free areas to the south housed arctic, arctic–alpine, and boreal elements that later formed the familiar modern northern patterns, and it is likely that these plants occurred at higher elevations in the montane parts of the refugial areas. The origins and histories of other temperate assemblages have also remained elusive. The full-glacial refugia of the temperate forests of Europe and of the Mediterranean vegetation have intrigued phytogeographers for over a century, but only the most rudimentary notions of their locations and extents have been established (Beug 1964 and 1975; Wright 1977; and Huntley and Birks 1983). The general supposition on both continents has been that, as the fossil record so far has proved to be fragmentary, refugial populations must have persisted in very localized situations in areas of topographical diversity. Graham (1986) suggests that intermingled biotas were widespread in the full glacial and he concludes that they "represent communities and environments without modern analogues." He then qualifies that statement by pointing out that the refugial environments of the full glacial were more heterogeneous than today's, being "composed of a fine-grained mosaic of habitats that today occur far apart."

Attempts to make palaeoclimatic inferences from the ice-marginal full-glacial plant assemblages recorded south of the Laurentide Ice Sheet are imprecise, as reviewed recently by King and Graham (1986), and in particular the nonarboreal pollen and macrofossil records with frequent *Oxyria*, *Saxifraga*, *Dryas*, and others are usually reconstructed as tundra in both vegetational and climatic terms, using the modern arctic as an analogue. This is probably misleading because one simple point is often overlooked. The annual and diurnal cycles of energy differ significantly between high and low latitudes (Table 2.1), and these differences, however modulated by circulation differences in the past or by the proximity of an ice sheet, would have existed also during full- and late-glacial times. In other words, modern arctic assemblages can never serve as precise climatic analogues for full-glacial fossils from sites at low latitudes. In fact, even if close similarity could be shown between modern arctic pollen spectra and fossil spectra from full-glacial deposits at approximately 40°N, palaeoclimatic reconstruction would have to be made with care. It is possible that the physiological climate experienced by a plant community could be derived from climates of quite different regional regime. For example, the increased day length in the high arctic summer (Table 2.1) compensates for the effect of the small angle of incidence of the solar rays, so that the mean June solar energy received at Resolute, Northwest Territories (latitude 74°N, 601 langleys/day), is slightly greater than that at Toronto (latitude 43°N, 516 langleys/day). At the ice margin in midcontinent during the full glacial, quite different compensatory effects might have offset the effects of a high solar angle and a potentially long growing season, in much the same way as present-day large ice caps influence the peripheral environment in summer. For example, Holmgren's (1971) studies of the Devon Island ice caps have shown that, in summer, downslope (katabatic) cold air off the ice and upslope (anabatic, due to warming of the ice-free land) air flows converge near the margin to produce frequent cloud cover, and cooler and sometimes wetter conditions than in the region as a whole. As Kutzbach and Wright (1986) suggest, a steep climatic gradient near the ice margin, a mosaic of topographic features, and the direct effects of the ice sheet on the marginal mesoclimate must have favoured the highly heterogeneous array of plant and animal communities recorded in the fossil record. On the other hand, a recent compilation of a full array of modern arctic pollen from northwest Canada has revealed similarities (quasianalogues) between high-latitude full-glacial samples and high-arctic modern spectra (Ritchie, Hadden, and Gajewski in press).

The general conclusion offered by Kutzbach and Wright (1986) is that a mosaïc of habitats of widely differing mesoclimate occurred along a steep macroclimatic gradient from the ice margin southward, with cooler summers (than modern) and decreased seasonality (Fig. 8.6).

Solomon and Shugart (1984) propose a very similar reconstruction based on the application of a model for ~~the~~ Appalachian forests to simulate the full-glacial vegetation on various slope and soil situations under different climatic scenarios. The pollen percentages for the interval 16,500 to 15,500 yr BP, recorded by Delcourt (1979) from Anderson Pond, Tennessee, were used as the measured vegetation in the simulation. They concluded that "decreased seasonal temperature extremes can indeed produce boreal forests that contain certain anomalous thermophilous tree admixtures." In other words, there is substantial compatibility between the simulations by general circulation models of cooler summers and winters during the full glacial with reduced seasonal contrast (Kutzbach and Guetter 1986) and both the simulated (Solomon, West, and Solomon 1981; Solomon and Shugart 1984) and the measured full-glacial (by pollen) vegetation at Anderson Pond, Tennessee.

Tentative bioclimatic reconstruction for full-glacial time is attempted in Fig. 8.6, with the important caveat that these are as much ideas to be tested by future proxy data sets and simulation as they are reasonable conclusions. I have determined the mean July and January temperatures and the length of the frost-free periods by comparing the fossil assemblages and the modern ranges of these parameters (Table 3.1). On the other hand, precipitation distributions have been based on the simulation model output (Kutzbach and Guetter 1986). In ice-free Beringia, a colder, drier climate with a seasonal pattern similar to modern supported a sparse herb tundra on uplands, with continuous sedge–grass–willow vegetation confined to coastal lowlands and interior valley bottoms. South of the ice, in the west, a cold, dry steppe environment prevailed in the Columbia Basin. In the central plains, the fossil record can be explained adequately by a steep climatic gradient from the ice margin southward, depicted speculatively by two palaeoclimate diagrams that provide macroclimates within which topographically controlled mesoclimates would have supported the heterogeneous or "intermingled" biota that are found in the fossil record.

Late glacial and Holocene

A warming trend began following 18,000 and culminated at 9,000 yr BP when the northern hemisphere summer solar radiation was 8 percent higher than at 0 and 18,000 yr BP. The land-surface July temperatures between 18,000 and 9,000 yr BP increased by about 7°C in the northern hemisphere according to the general circulation simulation experiments (Kutzbach and Guetter 1986), and the inferred vegetation changes correlate closely with such a reconstruction. *Picea*-dominated forests spread north into northern Iowa, Minnesota, Wisconsin, Michigan, Indiana, Ohio, Pennsylvania, New York, and southern New England, separated from the receding ice front by a belt of treeless vegetation that disappeared rapidly in the west-central sector, but persisted until the end of the late glacial east of the Great Lakes Basin. In the western mountains, elevation changes occurred in parklands, tundra, and steppes, but the record of the vegetation of the northern Great Plains is sparse. In the far northwest, vegetation responded initially to the warming trend by an increase in low shrub tundra (dwarf birch, ericads). Between 12,000 and 10,000 yr BP, a rapid recession of both Laurentide and Cordilleran Ice occurred and large areas of southern and west-central Canada became ice-free.

The movement northward of spruce-dominated forests, following an episode of variably treeless vegetation that occupied surfaces adjacent to the receding ice margin in eastern North America, is widely accepted as a response to climatic warming. The Conklin Quarry, Iowa, assemblage (Baker et al. 1986) provides convincing evidence for a cold climate at 18,000 yr BP, similar to modern subarctic conditions. The growing number of sites with reliable climatic indicators among the faunal remains strengthens the reconstruction that the climate warmed relatively rapidly between 14,000 and 10,000 yr BP. Péwé (1983) and Johnson (1986) have mapped the occurrence beyond the full-glacial ice margin of periglacial features, particularly patterned ground and solifluction features, and they cluster in southern Wisconsin and adjacent northern Iowa, Illinois, Indiana, and Pennsylvania, and at high elevations along the Appalachian Mountains. The palaeoclimatic implications of widespread permafrost, associated with fossil assemblages with arctic–subarctic modern analogues, led to the conclusion that a tundra bioclimate prevailed along the full-glacial ice margin in much of eastern North America. Johnson (1986) suggests that the occurrence of permafrost beyond the southern limit of the ice sheet, between 20,000 and 15,000 yr BP, "was continuous in a narrow zone adjacent to the ice margin, but became discontinuous to the south. . . ." He concludes that such an environment required a mean annual temperature of –5°C, cold winters, and low snowfall with strong winds. Ice wedge polygons form today in lowlands in the subarctic zone, for example, in the lower Mackenzie River valley, where the mean annual temperature is between –5 and –8°C. In these areas, permafrost is often absent from upland sites. Accordingly, the most conservative reconstruction of the full-glacial ice-marginal landscapes in the central plains and lower Great Lakes regions is that regional topographic diversity would have caused such local variations in the expression of the cold climate as the discontinuous occurrence of permafrost features, variations in active layer depth, and a vegetation mosaic of tundra, forest–tundra and woodland types. By 13,000 yr BP, farther north in southern Ontario, similar categories of evidence indicate a less cold, boreal climate, despite the immediate proximity of Laurentide Ice lobes (Terasmae 1981; Schwert et al. 1985).

As the late-glacial ice margin retreated in the Western Interior of Canada and adjacent United States, the northward expansion of the *Picea* forest complex was rapid, implying a significant climatic warming from 14,000 to 10,000 yr BP. The apparently abrupt replacement of conifer forest by prairie across the plains region in the early Holocene and the rapid extension eastward of the "Prairie Peninsula" by 8,000 yr BP have been interpreted as a transition from an early Holocene that "was slightly cooler and more moist than the late Holocene . . ." due to the influence of the wasting Laurentide Ice mass to the north, to a warmer, drier interior climate with steeper gradients across the interior than today (Webb, Cushing, and Wright 1983; Jacobson and Grimm 1986). Cooler, wetter climates similar to modern began in the late Holocene at about 4,000 yr BP. The major compositional change in the early Holocene forests of southeastern Canada, involving the addition of temperate species but no loss of boreal species, can be interpreted as a response to climatic change of the same order as the differences between stations on both sides of the modern boundary between boreal and temperate forests (Fig. 2.8) if we assume that the ranges of the dominant tree taxa are in equilibrium with the regional climate. The main differences between the stations in the boreal zone (Cameron Falls, Ontario; Earlton Airport, Ontario; and Taschereau, Quebec) and those in the temperate zone (Kenora, Combermere, and Mount Forest in Ontario; Bagotville, Quebec; and Woodstock, New Brunswick) are that, in the boreal zone, mean July temperatures are lower, the growing season is shorter, and the mean and absolute minimum temperatures in the coldest month are lower (>–22°C and >–45°C, respectively). The following speculative reconstruction of the early Holocene (9,000 yr BP) bioclimates of the interior and eastern part of Canada are offered simply as plausible conjecture to be tested against new proxy data. I am unable to suggest similar reconstructions for the Pacific–Cordilleran area because of the complexity of the area and its very spotty fossil record. The main features, all described with reference to the present day, are:

1. Warmer summers in the Northern Interior, promoting an extension eastward of tundra elements as the Laurentide Ice receded; an extension to the Beaufort Sea coast of a spruce-dominated assemblage, and an initial expansion westward into Interior Alaska and southern Yukon
2. Drier and warmer conditions in the Southern Interior, causing an expansion northward of the grassland–forest ecotone
3. A steep climatic gradient between the eastern sector of the Laurentide Ice Sheet and the ice-free Maritime regions, causing a slow expansion of boreal trees (spruce, poplar, birch, fir) into southern Quebec, Newfoundland, and Labrador. Farther west, white pine expanded rapidly westward, and the Appalachian region and adjacent lowlands was the centre of expansion into Canada of coniferous (hemlock) and deciduous (beech, maple, yellow birch) dominants of the modern northern mixed forest (Richard 1981b, in press). At the same time, deciduous trees (oak, elm, basswood, hickory) spread to replace the earlier conifer assemblage in southern Minnesota, Iowa, and Illinois

Mean July temperatures were 1 to 2°C higher in the Western Interior, but the effect of the Laurentide Ice depressed the hemispheric increased summer warmth in eastern Canada.

Subsequent Holocene vegetational changes were due to (1) lags in response to the Holocene warming by a few tree species because of slow rates of spread, e.g., *Carya*, *Castanea*; (2) edaphic changes, primarily responses to humus accumulation and the resulting nutrient deficiency and, in mesic–hydric habitats, paludification; (3) major disturbances, chiefly fire and large wind storms/hurricanes, that maintained a mosaic of successional responses within forested regions and that caused an extension of the width of the forest–tundra in Nouveau Québec; (4) climatic change involving trends in effective moisture that caused responses particularly, but not exclusively, in the mid-Holocene along the northwestern edge of the temperate zone, expressed in changes in the ranges of *Tsuga*, *Fagus*, and *Pinus strobus*; (5) a late-Holocene cooling that is undetected by most regional pollen diagrams at sites distant from ecotones or range boundaries of sensitive taxa, but that has been detected clearly, for example, in the White Mountain treeline, in southeastern Quebec, along the southern boreal forest boundary in the Western Interior, and at the arctic treeline in Quebec and elsewhere.

The later changes in range and relative abundance of boreal taxa provide some evidence for climatic change. At the northwestern Canadian extremity, on the Tuktoyaktuk Peninsula, spruce macrofossil and PAR data show that the boreal forest extended beyond its modern range, and Spear (1983) interprets these results as confirmation of similar indications of an early Holocene period of maximum summer warmth. A roughly contemporaneous movement northward of the southern ecotone in western Canada, by which boreal forest was replaced by grassland at roughly 9,000 yr BP, can be interpreted also in terms of increased summer warmth with soil-moisture deficits probably causing the demise of spruce and other boreal trees.

The late-Holocene extension southward of the western boundary of the boreal forest correlates reasonably with the widespread evidence in eastern Canada of increases in the relative abundance of spruce in the transitional region between the boreal and temperate forests in southern Quebec (Webb, Richard and Mott 1983). At the same time (2,000 to 3,000 yr BP), the northern limits of spruce forests and of shrub tundra in Ungava–Quebec receded southwards (see Chapter 5, pp. 90–2). All these events are interpreted as climatic cooling and/or increased moisture of unspecified magnitude. The northern treeline in Quebec has responded to the observed early twentieth-century warming (Payette and Filion 1985). Payette and Filion (1985) relate growth-form responses in white spruce at a rare site in southern Quebec to the Little Ice Age and subsequent climatic change (discussed more fully below).

The palaeoenvironmental history of the eastern temperate forest region of Canada remains imperfectly understood. General trends are apparent, but even their chronology and pattern depend on largely subjective interpretations of pollen data. Few disagree that, as the ice margin receded into southern Quebec and New Brunswick, a tundra assemblage of plants with wide-ranging arctic–boreal–alpine modern ranges formed the first vegetation, referred to in short as the *Oxyria* assemblage, enumerated at many sites, for example, in Table 5.1, pp. 73–4. Pollen spectra from the modern high-arctic herb tundra in western and eastern Canada and in northern Greenland (Figs. 3.16, and 3.18) are similar qualitatively and quantitatively to these late-glacial assemblages, so an interpretation as a herb tundra vegetation seems appropriate, as elaborated by Richard (1977a, 1985) for sites in southern Quebec. Though several of the most common late-glacial fossils are taxa with modern arctic–alpine distributions, they do not provide conclusive evidence of an arctic palaeoclimate. Several (*Dryas*, *Oxyria*, *Saxifraga*, and many of the mosses) are opportunistic species with tolerances of open habitats and unstable, immature, often calcareous soils, occurring today in disjunct open habitats in the boreal and northern temperate zones. However, ice wedge casts at sites scattered along the lower St. Lawrence Valley in southern Quebec are interpreted by Dionne (1975) and Péwé (1983) as evidence for permafrost conditions in the late glacial, coeval with the herb–tundra pollen and macrofossil assemblages. Tundra was replaced by arboreal vegetation at some sites with an intermediate shrub tundra phase. *Populus* and *Picea*, variably accompanied by *Betula papyrifera*, *Larix*, and small amounts of *Pinus banksiana*, followed, and many authors suggest that they formed an open parkland or forest–tundra, vaguely analogous to the modern subarctic vegetation. The evidence to support this reconstruction is twofold – low PAR values of *Picea* and other trees, and moderate frequencies of NAP. As a general proposition, this probably has some validity, but an accurate reconstruction must await results from many more sites, including macrofossil and bryophyte analyses. I noted in Chapter 5 how the latter sometimes provide convincing evidence for the presence, at least locally, of closed conifer forests in the late glacial, with modern analogues. Apparent nonanalogue assemblages from the late-glacial record have been highlighted by those who stress differences in migration patterns and community compositions between, for example, full-glacial and modern boreal assemblages, for example, by Davis (1981), who writes that "boreal species were associated very differently from the modern boreal forests of Canada." My own investigations of the western boreal forest led me to assert that the modern boreal forest was finally assembled and achieved regional stability only about 5,000 years ago (Ritchie 1976 and others). Likewise, Davis (1981) implies major differences between past and present assemblages when she points out that in the full glacial: "Trees that now characterize the temperate deciduous forest grew in relatively small populations" in the south and "In some areas they grew in close proximity to boreal species such as spruce." The close association of temperate forest and boreal species is not unusual – the southern part of the modern ranges of all of the boreal taxa except jackpine overlap completely with the eastern temperate mixed-forest region (Fig. 2.8). The relative proportions of these elements were different in the late glacial, but their presence together can be used to provide some general palaeoclimatic reconstructions. The spread of spruce northward was time-transgressive, rapid in southern Ontario but more gradual in the Maritimes. It is generally assumed to be a response to increased summer warmth. More precise palaeoclimatic reconstructions will no doubt be available when the calibration function technique has been further refined and, in particular, the use of multiple regression equations that relate climatic values to abundance values for individual taxa will overcome the apparent difficulty posed by the so-called nonanalogue late-glacial assemblages (Bartlein, Prentice, and Webb 1986). As a result of the sparseness and imprecision of the full- and late-glacial record of the early spruce assemblage, narrative description has predominated over analytical precision in communications on the topic. One unfortunate result has been a proliferation of images and metaphors that bear little relation to reality. The most common is that the early spruce-dominated assemblage was "a major biome" that experienced "an almost complete demise" and by 9,000 yr BP "had shrunk to a narrow band just in front of the retreating Laurentide Ice Sheet" (Webb 1981). I have already suggested that while "spruce biome" might be a reasonable term, it is highly unlikely that spruce was the major dominant of eastern forests on upland sites at the beginning of the Holocene. It was probably a local codominant with fir, larch, eastern cedar, birch, black ash, elm, and maple. However, putting aside the question of how "spruce biome" is defined, I estimate that the conifer-dominated assemblage at its late-glacial maximum occupied roughly one million square kilometers of east-central North America east of the Mississippi. At 9,000 yr BP, the same pollen assemblage was being recorded at sites east of the Manitoba–Ontario border to Labrador and Newfoundland, across an area of equal or greater size. By the early Holocene, the range of *Picea* (in the east) was roughly twice its full-glacial area, and, if the full Canadian range is included, the "spruce biome" had expanded almost to Alaska (Fig. 8.2). An ecologically more useful description, substantiated of course by the pollen and macrofossil record, is that at about 10,000 to 8,500 yr BP (time transgressive northward), the conifer-dominated assemblage was invaded by a number of trees dominant of the modern temperate mixed forests, and the relative abundances, and in some cases habitat ranges, of the boreal taxa decreased, but their geographic ranges in Canada and the adjacent United States did not change. With further ice recession, their ranges expanded northward round both flanks of the residual ice cap (Fig. 2.2). Their southern limits moved north from their full-glacial to modern positions (Fig. 3.1). The late-Pleistocene and Holocene pollen and macrofossil record of *Picea* at a site in the modern mixed forests (Chase Lake, Maine, by R. B. Davis and Kuhns, personal communication) illustrates this point clearly (Figs. 8.1a and 8.1b).

I emphasize the importance of precision in description because the coarse scale of the available isopoll maps (Webb 1981; Webb, Cushing, and Wright 1983) and migration maps (Davis 1981, 1983) appear to have generated misleading oversimplifications that influence palaeoenvironmental inferences. In particular, they exaggerate the problem of nonanalogues for late-glacial assemblages. It is of interest to note that Gordon (1985), discussing the postglacial history of spruce and the boreal forest in general from the viewpoint of a forest ecologist, has reached, quite independently, similar conclusions about the misinterpretation of the pollen record in terms of the lack of modern analogues.

By 9,000 yr BP, in southern Ontario, southern Quebec, New Brunswick, and Nova Scotia, the forest composition had changed to a mixed assemblage with abundant *Pinus strobus*, *Acer saccharum*, *Quercus* (probably chiefly *Q. rubra*), *Betula papyrifera*, and reduced amounts of *Picea*, *Abies*, and *Larix*. *Tsuga*, *Fraxinus americana*, and *Ostrya* were present in the southern areas. High early charcoal frequencies at several sites in the entire eastern area (Mirror Lake, New Hampshire, Davis 1983; Chase Lake, Maine, R. B. Davis and Kuhns, personal communication, see Fig. 8.1; Everitt Lake, Nova Scotia, Green 1982) and the maximum values of *Pinus strobus* are interpreted as an early Holocene climate with summers that were warmer and drier than modern. More compelling evidence has accrued from investigations in the White Mountains (New Hampshire) of elevational changes in tree limits during the Holocene (Davis, Spear, and Shane 1980). Pollen PAR and macrofossil analysis of *Tsuga* and *Pinus strobus* demonstrate an interval of warmer summers between 9,000 and 5,000 yr BP (Fig. 8.7) and "the occurrence of white pine above its present

Figure 8.7. Pine and hemlock PAR values plotted against radiocarbon age from Rogers Lake (Connecticut), two sites in southern New Hampshire, and an elevational transect of sites in the White Mountains from Davis, Spear, and Shane (1980).

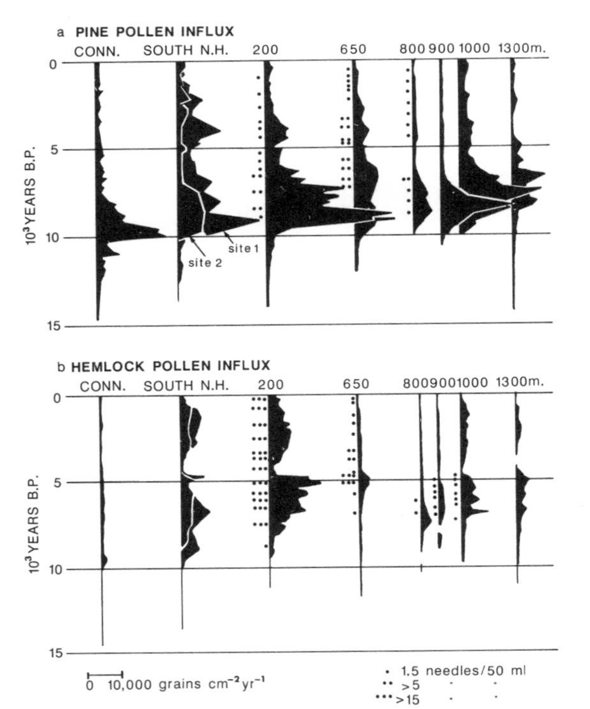

elevational limit in 9,000-year-old sediment means that the climate was already more favourable than today 9,000 yr BP, giving the earliest definitive evidence of a climate warmer than today." The range extension of white pine persisted until about 5,000 yr BP. Hemlock pollen and macrofossils indicate a similar response to that of white pine. Hemlock persisted at the 650-m site, slightly above its modern limit (600 m) until the Little Ice Age, in spite of the "synchronous decimation of northeastern hemlock populations" at 5,000 yr BP (Davis 1976). The authors conclude that the late glacial (14,000 to 10,000 yr BP) was colder than present, that from at least 9,000 to roughly 3,000 yr BP the summers (I presume) were warmer than modern, and that cooler conditions set in from about 3,000 to 2,000 yr BP. They estimate that the mean annual temperature during the early Holocene was 2°C warmer, and mean annual precipitation 125 mm lower, than at present. Similar inferences are made by Jackson (1986) in his interpretation of pollen and macrofossil data from the late-glacial and Holocene deposits in the High Peaks range of the Adirondack Mountains, New York. Elevational range extensions of white pine, hemlock, and yellow birch in the early Holocene (9,000 to 6,000 yr BP) are interpreted in terms of increases in annual temperature of 1 to 2°C and decreases in annual precipitation of up to 100 mm. Lowering of the elevational limits of these taxa between 6,000 and 4,000 yr BP is interpreted as a response to cooler climate. Webb, Richard, and Mott (1983) interpret increases in *Picea* and *Abies* in the St. Lawrence Lowlands of southern Quebec from 2,000 yr BP to the present as a response to a cooler climate.

While the area lies well to the south of Canada, it is of interest to consider the evidence for past climates along the southwestern margin of the temperate forest zone. This boundary extends in a wedge shape from southeastern Manitoba to near the southern tip of Lake Michigan, into southwestern Indiana and thence to the southwest (Fig. 2.8). A recent detailed investigation of sites in Wisconsin has filled in what has been a conspicuous gap in the Holocene palaeoenvironmental record, particularly as the prairie–forest border lies in the area (Winkler 1985a). The pollen, charcoal, and sedimentological evidence from one site (*Lake Mendota*) provides compelling evidence of a dry period between 6,500 and 3,500 yr BP with an estimated 10 percent decrease in precipitation, and 0.5°C warmer mean July temperature (Winkler, Swain, and Kutzbach 1986). The authors also note that the fossil data agree closely with the climate model results, using Milankovitch forcing factors (discussed below) and the proximity of the early Holocene Laurentide Ice. The magnitude of these changes can be visualized by reference to Figure 2.8, where climate diagrams from sites on both sides of the temperate forest boundary are shown.

The relevance of these midwestern Holocene palaeoclimates to considerations of climatic versus migrational factors in determining the pollen stratigraphy is clear, and Webb, Cushing, and Wright (1983) and Davis et al. (1986b) discuss the questions fully. It is evident that too few sites have been investigated to resolve the problem. The southern counties of Wisconsin are of great interest and potential, as Winkler (1985a) has demonstrated. Her pollen record from *Washburn Bog* in Sauk County is tantalizingly suggestive, but probably inconclusive. It shows the basic sequence for Wisconsin, from a mixed mesophytic forest that succeeded the early *Picea* forests at about 9,000 yr BP, replaced in turn by *Quercus* savanna between 6,500 and 3,500, with a return to mixed mesophytic forests in the late Holocene. The Washburn Bog diagram shows pollen occurrences of both *Fagus* and *Tsuga* in the early mesophytic forest at approximately 8,000 yr BP. The hypothesis that small populations of either or both trees were present in the early Holocene could be explained, if not tested as Davis et al. (1986a) imply, by sampling additional sites, and the question of whether small populations of these trees survived the warm, dry period might also be resolved from investigations of a large number of very small sites.

An investigation by Gajewski (1983) of the palaeoclimate of eastern North America during the past 2,000 years has provided an important demonstration of the sensitivity of the pollen record from lakes with annually varved sediments. A network of seven such sites in Maine (2), New York (1), Pennsylvania (1), Wisconsin (2), and Minnesota (1) was used, and 20-, 40-, or 80-year interval pollen samples analyzed for each lake core. Gajewski (1983) used principal components analysis to show a long-term trend over the 2,000-year period in all diagrams that he associated with large-scale changes in macroclimate: medium-frequency changes at the scale of a few centuries, also related to climate but expressed differently across the region, for example, as an expansion of hemlock and beech in the midwest and as decreases of these species in the northeast, responding differently to different regional climatic effects of the Little Ice Age (increased moisture in the west, cooling in the east): and high-frequency fluctuations that may be caused by local disturbance effect (=short-term vegetation dynamics). He then applied calibration functions from a network of modern pollen and climate data to three of the profiles (in the eastern sector) to reconstruct past temperature and precipitation. He showed a cooling trend of about 1.2°C in Maine and 1°C in New York over the past 1,800 years. Obviously, the results of these investigations have relevance to adjacent areas in Canada, and, as I will elaborate below, they provide a clue to both understanding the record of Holocene vegetation and devising ways of testing alternative explanations.

Several authorities in the southern Pacific–Cordilleran region as a whole, including sectors to the south and east beyond the scope of this book, have offered traditional, qualitative palaeoclimatic reconstructions with a large measure of agreement between them (Hebda 1983b; Barnosky 1984, 1985a, 1985b, 1986; Heusser 1985, and Mehringer 1985). Others have attempted to derive past temperature and precipitation values by applying calibration functions to pollen data from a lake site in southern British Columbia (Marion Lake; Mathewes and Heusser 1981) and from Hoh Bog on the Olympic Peninsula. The consensus is that, while many details and regional peculiarities remain to be elucidated by future investigations, four past climate regimes differing from modern can be recognized as follows: full-glacial conditions were significantly colder and drier; the late glacial (roughly, 13,000 to 10,000 yr BP) was cooler than present, but warmer than the full glacial, possibly as moist as present day in the Pacific zone, but moister than modern conditions in the interior; the early Holocene (10,000 to 8,000 yr BP) was both warmer and drier; and the mid- to late-Holocene cooling transition to modern conditions occurred at various times between 6,000 and 3,000 yr BP.

Palaeoclimatic reconstructions for the record from the northern Pacific are difficult because of the regional complexity resulting from the interplay of great orographic diversity, oceanic influences, valley glaciers and varied arrival and expansion times among the main taxa (Heusser 1985). Recently Heusser, Heusser, and Peteet (1985) derived mean July temperature and annual precipitation estimates from modern pollen–climate transfer functions for the coastal region of southern Alaska, British Columbia, and western Washington. They show an early Holocene warm, dry period followed by a cooler and wetter trend from about 5,000 yr ago to the present day. On the other hand, the record from the northern interior is relatively straightforward and consistent between sites. The full-glacial herb tundra is interpreted as indicating colder and drier conditions than modern; the slow increase of dwarf shrubs, particularly dwarf birch and ericads, is regarded as an indication of both a warming (summer) trend and slow development of humified soils; sites in eastern Alaska (Edwards and Brubaker 1986) and adjacent northern Yukon show a maximum summer warmth in the early Holocene that has been interpreted as a direct response to the increased high-latitude summer radiation levels expected according to the Milankovitch theory (Ritchie, Cwynar, and Spear 1983). A cooling trend began in the mid-Holocene and modern conditions were established by about 5,000 yr BP.

Although the early to mid-Holocene pollen records from arctic Canada and adjacent Greenland appear to be uninformative with respect to palaeoclimate, it is of interest that Fredskild (1985) found the remains of a beetle species (*Colymbetes dolabratus*) and the stickleback (*Gasterosteus aculeatus*) in 6,700- to 8,000-yr-old sediment from a small lake at 75°N in northwest Greenland. Both species are absent so far north today, indicating warmer summers than modern, "undoubtedly connected with a longer ice-free summer." It is likely that a clear pollen registration of an early Holocene warming in the high eastern arctic is lacking because of migrational delays, as Hyvärinen (1985) suggests. A more sensitive response is found in the palaeolimnological record (e.g., Smol, 1983). Funder (1979) arrives at essentially the same conclusion for the origin of the Greenland flora and vegetation as I suggested above for that of eastern Canada – a post-10,000-year immigration from North America and, for the east Greenland plants, from the North

Atlantic islands. In striking contrast, the western Canadian arctic has a long history of deglaciation, or nonglaciation, with a resulting long history of vegetation. Predominantly herb tundras prevailed until the late glacial when dwarf shrubs expanded (dwarf birch, heaths, willows) apparently in response to climatic warming. Cwynar (1982) notes that his Hanging Lake site gives evidence for "an early warm interval dating from 11,100 to 8,900 yr BP," and when these results were combined with pollen and macrofossil evidence of treeline changes, a secure record emerged of an early Holocene summer thermal maximum (Ritchie, Cwynar, and Spear 1983). It is probable, and both Funder (1979) and Hyvärinen (1985) make the point, that the persistence of a large part of the Laurentide Ice Sheet in eastern North America between 12,000 and 8,000 yr BP (Fig. 2.2) influenced the climate of the eastern arctic, including Greenland, so as to repress the direct expression of increased summer radiation.

Fire

A large literature has appeared during the past decade on the role of natural fires in boreal and other forest regions, so no attempt is made here to provide a detailed review. The interested reader can get access to the literature by such recent reviews as those of Heinselman (1981), Wein and Maclean (1982), and Clark (1983). However, it is relevant here to appreciate in general terms the frequency and areal extent of wildfire in Canada. The following remarks refer to all forested lands in Canada, although the boreal forest makes up the largest portion, roughly 75 percent. Lightning is the main cause of fire. For example, 87 percent of the total area (annual average 12,841 km^2) burned during the period 1970–9 was the result of lightning, although that amounted to only an annual average of 32 percent (2,908) of the number of fires (Ramsey and Higgins 1982). The number of fires and total area of forested land burned during the ten years from 1972 to 1981 (Table 8.1) shows an apparent trend of increase, but Harrington (1982) advises caution in drawing any conclusions as "in the short period of accurate records available, many apparent trends may reflect nothing more than random occurrences." He does note, however, "a generally negative correlation between area burned in the far west (excluding British Columbia) and that in the east, suggesting a wavelength of high-pressure ridges and low-pressure troughs that is approximately a continent wide." During the period 1972 to 1981, the mean annual area burned was 20,092 km^2 (Table 8.1), which can be envisioned as a strip 3.9 km wide across the maximum latitudinal dimension of Canada (roughly 5,000 km in length). This is equivalent to roughly 0.5 percent of all forested lands and yields an estimate of the fire cycle average of about 200 years (after Clark 1983).

Fire cycles in the boreal forest are short in the drier, western sector (50 to 100 yr) and longer in eastern Canada (100 to 200 yr). However, the cycle period varies greatly with landform and soil patterns, and periods of high-intensity fires have been associated with successive drought years. Cogbill (1985) notes that for the eastern boreal forest "a natural mean rotation time of about 100 years and a maximum stand age of 250 years are the essential parameters describing the age structure." Many of the dominant taxa of boreal forests have reproductive and structural attributes that are apparently effective adaptations to fires of varied intensities and cycle lengths (Heinselman 1981). For example, Jeffrey (1961) concluded from an age analysis of white spruce forests on alluvial soils in northeastern Alberta that most stands originated from fire and rela-

Table 8.1 *Yearly fire totals, areas burned, and approximate percentage of total forested lands of Canada*

Year	Total number of fires	Total area burned (km^2)	Approximate % of all forested lands
1981	9,990	51,377	1.3
1980	8,973	48,222	1.2
1979	9,793	27,008	0.7
1978	7,928	2,894	0.1
1977	8,888	14,832	0.4
1976	10,161	18,139	0.5
1975	10,995	10,320	0.3
1974	8,035	8,487	0.2
1973	7,503	11,843	0.3
1972	8,153	7,800	0.2

Source: Derived from Ramsay and Higgins (1982).

tively few, with longer histories of no disturbance, showed long-term seral change through *Salix*, *Alnus*, and *Populus balsamifera* stages. In theory, successional sequences of variably adapted plants should be expected, from the aggressive, r-selected colonizers (herbs, willows, poplars, birch) to the shade-tolerant K-selected, longer-lived trees. In practice, the fire cycle may be so short and fire types vary so greatly that simple successional patterns are not common. Cogbill (1985) recognizes a few such chronosequences in his detailed age analysis of many stands in Quebec, but in general he finds that few sites survive without further fire to permit later stages to develop. "The dynamics of these boreal forests is simply the development and aging of the contemporaneously initiated trees." Farther east however, quite different conclusions were reached from an equally detailed investigation, in the boreal forests of southeastern Labrador, where the cooler, wetter climate determines extremely long intervals between fires (Foster 1983a, 1983b, 1985; Foster and King 1986). The result is a "multiaged structure of the conifer forests . . . in sharp contrast to the even-aged structure from black spruce forests in the central and western boreal forest." He also shows that birch forests have a successional role after fire, but, because of their lesser longevity, there is "a gradual conversion of birch stands to forests dominated by fir and black spruce" (Foster 1983a, 1983b).

In the lower foothills region of Alberta, in the mixed-woods section of the boreal forest, it has been long recognized that "very few stands have escaped fire for more than 100 years, which means that succession is unimportant" (Horton 1956). The same author notes that at higher elevations in the foothills, rougher, broken terrain appears to restrict the size and spread of fires so that alder stands persist with a consequent succession from lodgepole pine to Engelmann spruce and subalpine fir at higher elevations and to Douglas-fir and white spruce at middle elevations. In his review of forest responses (North American) to such factors as wildfire, Oliver (1981) points out that the frequency and intensity of the "disturbance" factor determines which species will dominate and that these species "can often dominate for a long time."

On the other hand, near the northern boundary of the boreal forest, evidence has been assembled to show that the interaction of climatic change and fire has caused a significant deforestation (Payette and Gagnon 1985), as described in Chapter 5. Recently, Millet and Payette (in press) have calculated fire cycles and mapped fire sizes across a wide transect in western Nouveau Québec, from the northern edge of the boreal forest to the shrub tundra zone. They further develop the hypothesis of a maximum deforestation in what is now the forest subzone of the forest–tundra caused by frequent fires in a regional climate unfavourable to rapid regeneration, as shown in Figure 8.8. The hypothesis is testable by pollen, charcoal, and geochemical analyses of sediments from carefully chosen lakes along this transect. Fire frequency declines sharply in the shrub subzone of the forest–tundra and in the shrub tundra zone.

Figure 8.8. Recovery of boreal forest after fire is often very slow near the northern limits of the zone. Here an open stand of *Picea glauca* on an exposed ridge summit in the north Yukon was destroyed by fire in the early 1960's and, almost 20 years later, tree regeneration is absent.

Several recent investigations have shown that wildfire has played a major role in determining the structure and composition of forests during the Holocene. In eastern Canada and the adjacent United States, Green (1981, 1982, in press) and Davis et al. (1986) demonstrate that fire has varied in frequency during the Holocene. Green (in press) interprets both charcoal and pollen concentration data in terms of fire as the factor that triggered community change in the early Holocene at several sites in Nova Scotia and New Brunswick. Davis et al. (1986) identify a period of maximum charcoal frequency between 11,000 and 8,000 yr BP at several sites in Maine and suggest that it was characterized by "relatively unstable and dry conditions with frequent fire." Green (in press) concludes that the frequency and spread of fire has been influenced by the roughness of the landscape, a point demonstrated also by Grimm (1984) in his analysis of the postglacial history of the prairie–forest border in southern Minnesota.

Fire has been identified as a significant factor throughout the Holocene history of the boreal forest of Canada (e.g., Cwynar 1978; MacDonald 1984) as it has in adjacent areas of the United States (Swain 1973, 1980).

The role of fire in the early Holocene of the southern Pacific zone is alluded to frequently in the early literature, but it was only when Tsukada and Sugita (1982) conducted a systematic analysis of charcoal along with PAR estimates for the Mineral Lake site (Fig. 7.2) that firm evidence came to hand. They show a relationship between a decline in *Pseudotsuga* PAR at 9,700 yr BP and increases of *Pteridium* spores and charcoal fragments. They also point out that *Alnus* shows maximum development during this period and suggest that forest fires close to the site were frequent, attested by the presence of large (>250 microns) charcoal fragments. Mathewes (in press) appears to have been unaware of this investigation when he speculated on the role of fire in the same forest zone. The Kirk Lake record (Cwynar in press), described in Chapter 7, greatly strengthens the proposition that the warm dry summers of the early Holocene promoted frequent wildfires that maintained a mosaic, successional structure in the vegetation that would account for the high alder and bracken PAR values associated with Douglas-fir.

Pathogens

It is not my intention to attempt a review of the role of forest pathogens in Canada, a field outside my range of competence, but only to describe concisely those cases where a pathogenic organism has a major effect on the composition and structure of the vegetation and therefore could have played a significant role in the history of the forests. A recent review by Martineau (1984) shows that most of the dominant trees of the forests of Canada are each attacked by fifteen to thirty insect species, but only a few have fatal effects and, of these, fewer cause large-scale epidemics. But it is clear that these few can have major destructive effects on the forests, so I will review them briefly here, along with a few important plant pathogens.

The spruce budworm, *Choristoneura fumiferana* (Clem.), is one of the most widespread and destructive native insect pests in Canadian forest vegetation, and because of its significant economic effect, it has been surveyed and investigated in some detail. Blais (1983) has collated the scattered literature into a coherent statement of the current state of knowledge, and the following comments draw heavily on his paper. Budworms attack the leaves of *Abies balsamea* and *Picea glauca*. Its area of concentration is eastern Canada, from central Ontario to Newfoundland. The number of infestations has increased significantly, and the areas of infestation have become larger since the last century. The estimated total areas of defoliation in Ontario, Quebec, and the Maritimes were 100,000, 250,000 and 550,000 km^2 in the 1930s, 1940s, and the 1970s, respectively. Blais (1983) suggests that increased anthropogenic influence has caused the change, in particular by increasing greatly the relative area of the forests dominated by *Abies*. The initial cultural effect was the intensive logging of *Pinus strobus* from the early nineteenth century until the beginning of this century. Then, during most of this century, clear-cutting of fir–spruce and mixed stands has accelerated regeneration of fir. The result has been to produce large areas of balsam fir, promoting massive infestations of budworm. A further influence has been the increased role of fire control and protection activities, particularly in accessible areas of merchantable timber, so that, as Blais (1983) puts it: "fire protection means that our forest is progressing toward the climax stage of succession. It has long been recognized that budworm outbreaks are associated with the climax (fir–spruce stands) of the boreal forest."

The spread of birch dieback in both *Betula alleghaniensis* and *B. papyrifera* (Fig. 8.9), which has increased since the 1930s, has further promoted the increase of fir and spruce, which in turn favours more frequent and extensive outbreaks of spruce budworm. In general, according to Blais (1983), the less the influence of anthropogenic factors, the less frequent and extensive are the outbreaks of the pathogen, implying that in presettlement times the budworm functioned to "prevent the perpetuation of decadent stands and bring about a rejuvenation of the forest." It appears likely that, while the modern scale of infestation and its effects are significant, past changes in forest composition would have been small, detectable only at the individual stand level and therefore unlikely to be registered in the pollen record.

The spruce beetle, *Dendroctonus rufipennis* (Kby.), is a bark borer that attacks all spruce species, and it has been recorded as effecting widespread stand destruction in local river valleys and catchments, such as in certain river valleys of Gaspé, Quebec, in 1928–34 (Gobeil 1938).

The larch sawfly, *Pristiphora erichsonii* (Htg.), is an epidemic pest of *Larix laricina* and *L. occidentalis*, causing both damage to the leaves and shoots with consequent growth reductions, and mortality (Nairn et al. 1962). However, larch is shade-intolerant and does not readily regenerate in treed communities, so the effect of the sawfly might be to accelerate the eventual replacement by such species as *Picea mariana* and *Fraxinus nigra*.

Figure 8.9. The composition and structure of forest stands can be altered significantly by pathogens. Here a stand of *Betula papyrifera* in the La Verendry Provincial Park, Quebec, has been killed by birch dieback disease and is being replaced rapidly by *Abies balsamea* and *Picea glauca*.

The hemlock looper, *Lambdina fiscellaria fiscellaria* (Gven.), is a native defoliator that attacks mainly *Abies*, but feeds on many other conifers and several hardwood species (birch, maple, poplar). The most serious recorded outbreaks were in Newfoundland, where five epidemic waves, each of 3- to 6-yr duration, occurred at 10- to 15-yr intervals between 1912 and 1970. Large areas of balsam fir were destroyed (Martineau 1984). Attacks on hemlock have been very localized. For example, 90 percent of hemlock was destroyed in 1953–6 on small islands in Ontario (Martineau 1984).

Pathogen attacks that result in extensive tree mortality can be expected to be recorded in the regional pollen rain, and one well-documented and several putative instances are familiar. The chestnut, *Castanea dentata*, has been virtually eliminated from Canada and the adjacent United States by a fungal disease, *Endothia parasitica* (Murr.) A. and A., that appeared first in 1904 in New York City and spread rapidly throughout the range of the tree until by the late 1930s virtually all mature trees had been killed, and a survey of chestnuts in Ontario reported no surviving mature trees (Fox 1949). The pollen registration of this event has been documented, and the "chestnut decline" has become a useful marker (Anderson 1974). Before the blight, chestnut was a major codominant in the deciduous forest belt of Appalachia.

Other pathogens, introduced from Europe, have caused less complete but still serious decimation of trees in eastern North America – the Dutch elm disease, a fungus (*Ceratocystis ulmi*) transmitted by a beetle (*Scolytis*), and a similar pathogen in beech. Davis (1981) proposed a similar pathogenic attack and response mechanism to the chestnut blight to account for the apparently synchronous (Webb 1982) hemlock decline in many (but not all) pollen diagrams from eastern North America. The decline is more conspicuous in diagrams from the northern part of the range of hemlock (Gaudreau and Webb 1985). As Birks (1986) points out, adequate testing of an hypothesis of pathogenic cause is extremely difficult. However, Allison, Davis and Moeller (1984) compared the decrease in PAR of chestnut corresponding to its decline in the landscape with the PAR decrease of hemlock in the mid-Holocene. Both sample sets came from the same pond in New Hampshire, and the analysis was based on carbowax-embedded samples of annually laminated sediment. The chestnut pollen decline occurred in 7 to 8 yr and the hemlock decrease spanned 6 to 7 yr, strong evidence in support of a pathogen-caused response. However, no evidence has been offered of forest destruction by an insect pathogen on the scale of the hemlock decline, and the modern depredations of the hemlock looper, suggested by Allison, Davis, and Moeller (1984) as the possible cause, are on very local scales. The problem will not be resolved further until a large number of sites is at hand with reliable chronologies and comparable pollen source areas, so that the spatial and temporal ranges of the hemlock decline can be made more precise.

Finally, the beech bark disease is caused by the combined activity of an insect, *Cryptococcus fagi* Baer., which feeds on the bark and cambium of the beech (*Fagus grandifolia*) and introduces an Ascomycete fungus, *Nectria coccinea* var. *faginata* Lohman, Watson, and Ayers, which destroys the tissues initially attacked by the insect. Severely attacked trees become deformed and stunted, and epidemics causing extensive mortality have been reported from maritime Canada and adjacent Maine (Ehrlich 1934). Bormann et al. (1970) speculate that local declines of beech might be related to this pathogenic combination. Both fungus and insect were probably introduced accidentally from Europe early in this century (Boyce 1961). The current status of the disease is poorly known. It was originally reported in North America in 1920 near Halifax, Nova Scotia, and its subsequent spread in the Maritime Provinces and adjacent New England was documented by

Ehrlich (1934). It appears to be confined to the Atlantic region and may be climatically limited.

Paludification

The pollen record from the northwest provides increasing evidence, as more investigators make the distinction between white and black spruce pollen, that *Picea glauca* was the initial conifer dominant in the early Holocene (e.g., Brubaker, Garfinkel, and Edwards 1983; MacDonald 1984; Anderson 1985; Edwards et al. 1985; Ritchie 1985b; Edwards and Brubaker 1986). After a short period of dominance, white spruce PAR values declined in the early Holocene at all Canadian sites. The chronology of replacement of white spruce by black spruce changes westward into central Alaska because white spruce appears to have spread more slowly through Alaska between 9,000 and 5,000 yr BP than it did into northwest Canada (Ager 1983; Ritchie and MacDonald 1986). At sites near or north of the modern treeline, the mid-Holocene white-spruce pollen decline is a reflection of a movement southward of the limit of trees (e.g., Spear 1983), and the primary controlling factor is thought to have been a cooling of summer temperatures after a maximum of summer warmth in the early Holocene (Ritchie, Cwynar, and Spear 1983). However, within the boreal forest, the decline of white spruce was accompanied by an increase in other boreal trees, particularly black spruce and white birch, and some factor other than climatic change must be adduced to explain the decrease. Recent studies on the autecology of white spruce are suggestive. Juday and Zasada (1984) have shown from an analysis of the age of the land surfaces of Alaskan floodplains and the age structure of the trees that, while white spruce replaces balsam poplar in a successional process and can replace itself for one generation, ultimately old-growth white spruce degenerates and is replaced by black spruce. The soil changes that accompany and in part cause this change in species dominance are the accumulation of a thick moss mat and organic layer, increasingly moist soils, and the development of permafrost (Viereck 1970; Van Cleve, Dyrness, and Viereck 1980). Permafrost aggradation in response to the accumulation of humus and peat layers might have been a general trend in the early Holocene, responsible in part for the changes in species composition and abundance of the early forests. Zoltai and Tarnocai (1975), while they correctly emphasize that their data are not adequate for firm conclusions, show that permafrost development in poorly drained sites along the Mackenzie River valley resulted from the accumulation of an initial layer of organic debris or peat, and their stratigraphic studies depict transitions from initial moist mineral soil to peat-capped surfaces with permafrost. The radiocarbon ages of the basal peats in the Lower Mackenzie region cluster in the early Holocene. It is possible, though adequate macrofossil evidence is still lacking to test the idea, that *Picea glauca* was common on these early Holocene, moist mineral soils, and was later replaced by *P. mariana* as peat and permafrost developed (MacDonald 1983; Ritchie 1984a). The modern distribution of the two species in the northern part of the boreal forest appears to be related to the microtopography as it regulates the thickness of the active layer and the resulting soil depth, moisture, and temperature (Sakai et al. 1979; Viereck and Dyrness 1980; Van Cleve and Viereck 1981). Black spruce occurs on the cooler, moister soils with shallower active layers.

An investigation of forest history in southeast Labrador, using palynology and geochemistry, led to essentially similar conclusions about the long-term dynamics of white and black spruce. Engstrom and Hansen (1985) presented detailed diagrams, in addition to the pollen diagram I have made use of earlier (Fig. 5.2), showing the changes in stratigraphic pattern of elemental concentration with time. They used their analyses to test and confirm the idea of Lamb (1980), from pollen studies in the same area, that the change in forest dominance from *Picea glauca* followed by *Abies balsamea* and then by *Picea mariana* "resulted from the progressive deterioration of soil conditions (the gradual buildup of waterlogged forest peat) which ultimately favoured the edaphically tolerant black spruce" (Engstrom and Hansen 1985, p. 558). The most striking geochemical evidence from the lake sediments is the increase in authigenic Fe and Mn together with humic fractions at about 7,000 yr BP, coinciding with the transition from white spruce to black spruce pollen predominance (*authigenic minerals* accumulate in sediments by diagenetic processes and by solution and precipitation; *allogenic minerals* are derived from soil erosion in the lake catchment area). On the other hand, they recorded "no stratigraphic changes in allogenic minerals that might indicate intensive soil weathering," which led them to conclude that the vegetational change "resulted from the sequestering of nutrients in a cold waterlogged substrate." They offer the interesting generalization, which should be tested by the application of their own elegant methods elsewhere in the boreal forest, that: "Autogenic changes of vegetation have occurred (in Labrador) because of the buildup of deep forest peats, a process taking perhaps 2,000 years to complete. Given the normal length of interglacial cycles, ecosystems in the boreal region are more likely to be affected by vegetationally mediated soil changes than the progressive loss of fertility through weathering."

Forest ecologists who still take seriously the

tendentious and heavily semantic discussions of autogenic/holistic versus reductionist/individualist concepts of plant succession, reviewed recently by McIntosh (1985), might usefully pay attention to such investigations of boreal forest history as that of Engstrom and Hansen (1985) as sources of what is usually lacking in their deliberations – chronologically secure data on both vegetational and environmental change over significantly long time periods. The changes recorded in the above example occurred over periods involving relatively few generations of the dominant trees; *Picea glauca* abundances had declined significantly, followed by a short-lived increase in *Abies*, then an increase in *Picea mariana*, all within 1,500 years. The operative factor in this example of long-term dynamics (as defined by Johnson 1981) is edaphic change – an autogenic factor – and the response is change in the relative abundances of the three dominant trees.

Recent investigations of white spruce regeneration on upland sites following disturbances provide a clue to explain in part the early Holocene decrease in white spruce followed by increases in the other boreal trees and, later, alder. White spruce in the northwest produces large seed crops irregularly, as seldom as once every 10 yr (Zasada and Viereck 1979). It rarely reproduces vegetatively, and it requires an open, mineral soil lacking competition from shrubs and herbs for successful germination and establishment. Fox et al. (1984) suggest on the basis of both field observations and simulation studies that "the probability that seed availability coincides with the brief period between initial disturbances (logging and fires) and complete colonization by other species can be very low" and they suggest that "it is not an unlikely proposition that periodic disturbance of an area initially fully stocked with (white) spruce will result in a progressive decline or exclusion of spruce from the area."

Problems for the future

Climate disequilibrium

One of the familiar problems of science is that when a quantitative factual record is inadequate, discursive narrative predominates from which various subjective emphases emanate and often have inordinate influence on the development of the science. For example, when West (1964, 1970) concluded from his detailed analysis of interglacial floras in Britain that successive interglacials supported different aggregations of taxa "with no long history in the Quaternary . . . merely temporary aggregations under given conditions of climate, other environmental factors, and historical factors," the apparent congruence of that conclusion with Gleasonian ideas of individualistic plant communities assembled by chance, led several North American investigators, including the influential M. Davis (1976, 1981), to interpret the eastern North American Holocene record as a predominantly chance sequence of communities assembled primarily as a function of time and distance from a full-glacial refugium. The emphasis was that "during much of the Holocene the distributions of many species were in disequilibrium with climate." She used migration maps to support these notions and, while the apparently different migration routes of some taxa (*Picea*, *Larix*, *Abies*, *Pinus*) now can be shown to be similar as the fossil record expands, few would dispute that the routes and timing of chestnut and hickory seem to demonstrate the disequilibrium notion. In interpreting the pollen record from the Lower Mackenzie basin, I also subscribed to the disequilibrium–chance aggregation concept (Ritchie 1977).

However, the British interglacial record might not be representative of Europe as a whole; it is not surprising that Britain, the extreme distal, periodically insular fragment of Eurasian migration routes, might show a maximum of variability in species composition from one interglacial to the next. But the long sequences in southern Europe are remarkable for the consistency with which particular assemblages recur (Wijmstra 1969; van der Hammen, Wijmstra, and Zagwijn 1971; Woillard 1978; de Beaulieu and Reille 1983; Suc and Zagwijn 1983). Likewise, a long record from South America, with at least fifteen cycles, shows a pattern of consistency of recurring ecological–physiognomic assemblages, although they differ floristically from one cycle to the next (Hooghiemstra 1984).

Long, continuous pollen records spanning several glacial–nonglacial cycles are lacking in North America, so a systematic examination of the question of different, temporary aggregations versus recurring assemblages is not possible. However, roughly twenty sites have been reported from eastern Canada with palynological evidence from the last interglacial or Sangamon Stage, "considered by Atlantic Canada workers to encompass all of oxygen isotope stage 5" (Mott and Grant 1985). Several records are detailed, showing up to 100 different plant taxa from both pollen and macrofossils. Direct inspection of the diagrams suggests that the five main pollen assemblages identified so far are indistinguishable from Holocene assemblages. They are tundra, dominated by grasses, NAP, and *Betula*; forest–tundra dominated by *Picea*, *Betula*, ericads, Cyperaceae, and NAP; conifer forest dominated by *Picea*, *Pinus*, *Abies*, *Larix*, and *Betula*; mixed forest dominated by *Pinus*, *Abies*, *Tsuga*, *Fagus*, *Betula* (*alleghaniensis*), and *Acer*; and hardwood forest dominated by *Tilia*, *Acer*, *Fagus*, and *Fraxinus* with lesser amounts of *Juglans*, *Carpinus*, *Liquidambar*, and *Castanea* (Brookes, McAndrews, and von Bitter

1982; Mott, Anderson, and Matthews 1982; deVernal, Richard and Occhietti 1983; deVernal et al. 1986). These interglacial regional pollen assemblages are very similar also in sequence to the Holocene regional patterns described in Chapter 5 for eastern Canada. It might well be, of course, that the composition and structure of the vegetation at the local community level was significantly different during, say, the mixed-forest assemblage zones of the Sangamon and those of the Holocene, but so far pollen analysis has been unable to resolve differences in vegetation at scales finer than the regional level. All of the common pollen taxa recorded in Holocene assemblages from eastern Canada have been found in sediments from the last interglacial, in the same combinations, proportions, and sequences. However, it would be premature to draw firm conclusions. The recent assessment by Birks (1986), in which he develops the glacial–interglacial cycle, described by Iversen (1958), to describe the Quaternary vegetation pattern and process of northwest Europe, is a reasonable statement of our current understanding. He suggests that "broadly similar ecological–physiognomic assemblages recur in interglacials at the broad scales of 10^6 to 10^{12} m^2 and 10 to 12 $\times$ 10^4 years in response to regional climatic and edaphic change and competition" and "within these broad assemblages . . . the actual floristic and vegetational composition varies from interglacial to interglacial." (Although the elaboration of the cryocratic–protocratic–mesocratic–telocratic scheme of Iversen (1958) to describe vegetational and ecological patterns and processes in the late Quaternary of northwest Europe is a very effective and systematic method of organizing information, I have not attempted it here because the database is too variable and the schema would require extensive regional adaptation to fit the much broader range of Canadian ecosystems.)

In the same review of the northwest European literature, Birks (1986) draws attention to two lines of evidence that provide compelling support for the disequilibrium concept. First, the abundance of "thermophilous beetles and marsh and aquatic plants . . . indicates that climate was warmer than is suggested by the terrestrial vegetation . . ." during the Allerød interstadial. Second, at 10,000 yr BP, while temperate tree taxa were present, thermophilous aquatic and marsh plant taxa were absent, suggesting that "tree distributions (were) not in equilibrium with climate for at least 10^3 years."

In the absence in North America of conclusive, precise data, the long and interesting exchange of subjective opinions will continue to enliven palaeoecology. For example, an equally plausible alternative to the disequilibrium emphasis of Davis (1981, 1986) is one that gives primacy to climatic control and has conceptual affinities with a Clementsian view of community development (Solomon, West, and Solomon 1981). Such a viewpoint would contend that because of the great topographic and, therefore, mesoclimatic diversity of the refugial areas along and adjacent to the central and southern Appalachian axis, most of the taxa that dominate the modern arctic, boreal, and temperate plant communities were within reasonable dispersal distance to occupy the expanding deglaciated landscape to the north. The actual sequence was a response to changing climate from conditions with a short growing season, cool summers but relatively mild winters (Fig. 8.6) prevailing on an open landscape with immature soils and widespread surface instability, to warmer summers and greater diversity of soil conditions. Such an environment selected out from the large available pool a mixture of life-cycle strategies, primarily mobile taxa with relatively undemanding requirements for establishment and spread, but all species adapted to short growing seasons with cool summers. In other words, an assemblage of taxa that today are arranged into roughly zonal patterns of herb tundra, shrub tundra, forest–tundra, and various subzones of the boreal forest moved en masse into the deglaciated landscapes at the beginning of ice recession, with only such minor sorting effects of different migration rates on the pattern as the extreme mobility of *Populus* species enabling them, locally, to provide the first arboreal presence. On the other hand, the equally available pool of trees that today are more or less confined to the temperate forest region were held in their refugia until a few millennia later when the climate changed to warmer summers with longer growing seasons. To the north near the receding ice, forest communities of boreal trees (spruce, fir, pine, birch, with more local communities of black ash, elm, and other hardwoods) were replacing treeless vegetation, and species referred to here as arctic–montane outliers were shrinking in range to montane or locally unshaded habitats. Progressive climate change towards longer, warmer growing seasons and milder winters both eliminated the cold, short-season taxa and enabled the pool of temperate taxa to spread north. The consistent pattern involving white pine, sugar maple, hemlock, beech, and yellow birch is then explained as a response to a series of climatic changes involving precipitation, length of growing season, summer and winter temperature extremes, and other climatic factors. Some of the changes involved expansion or recession of the boundary of the vegetation dominated by these temperate species, and the extension northward of *Pinus strobus* is an example. The primary changes in the pollen record were due to shifts in the dominance of the major taxa on mesic sites in response to climate change – for example, increased effective precipitation promoted hemlock over white pine, maple, beech, and yellow birch.

Spatial resolution

A continuing challenge to the palaeoecologist is the commonly held view among ecologists (e.g., Connell and Sousa 1983) that pollen analysis is ineffective in addressing problems of vegetation dynamics because of its broad spatial and temporal scales. Bridging the gap between long-term regional vegetational history, to which almost all available pollen data have exclusive relevance, and changes at the plant community level, measurable on decadal time scales, is not going to be easily accomplished. Studies of very small basins (<1 ha) and annually laminated sediments hold out some promise. Detailed analyses of plant macrofossils can be useful, but they too have limitations. The cautionary comments of Jackson (1986), after completing what is probably the most thorough investigation of North American plant macrofossils, are pertinent. He writes: "Macrofossil assemblages do not provide an unbiased sample of watershed vegetation. Many important forest taxa (maples, beech, mountain ash) are extremely rare in surface macrofossil assemblages, despite their abundance in watershed vegetation. . . . In contrast, conifers and birches are abundant in macrofossil assemblages, even when present only in small numbers in the vegetation." He also notes a strong bias towards streamside and lakeshore vegetation. Modern studies indicate that "presence/absence data are reliable for detecting significant populations of coniferous taxa and birches. . . . Presence/absence data are less effective at detecting very small populations of conifers and birches, and completely ineffective for other hardwoods (maples, beech)."

Earlier in this chapter, it was noted that most of the variation in community composition and structure in forested regions of Canada is due to wildfire and pathogens. Detailed ecological mapping has shown that the smallest units of forest community are detectable at scales of 1:500 to 1:10,000 (e.g., Jurdant et al. 1972); and Damman (1979) has concluded that depending on the topography of the landscape, "it is impossible to map individual site types or plant communities at scales over 1:20,000." In forested areas of moderate topographic diversity with average conditions of additional heterogeneity due to disturbance factors (fire, pathogens), scales less than 1:10,000 are necessary to distinguish the plant community units that are important from ecological or dynamic viewpoints. As Birks and Gordon (1985) and Birks (1986) point out in reviews of the literature on spatial scale and pollen-representation factors, reconstructions of past communities from pollen analysis at this local scale are feasible only when sampling sites are chosen that register predominantly a source area of 20- to 30-m radius. Small hollows or ponds, 10 to 30 m in diameter, are appropriate. By contrast, the entire Canadian, and North American, pollen record is based on sites that register a regional or larger scale signal and therefore contain no information that can be related to those small-scale spatial and temporal changes that are referred to generally as "community dynamics."

Now that the regional sequences of pollen assemblages are established, particularly for eastern Canada, specific problems of spatial resolution could be addressed by making imaginative use of information on modern communities and environments. For example, Bergeron et al. (1983) and Bergeron, Bouchard, and Massicotte (1985) have provided detailed plant community gradient analyses in the Abitibi region of the boreal forest, and they have identified the chief environmental variables that control the mosaic pattern of the forest communities. Carefully designed investigations, using both pollen and macrofossil analyses, of sediments in basins of varied size might provide discernible records of variation due to each of climatic stress, competitive adaptations to drainage and trophic regime, and fire history. A useful review by Oliver (1981) of the status of understanding of forest development following "disturbance" at various scales would help to indicate the potential and limitations of such investigations. He defines "major disturbances" as factors that "knock over or kill all living tree stems in an area large enough to ensure that most trees beginning growth after the disturbance do not encounter competition from surrounding, undisturbed trees," and he lists as examples crown fires, severe windstorms, snow avalanches, landslides, mud flows, and other severe soil erosion or deposition effects and forest clear-cutting. By contrast, he defines "minor disturbances" as factors that affect only individual trees – windthrow, lightning strike, local pathogen effects, and partial cutting or thinning. Changes at the stand or community level (major), by contrast with individual (minor) gap-filling processes, can be studied by pollen analysis of small (<0.5 ha) basins, as Andersen (1978) and several others have shown so effectively. Oliver assembles North American examples to show that the frequency and intensity of the disturbances determine which species assume dominance, and he emphasizes that different species can dominate a stand for a long time following the change, so that, contrary to the succession-to-climax view, "different forest communities could potentially inhabit the same area for an indefinite period" (1981, p. 155). In addition, the results of forest succession model simulations can provide useful insights to facilitate the correct choice of stand size and time interval, and the appropriate variables to include

(Shugart, Crow, and Hett 1973; Davis and Botkin 1985; Dale, Hemstrom, and Franklin 1986). For example, the apparent contradictions between the views noted above (Oliver 1981) and the modeling results of Dale, Hemstrom, and Franklin (1986) regarding succession-to-climax theory relate to such critical factors as the availability of seed sources at the time of disturbance, the age structure and composition of the initial stand, and the intensity of the disturbance factor. Pollen and macrofossil analysis of the sediments in small hollows could test these ideas in particular regions and the western hemlock–silver fir–Douglas-fir forests of the Pacific northwest would provide scope for such fine-scale investigation.

Such data sets could also be used effectively to analyze plant taxa and community responses to climatic change, a topic of current interest and controversy (Birks 1981c; Wright 1984; Cole 1985; Markgraf 1986). Many areas in southern Canada provide appropriate landscapes and potential fossil sites to explore the questions of response and sensitivity of vegetation to climatic change. One of the most promising lines of research would be to choose small sites in areas of differing topographical complexity within the same biogeographic zone and to measure and compare the pollen responses. Jacobson and Grimm (1986) assemble convincing evidence that the regional vegetation response, and therefore the regional pollen sequence, can vary according to the physiographic complexity of an area. They show that a site in central Minnesota surrounded by relatively heterogeneous topography yielded a pollen sequence of continual, gradual change over time whereas sites elsewhere in the same biogeographic region (prairie–forest border) revealed abrupt changes ascribed to the response of relatively uniform vegetation in areas of subdued topography (Grimm 1983). Topographically diverse regions can house a wide range of ecological conditions and communities, illustrated by the familiar vegetational contrasts of slopes with different aspects.

Pollen source area

One corollary of the problem of spatial resolution is that the source areas of pollen registered at fossil sites are poorly known (Birks 1981b). As a result, largely untested theoretical assertions about pollen dispersal behaviour have unduly influenced some investigators (e.g., Tauber 1965). A modest beginning has been made in Canada to accumulate the necessary long-term data on patterns of atmospheric pollen. However, nothing conclusive will emerge until a data bank is available comparable to the Montpellier-based network of monitoring stations, which has recorded for over 10 yr the year-round weekly atmospheric pollen from the subarctic of Sweden to North Africa (Cour et al. 1980).

None of the above problems will be solved easily or quickly. The progress of palaeoecology remains dependent on the adequacy of its database, and as most types of data analysis (pollen, macrofossils, diatoms, etc.) require necessarily laborious and meticulous identification and enumeration, the rate of advance of theoretical developments remains frustratingly slow. As Birks and Gordon (1985) emphasize, even the availability of computer-based manipulative capacities combined with various numerical methods will make little impact on problems of theoretical ecology in the absence of an adequate database.

Concluding comments

The patterns of vegetational change in Canada since deglaciation have been described in as much detail as the fossil data permit, and it has been emphasized that even in regions with the most detailed networks of sites, the record refers to vegetation only at the regional scale in both space and time. Several important categories of proximal causal factors have been identified to explain these changes, of which climate, soil development, wildfire, windstorms, erosion, epidemic pathogens, and cultural effects are the most important.

A broader temporal perspective has not been attempted because of the sparse fossil record from older sediments. However, that does not imply that I do not share the fundamental motivation of palaeoecologists to illuminate the ultimate causes that have moulded the Quaternary and modern vegetation of Canada. These have been essentially evolutionary responses to major environmental changes in the late Tertiary and early Quaternary. The uplifting of the Cordillera and other large-scale changes in the physical geography of Canada produced an entirely new macroclimate that, over several millions of years, differentiated into arctic, boreal, temperate, and arid regional patterns, which led in turn to the evolution of several distinctive responses in the Angiosperm flora, including morphological and physiological tolerance to extremes of winter cold, short growing seasons, and increased seasonality; the emergence of herbs and grasses adapted to summer drought, wildfire, and a coevolved, herbivorous, plains fauna; life-cycle adaptations to wildfire in forest tree species; and other trends. This book has attempted to reconstruct the vegetation during only the latest of a large number of cycles of glacial/interglacial climate, but the long-term use of such a compilation will be found in terms of understanding of the ultimate evolutionary processes that have governed and determined the patterns of change.

Notable progress has been made in Canada and all of North America in elaborating the history of "biotic communities evolved in adjustment to

. . . the *climatic complexes* present in their region" (Bryson and Wendland 1967 in what is now regarded as a seminal contribution, although fifteen years ago it was greeted with silence or scepticism). The record reviewed in this book, and the much richer documentation from Eurasia, reveals that the response of plants to macroclimatic change at the scale of 10^3 years and 100 km^2 can be comprehended in a deterministic framework, in the sense that the Quaternary vegetational complexes consist of a number of dominant taxa with discrete ecological amplitudes that have evolved in response to the late Tertiary and Quaternary change in climate toward increased continentality, including much colder winters and warmer, drier summers, high but regionally variable frequencies of wildfire, such regional edaphic trends at higher latitudes as podzolization, paludification, and permafrost aggradation, and the influence of coevolved or immigrant fauna, including pathogens and browsers. In this perspective of ultimate evolutionary causes, the major vegetational sequences recorded through the more than fifteen glacial–interglacial cycles of the later Quaternary can be expected to have regional consistency, as the still meagre record from Atlantic Canada shows. It is likely that the same dominant taxa will have persisted throughout most of the Quaternary and that measurable change as a function of evolutionary process will be registered only in older sediments. For example, Suc and Zagwijn (1983) and Suc (1984), drawing on a rich fossil record from southern Europe, describe the appearance and evolution of the Mediterranean vegetation about 4 million years ago as a response to the development of increased climatic seasonality with dry, warm summers. This was followed at 2.3 million years ago by cyclical variations in climate and vegetation throughout the Quaternary with little or no change in the floristic composition of the recurring assemblages. New discoveries from the high arctic of both western Canada and Greenland provide tentative intimations of similar patterns. Late-Tertiary records show a forest–tundra assemblage dominated by *Larix* (Kuc 1974; Funder et al. 1985; Matthews, Mott, and Vincent 1986); regional modern analogues are not found in North America, though they might occur in Siberia, and it is likely that the apparent shift from a regional *Larix* forest–tundra to a *Picea* forest–tundra in the Quaternary reflects an evolutionary response.

Such a deterministic view need not be seen as a contradiction of the so-called "individualistic" theory, which is often interpreted as a chance or stochastic model of vegetation dynamics. In my view, the differences can be attributed largely to the differing spatial and temporal scales employed. The individualistic concept rests on the "assertions that species had individualistic ecological characteristics and were assembled into local communities with a large stochastic component according to the vagaries of dispersal and the available environments" (McIntosh 1985). An example will illustrate what I regard as a sensible view of this largely spurious controversy. The pollen record shows that the response of vegetation to Holocene climatic change in the Pacific northwest of North America has resulted in a low-elevation forest dominated by western hemlock, Douglas-fir, and silver fir, and the many hundreds of stands described in the botanical and forestry literature show a high degree of constancy in species composition and relative abundance (Krajina 1969; Franklin and Dyrness 1973). Interstand variation is common, however, and it can be ascribed to variations in the availability of seed after partial or complete destruction by fire, erosion, or pathogens, the composition of the stand at the time of change, and local microclimate, and the so-called "vagaries of seed dispersal" are modal responses to the interaction of genotypes and environment.

Appendix

Sites used for modern pollen spectra

Map no.	Site name or no.	Author(s)
	Pacific–Cordilleran	
1	Pinecrest	Mathewes and Rouse (1975)
2	Marion Lake	Mathewes (1973)
3	Hoh Bog	Heusser (1974)
4	Kirk Lake	Cwynar (in press)
5	Hall Lake	Sugita and Tsukada (1982)
6	Davis Lake	Barnosky (1981)
7	Mineral Lake	Sugita and Tsukada (1982)
8	Battle Ground Lake	Barnosky (1985a)
9	Fiddler's Pond	White and Mathewes (1982)
10	Fairfax	Schweger, Habgood, and Hickman (1981)
11	Wedge Lake	MacDonald (1982)
12	Yamnuska Bog	MacDonald (1982)
13	Chalmers Bog	Mott and Jackson (1982)
14	2	Hazell (1979)
15	4	Hazell (1979)
16	7	Hazell (1979)
17	5	Kearney (1983)
18	9	Hazell (1979)
19	10	Hazell (1979)
20	12	Hazell (1979)
21	29	Kearney (1983)
22	44	Mack, Bryant, and Pell (1978)
23	36	Mack, Bryant, and Pell (1978)
24	Waits Lake	Mac et al. (1978a)
25	Big Meadow	Mack et al. (1978b)
26	29	Mack, Bryant, and Pell (1978)
27	16	Mack, Bryant, and Pell (1978)
28	Kelowna Bog	Alley (1976)
29	Mud Lake	Mack, Rutter, and Valastro (1979)
30	Bonaparte	Mack, Rutter, and Valastro (1979)
31	28	Mack and Bryant (1974)
32	75	Mack and Bryant (1974)
33	40	Mack and Bryant (1974)
34	42	Mack and Bryant (1974)
35	13	Mack and Bryant (1974)
36	Carp Lake	Barnosky (1985b)
37	Triangle Lake	Birks (1980)
38	K17	Birks (1977)
39	Gull Lake	Birks (1980)
40	Natla Bog	MacDonald (1983)
41	Lateral Pond	Ritchie (1982)
42	Antifreeze Pond	Rampton (1971)
43	Monkshood	Cwynar (personal communication)
44	99	Birks (1977)
45	111	Birks (1977)
46	Grayday	Cwynar (personal communication)
47	Hanging Lake	Cwynar (1982)
48	Kettlehole	Cwynar (personal communication)
49	Waterdevil	Cwynar (personal communication)
50	Drizzle Lake	Cwynar (personal communication)
51	Lily Lake	Cwynar (personal communication)
52	Cape Ball	Warner, Clague, and Mathewes (1984)
	Western Interior	
1	39	Ritchie, Hadden, and Gajewski (in press)
2	37	Ritchie, Hadden, and Gajewski (in press)

Map no.	Site name or no.	Author(s)
3	35	Ritchie, Hadden, and Gajewski (in press)
4	33	Ritchie, Hadden, and Gajewski (in press)
5	31	Ritchie, Hadden, and Gajewski (in press)
6	28	Ritchie, Hadden, and Gajewski (in press)
7	26	Ritchie, Hadden, and Gajewski (in press)
8	23	Ritchie, Hadden, and Gajewski (in press)
9	43	Ritchie (1974)
10	52	Ritchie (1974)
11	34	Ritchie (1974)
12	41	Ritchie, Hadden, and Gajewski (in press)
13	48	Ritchie, Hadden, and Gajewski (in press)
14	55	Ritchie, Hadden, and Gajewski (in press)
15	5	Lichti-Federovich and Ritchie (1968)
16	9	Lichti-Federovich and Ritchie (1968)
17	4	Ritchie (1974)
18	38	Ritchie (1974)
19	31	Ritchie (1974)
20	Small Tree Lake	Mott (1969)
21	19	Lichti-Federovich and Ritchie (1968)
22	17	Lichti-Federovich and Ritchie (1968)
23	20	Lichti-Federovich and Ritchie (1968)
24	16	Lichti-Federovich and Ritchie (1968)
25	Twin Tamarack	Ritchie (1985b)
26	Lac Mèléze	MacDonald (1984)
27	Lac Demain	MacDonald (1984)
28	67	MacDonald (1984)
29	73	MacDonald (1984)
30	Wild Spear	MacDonald (1984)
31	Porter Lake	Ritchie (1980)
32	111	Lichti-Federovich and Ritchie (1968)
33	Cree Lake	Mott (1969)
34	Reindeer Lake	Ritchie (1976)
35	Thompson	Ritchie (1976)
36	Lofty Lake	Lichti-Federovich (1972)
37	Cycloid Lake	Mott (1969)
38	Flin Flon	Ritchie (1976)
39	Grand Rapids	Ritchie and Hadden (1975)
40	120	Lichti-Federovich and Ritchie (1968)
41	102	MacDonald (1984)
42	104	MacDonald (194)
43	Midnight Lake	Mott (1969)
44	98	Lichti-Federovich and Ritchie (1968)
45	99	Lichti-Federovich and Ritchie (1968)
46	100	Lichti-Federovich and Ritchie (1968)
47	103	Lichti-Federovich and Ritchie (1968)
48	104	Lichti-Federovich and Ritchie (1968)
49	122	MacDonald (1984)
50	123	MacDonald (1984)
51	119	MacDonald (1984)
52	117	MacDonald (1984)
53	120	MacDonald (1984)
54	Clearwater Lake	Mott (1969)
55	109	Lichti-Federovich and Ritchie (1968)
56	110	Lichti-Federovich and Ritchie (1968

Eastern

Map no.	Site name or no.	Author(s)
1	Rock Basin Lake	Hyvärinen (1985)
2	Moraine Lake	Hyvärinen (1985)
3	Proteus Lake	Hyvärinen (1985)
4	Klaresø	Fredskild (1969)
5	Iglutalik Lake	Davis (1980)
6	Söndre 3	Pennington (1980)
7	Disko 5	Pennington (1980)
8	Terte	Fredskild (1983)
9	21	Richard (1981a)
10	19	Richard (1981a)
11	18	Richard (1981a)
12	43	Lamb (1984)
13	46	Lamb (1984)
14	52	Lamb (1984)
15	10	Richard (1981a)
16	Chism I	Richard (1979)
17	Chism II	Richard (1979)
18	Delorme II	Richard, Larouche, and Bouchard (1982)
19	Daumont	Richard, Larouche, and Bouchard (1982)
20	Coghill	King (1984)
21	Lac Gras	King (1984)
22	53	Lamb (1984)
23	38	Lamb (1984)
24	72	Lamb (1984)
25	Rattle Lake	Björck (1985)
26	Sioux Lake	Björck (1985)
27	Indian Lake	Björck (1985)
28	Crates Lake	Liu (1982)
29	Yelle	Richard (1980a)
30	Antoine	Saarnisto (1974)
31	Jack Lake	Liu (1982)
32	Lac La Fourche	Richard (1981b)
33	Lac Turcotte	Labelle and Richard (1984)

Map no.	Site name or no.	Author(s)
34	Sable	King (1984)
35	22	Lamb (1984)
36	7	Lamb (1984)
37	N5	Railton (1972)
38	Sugarloaf	Macpherson (1982)
39	Hayes Lake	McAndrews (1982)
40	Upper Twin Lake	Saarnisto (1974)
41	Nina Lake	Liu (1982)
42	Lac Bastien	Bennett (personal communication)
43	Nutt Lake	Bennett (personal communication)
44	Ramsay Lake	Mott and Farley-Gill (1981)
45	Sav. 1	Savoie and Richard (1979)
46	Mont Shefford	Richard (1978)
47	Lotbiniere	Richard (1975a)
48	Lac Colin	Mott (1977)
49	Hams Lake	Bennett (personal communication)
50	Lake Hunger	Bennett (personal communication)
51	Basswood Road Lake	Mott (1975)
52	NS 4	Railton (1972)
53	Canoran Lake	Railton (1972)
54	Bluff Lake	Livingstone (1968)
55	NS15	Railton (1972)
56	Upper Gillies Lake	Livingstone (1968)

References

Adams, M.S., & Loucks, O.L. (1971). Summer air temperatures as a factor affecting net photosynthesis and distribution of eastern hemlock (*Tsuga canadensis* (L.) Carriere) in southwestern Wisconsin. *The American Midland Naturalist* 85(1):1–10.

Ager, T.A. (1975). *Late Quaternary Environmental History of the Tanana Valley, Alaska*, Report 54. Ohio State University Institute of Polar Studies, Columbus.

Ager, T.A. (1982). Vegetational history of western Alaska during the Wisconsin glacial interval and the Holocene. In D.M. Hopkins, J.V. Matthews, C.E. Schweger, & S.B. Young (eds.), *Paleoecology of Beringia*, pp. 75–93. Academic Press, New York.

Ager, T.A. (1983). Holocene vegetational history of Alaska. In H.E. Wright, Jr. (ed.), *Late-Quaternary Environments of the United States*, pp. 128–141. University of Minnesota Press, Minneapolis.

Ager, T.A. (1986). Ice-marginal vegetation development in southern Alaska during the late Pleistocene and early Holocene: Pollen evidence from Cook Inlet Region. *American Quaternary Association Program and Abstracts*, p. 11.

Ager, T.A., & Brubaker, L.B. (1985). Quaternary palynology and vegetational history of Alaska. In V.M. Bryant Jr. & R.G. Holloway (eds.), *Pollen Records of Late-Quaternary North American Sediments*, pp. 353–384. American Association of Stratigraphic Palynologists Foundation, Dallas.

Alley, N.F. (1976). The palynology and palaeoclimatic significance of a dated core of Holocene peat, Okanagan Valley, southern British Columbia. *Canadian Journal of Earth Sciences* 13:1131–1144.

Allison, T.D., Davis, M.B., & Moeller, R.E. (1984). Chestnut and hemlock pollen in laminated sediments: evidence for a pathogen-induced decline. *Abstracts: Sixth International Palynological Conference, Calgary*, p. 2.

Ammann, B.R. (1977). A pollenmorphological distinction between *Pinus banksiana* Lam. and *P. resinosa* Ait. *Pollen et Spores* 19(4):521–529.

Andersen, S.Th. (1978). Local and regional vegetational development in eastern Denmark in the Holocene. *Danmarks Geologiske Undersøgelse Arbog* 5:27.

Anderson, P.M. (1985). Late Quaternary vegetational change in the Kotzebue Sound area, northwest Alaska. *Quaternary Research* 24:307–321.

Anderson, P.M., & Brubaker, L.B. (1986). Modern pollen assemblages from Northern Alaska. *Review of Palaeobotany and Palynology* 46:273–291.

Anderson, R.C., & Loucks, O.L. (1979). White-tail deer (*Odocoileus Virginianus*) influence on structure and composition of *Tsuga canadensis* forests. *Journal of Applied Ecology* 16:855–861.

Anderson, T.W. (1974). The chestnut pollen decline as a time horizon in lake sediments in eastern North America. *Canadian Journal of Earth Sciences* 11(5):678–685.

Anderson, T.W. (1980). Holocene vegetation and climatic history of Prince Edward Island, Canada. *Canadian Journal of Earth Sciences* 17(9):1152–1165.

Anderson, T.W. (1983). Preliminary evidence for Late Wisconsinan climatic fluctuations from pollen stratigraphy in Burin Peninsula, Newfoundland. In *Current Research, Part B*, paper 83–1B, pp. 185–188. Geological Survey of Canada, Ottawa.

Anderson, T.W. (1985). Late-Quaternary pollen records from Eastern Ontario, Quebec, and Atlantic Canada. In V.M. Bryant Jr. & R.G. Holloway (eds.), *Pollen Records of Late-Quaternary North American Sediments*, pp. 281–326. American Association of Stratigraphic Palynologists Foundation, Dallas.

Anonymous. (1981). *Weather of U.S. Cities*, 2 volumes. Gale Research Company, Detroit.

Ashworth, A.C., & Cvancara, A.M. (1983). Paleoecology of the southern part of the Lake Agassiz Basin. In J.T. Teller & L. Clayton (eds.), *Glacial Lake Agassiz*, Special Paper 26, pp. 133–156. Geological Association of Canada, Toronto.

Ashworth, A.C., & Schwert, D.P. (1986). The effects of late Wisconsinan extinction on the postglacial development of the northern North American beetle fauna. *American Quaternary Association Program and Abstracts*, pp. 46–47.

Bailey, R.G. (1980). *Description of the Ecoregions of the*

United States, Miscellaneous Publications No. 1391. United States Department of Agriculture Forest Service, Ogden, Utah.

Baker, R.G. (1983). Holocene vegetational history of the western United States. In H.E. Wright Jr. (ed.), *Late-Quaternary Environments of the United States*, vol. 2, *The Holocene*, pp. 109–127. University of Minnesota Press, Minneapolis.

Baker, R.G., Schwert, D.P., Frest, T.J., Rhodes, R.S. II, Hallberg, G.R., Ashworth, A.C., & Janssens, J.A. (1986). A full-glacial biota from southeastern Iowa. *Journal of Quaternary Science* 1:91–107.

Baker, R.G., & Waln, K. (1985). Quaternary pollen records from the Great Plains and Central United States. In V.M. Bryant Jr. & R.G. Holloway (eds.), *Pollen Records of Late-Quaternary North American Sediments*, pp. 191–204. American Association of Stratigraphic Palynologists Foundation, Dallas.

Bakuzis, E.V., & Hansen, H.L. (1965). *Balsam fir: Abies balsamea (Linnaeus) Miller – A Monographic Review*. University of Minnesota Press, Minneapolis.

Barnosky, C.W. (1981). A record of late Quaternary vegetation from Davis Lake, Southern Puget Lowland, Washington. *Quaternary Research* 16:221–239.

Barnosky, C.W. (1984). Late Pleistocene and early Holocene environmental history of southwestern Washington State, U.S.A. *Canadian Journal of Earth Sciences* 21:619–629.

Barnosky, C.W. (1985a). Late Quaternary vegetation near Battle Ground Lake, Southern Puget Trough, Washington. *Geological Society of America Bulletin* 96:263–271.

Barnosky, C.W. (1985b). Late Quaternary vegetation in the southwestern Columbia Basin, Washington. *Quaternary Research* 23:109–122.

Barnosky, C.W. (1986). Vegetation and climate of the Puget Trough during the advance and retreat of the Puget lobe. *American Quaternary Association Program and Abstracts*, p. 26.

Bartlein, P.J., Prentice, I.C., & Webb, T. III. (1986). Climatic response surfaces based on pollen from some eastern North America taxa. *Journal of Biogeography* 13:35–57.

Bennett, K.D. (1985). The spread of *Fagus grandifolia* across eastern North America during the last 18,000 years. *Journal of Biogeography* 12:147–164.

Bennett, K.D. (In press). Holocene history of forest trees in Southern Ontario. *Canadian Journal of Botany*.

Berger, A. (1981). Astronomical theory of paleoclimates. In A. Berger (ed.), *Climatic Variations and Variability: Facts and Theories*, pp. 501–526. Reidel, London.

Berger, A., Imbrie, J., Hays, J., Kukla, G., & Saltzman, B. (1984). *Milankovitch and Climate, Understanding the Response to Astronomical Forcing*, 2 vols. Reidel, Boston.

Bergeron, Y., Bouchard, A., Gangloff, P., & Camite, C. (1983). *La classification ecologique des milieux forestiers de la partie ouest des cantons d'Hebecourt et de Roquemaure, Abitibi, Québec*. Laboratoire d'ècologie forestière, Laval University No. 9, Quebec.

Bergeron, Y., Bouchard, A., & Massicotte, G. (1985). Gradient analysis in assessing differences in community pattern of three adjacent sectors within Abitibi, Quebec. *Vegetatio* 64:55–65.

Berglund, B.E. (ed). (1986). *Handbook of Holocene Palaeoecology and Palaeohydrology*. Wiley, New York.

Berglund, B.E. (ed). (1986). *Palaeohydrological Changes in the Temperate Zone in the last 15,000 Years*. Project Guide, vols. 1 and 2, pp. 11–22. Department of Quaternary Geography, Lund University, Lund, Sweden.

Berglund, B.E., & Digerfeldt, G. (1976). Environmental changes during the Holocene – a geological correlation project on a Nordic basis. *Newsletter of Stratigraphy* 5:80–85.

Beug, H.J. (1964). Untersuchungen zur spat-und postglazialen Vegetationsgeschichte im Gardaseegebiet unter besonderer Berucksichtigung der mediterranen Arten. *Flora* 154:401–444.

Beug, H.J. (1975). Changes of climate and vegetation belts in the mountains of Mediterranean Europe during the Holocene. *Bulletin of Geology*, Warszawa, Habditka.

Bird, J.B. (1980) *The Natural Landscapes of Canada*, 2d ed. Wiley, Toronto.

Bird, R.D. (1961). *Ecology of the Aspen Parkland of Western Canada*, Publication 1066. Canadian Department of Agriculture, Ottawa.

Birks, H.J.B. (1973). Modern pollen rain studies in some arctic and alpine environments. In H.J.B. Birks, & R.G. West (eds.), *Quaternary Plant Ecology*, pp. 143–168. Blackwell, Oxford.

Birks, H.J.B. (1976). Late-Wisconsinan vegetational history at Wolf Creek, Central Minnesota. *Ecological Monographs* 46:395–429.

Birks, H.J.B. (1977). Modern pollen rain and vegetation of the St. Elias Mountains, Yukon Territory. *Canadian Journal of Botany* 55(18):2367–2382.

Birks, H.J.B. (1980). Modern pollen assemblages and vegetational history of the moraines of the Klutlan Glacier and its surroundings, Yukon Territory, Canada. *Quaternary Research* 14:101–129.

Birks, H.J.B. (1981a). Late Wisconsin vegetational and climatic history at Kylen Lake, northeastern Minnesota. *Quaternary Research* 16:322–355.

Birks, H.J.B. (1981b). Long-distance pollen in Late Wisconsin sediments of Minnesota, U.S.A.: A quantitative analysis. *New Phytologist* 87:630–661.

Birks, H.J.B. (1981c). The use of pollen analysis in the reconstruction of past climates: A review. In T.M.L. Wigley, M.J. Ingram, & G. Farmer (eds.), *Climate and History*, pp. 111–138. Cambridge University Press, New York.

Birks, H.J.B. (1985). Recent and possible future mathematical developments in quantitative palaeoecology. *Palaeogeography, Palaeoclimatology, Palaeoecology* 50:107–147.

Birks, H.J.B. (1986). Late-Quaternary biotic changes in terrestrial and lacustrine environments, with particular reference to north-west Europe. In B.E. Berglund (ed.), *Handbook of Holocene Palaeoecology and Palaeohydrology*, pp. 3–65. Wiley, New York.

Birks, H.J.B., & Birks, H.H. (1980). *Quaternary Palaeoecology*. Arnold, London.

Birks, H.J.B., & Gordon, A.D. (1985). *Numerical Methods in Quaternary Pollen Analysis*. Academic Press, London.

Birks, H.J.B., & Peglar, S.M. (1980). Identification of *Picea* pollen of late-Quaternary age in eastern North

America: a numerical approach. *Canadian Journal of Botany* 58:2043–2058.

Björck, S. (1985). Deglaciation chronology and revegetation in northwestern Ontario. *Canadian Journal of Earth Sciences* 22:850–871.

Black, R.A., & Bliss, L.C. (1978). Recovery sequence of *Picea mariana/Vaccinium uliginosum* forests after burning near Inuvik, Northwest Territories, Canada. *Canadian Journal of Botany* 56:2020–2023.

Black, R.A., & Bliss, L.C. (1980) Reproductive ecology of *Picea mariana* (Mill.) BSP, at tree line near Inuvik, Northwest Territories, Canada. *Ecological Monographs* 50:331–354

Black, R.A., & Mack, R.N. (1976). *Tsuga canadensis* in Ohio: Synecological and phytogeographical relationships. *Vegetatio* 32(1):11–19.

Blais, J.R. (1983). Trends in the frequency, extent, and severity of spruce budworm outbreaks in eastern Canada. *Canadian Journal of Forest Research* 13:539–547.

Borman, F.H., Siccama, T.G., Likens, G.E., & Whittaker, R.H. (1970). The Hubbard Brook ecosystem study: composition and dynamics of the tree stratum. *Ecological Monographs* 40(4):373–388.

Boyce, J.S. (1961). *Forest Pathology*. McGraw-Hill, New York.

Boyko, M. (1973). *European Impact on the Vegetation Around Crawford Lake in Southern Ontario*. M.Sc. thesis. University of Toronto, Toronto.

Bradley, R.S. (1985). *Quaternary Paleoclimatology*. Allen & Unwin, Winchester, Massachusetts.

Bradshaw, R.H.W., & Webb, T. III. (1985). Relationships between contemporary pollen and vegetation data from Wisconsin and Michigan, U.S.A. *Ecology* 66:721–737.

Brassard, G.R. (1984). The bryogeographical isolation of the Island of Newfoundland, Canada. *The Bryologist* 87(1):56–65.

Brookes, I. (1972). The physical geography of the Atlantic Provinces. In A. Macpherson (ed.), *Studies in Canadian Geography, the Atlantic Provinces*, pp. 1–45. University of Toronto Press, Toronto.

Brookes, I.A., McAndrews, J.H., & von Bitter, P.H. (1982). Quaternary interglacial and associated deposits in southwest Newfoundland. *Canadian Journal of Earth Sciences* 19(3):410–423.

Brubaker, L.B., Garfinkel, H.L., & Edwards, M.E. (1983). A late-Wisconsin and Holocene vegetation history from the Central Brooks Range: Implications for Alaskan palaeoecology. *Quaternary Research* 20:194–214.

Brubaker, L.B., Graumlich, L.J., & Anderson, P.M. (1987). An evaluation of statistical techniques for discriminating *Picea glauca* from *Picea mariana* pollen in northern Alaska. *Canadian Journal of Botany* 65:899–906.

Bryant, V.M. Jr., & Holloway, R.G. (eds). (1985). *Pollen Records of Late-Quaternary North American Sediments*. American Association of Stratigraphic Palynologists Foundation, Dallas.

Bryson, R.A. (1966). Air masses, streamlines and the Boreal Forest. *Geographical Bulletin* 8(3):228–269.

Bryson, R.A., & Hare, F.K. (1974). The Climates of North America. Volume VII. In H. Landsberg (ed.), *World Survey of Climatology*. Elsevier, New York.

Bryson, R.A., & Wendland, W.M. (1967). Tentative climatic patterns for some late glacial and post-glacial episodes in central North America. In W.J. Mayer-Oakes (ed.), *Life, Land and Water*, Occasional Papers No. 1, pp. 271–298. University of Manitoba, Department of Anthropology, Winnipeg.

Burden, E.T., McAndrews, J.H., & Norris, G. (1986). Palynology of Indian and European forest clearance and farming in lake sediment cores from Awenda Provincial Park, Ontario. *Canadian Journal of Earth Sciences* 23:43–54.

Burden, E.T., Norris, G. & McAndrews, J.H. (1986). Geochemical indicators in lake sediment of upland erosion caused by Indian and European farming, Awenda Provincial Park, Ontario. *Canadian Journal of Earth Science* 23:55–65.

Burke, M.J., Gusta, L.V., Quamme, H.A., Weiser, C.J., & Li, P.H. (1976). Freezing and injury in plants. *Annual Review of Plant Physiology* 27:507–528.

Buxton, G.F., Cyr, D.R., Dumbroff, E.B., & Webb, D.P. (1984). Physiological responses of three northern conifers to rapid and slow induction of moisture stress. *Canadian Journal of Botany* 63:1171–1176.

Canada Department of Agriculture (1970). *The System of Soil Classificiation for Canada*. Queen's Printer, Ottawa.

Charney, J.D. (1980). Hemlock-hardwood community relationships in the Highlands of southeastern New York. *Bulletin of the Torrey Botanical Club* 107(2):249–257.

Clark, W.R. (1983). *Forest Depletion by Wildland Fire in Canada, 1977–1981*. Environment Canada Information Report P1-X-21, Department of the Environment, Ottawa.

Clayton, L., Teller, J.T., & Attig, J.W. (1985). Surging of the southwestern part of the Laurentide Ice Sheet. *Boreas* 14:235–241.

Cogbill, C.V. (1985). Dynamics of the boreal forests of the Laurentian Highlands, Canada. *Canadian Journal of Forest Research* 15:252–261.

Cole, K. (1985). Past rates of change, species richness, and a model of vegetational inertia in the Grand Canyon, Arizona. *American Naturalist* 125:289–303.

Colinvaux, P.A. (1967). Quaternary vegetational history of arctic Alaska. In D.M. Hopkins (ed.), *The Bering Land Bridge*, pp. 207–231. Stanford University Press, Stanford.

Comtois, P., & Simon, J.P. (1985). *Inheritance of Allozyme Traits in Northern Populations of Balsam Poplar, and Future Directions for Genetic Studies in Nouveau-Québec*. University of Montreal, Montreal.

Connell, J.H., & Sousa, W.P. (1983). On the evidence needed to judge ecological stability or persistence. *The American Naturalist* 121(6):789–824.

Corns, I.G.W. (1983). Forest community types of west-central Alberta in relation to selected environmental factors. *Canadian Journal of Forest Research* 13(5):995–1010.

Cour, P., Seignalet, C., Guérin, B., Mayrand, L., Nilsson, S., & Michel, F.B. (1980). Inter-regional studies of pollen incidence from Lapland to North Africa. Proceedings of the 1st International Conference on Aerobiology, pp. 61–79, Munich 1978, Berlin 1980.

Curtis, J.T. (1959). *The Vegetation of Wisconsin*. University of Wisconsin Press, Madison.

Cwynar, L.C. (1977). The recent fire history of Barron Township, Algonquin Park. *Canadian Journal of Botany* 55(11):1524–1538.

Cwynar, L.C. (1978). Recent history of fire and vegetation from laminated sediment of Greenleaf Lake, Algonquin Park, Ontario. *Canadian Journal of Botany* 56(1):10–21.

Cwynar, L.C. (1982). A Late-Quaternary vegetation history from Hanging Lake, northern Yukon. *Ecological Monographs* 52(1):1–24.

Cwynar, L.C. (In press). The role of fire in the forest history of the western flank of the North Cascade Range. *Ecology* 68.

Cwynar, L.C., & Ritchie, J.C. (1980). Arctic steppe-tundra: A Yukon perspective. *Science* 208:1375–1377.

Dale, V.H., Hemstrom, M., & Franklin, J. (1986). Modeling the long-term effects of disturbances on forest succession, Olympic Peninsula, Washington. *Canadian Journal of Forest Research* 16:56–67.

Damman, A.W.H. (1976). Plant distribution in Newfoundland especially in relation to summer temperatures measured with sucrose inversion method. *Canadian Journal of Botany* 54:1561–1585.

Damman, A.W.H. (1979). The role of vegetation analysis in land classification. *Forestry Chronicle* 55:175–182.

Dansereau, P. (1959) *Phytogeographia Laurentiana II: The Principal Plant Associations of the Saint Lawrence Valley*, Contributions de l'Institut Botanique de l'Université de Montréal No. 25. University of Montreal, Montreal.

David, P.P. (1981). Stabilized dune ridges in northern Saskatchewan. *Canadian Journal of Earth Sciences* 18(2):286–310.

Davis, M.B. (1967). Late-glacial climate in northern United States: A comparison of New England and the Great Lakes region. In E.J. Cushing & H.E. Wright Jr. (eds.), *Quaternary Paleoecology*, pp. 11–44. Yale University Press, New Haven.

Davis, M.B. (1969). Palynology and environmental history during the Quaternary period. *American Scientist* 57(3):317–332.

Davis, M.B. (1976). Pleistocene biogeography of temperate deciduous forests. *Geoscience and Man* 13:13–26.

Davis, M.B. (1981). Quaternary history and the stability of forest communities. In D.C. West, H.H. Shugart & D.B. Botkin (eds.), *Forest Succession: Concepts and Application*, Chapter 10. Springer-Verlag, New York.

Davis, M.B. (1983). Holocene vegetational history of the eastern United States. In H.E. Wright Jr. (ed.), *Late-Quaternary Environments of the United States*, vol. 2, *The Holocene*, pp. 166–181. University of Minnesota Press, Minneapolis.

Davis, M.B. (1986). Climatic instability, time lags, and community disequilibrium. In J. Diamond & T.J. Case (eds.), *Community Ecology*, pp. 269–284. Harper and Row, New York.

Davis, M.B., & Botkin, D.B. (1985). Sensitivity of cool-temperate forests and their fossil pollen record to rapid temperature change. *Quaternary Research* 23:327–340.

Davis, M.B., & Goodlett, J.C. (1960). Comparison of the present vegetation with pollen-spectra in surface samples from Brownington Pond, Vermont. *Ecology* 41(2):346–357.

Davis, M.B., Schwartz, M.W., Woods, K.D., & Webb, S.L. (1986a). Detecting beech and hemlock species limits from pollen in sediment. *American Quaternary Association Program and Abstracts*, p. 76.

Davis, M.B., Spear, R.W., & Shane, L.C.K. (1980). Holocene climate of New England. *Quaternary Research* 14:240–250.

Davis, M.B., Woods, K.D., Webb, S.L., & Futyma, R.P. (1986b). Dispersal versus climate: Expansion of *Fagus* and *Tsuga* into the Upper Great Lakes region. *Vegetatio* 67:93–103.

Davis, P.T. (1980). *Holocene Vegetation and Climate Record from Iglutalik Lake, Cumberland Sound, Baffin Island, Northwest Territories, Canada*. Ph.D. thesis. University of Colorado, Boulder.

Davis, P.T., Nichols, H., & Andrews, J.T. (1980). Holocene vegetation and climate record from Iglutalik Lake, Baffin Island. *Abstracts, Fifth International Palynological Conference*, Cambridge.

Davis, R.B., Bradstreet, T.E., Stuckenrath, R., & Borns, H.W. (1975). Vegetation and associated environments during the past 14,000 years near Moulton Pond, Maine. *Quaternary Research* 5:435–465.

Davis, R.B., & Jacobson, G.L. Jr. (1985). Late Glacial and early Holocene landscapes in northern New England and adjacent areas of Canada. *Quaternary Research* 23:341–368.

Davis, R.B., Jacobson, G.L. Jr., Anderson R.S., & Tolonen, M. (1986). Macrofossils in Maine lake sediments confirm and extend paleoecological inferences based on pollen. *American Quaternary Association Program and Abstracts*, p. 77.

Davis, R.B., & Webb, T. III. (1975). The contemporary distribution of pollen from eastern North America: A comparison with the vegetation. *Quaternary Research* 5:395–434.

de Beaulieu, J.L., & Reille, M. (1983). A long Upper Pleistocene pollen record from Les Echets, near Lyon, France. *Boreas* 13:111–132.

Delage, M., Gangloff, P., Larouche, A., & Richard, P.J.H. (1985). Note sur un site à macrorestes végétaux tardiglaciaires au sud-ouest de Montréal, Québec. *Géographie physique et Quaternaire* 34(1):85–90.

Delcourt, H.R. (1979). Late Quaternary vegetation history of the eastern highland rim and adjacent Cumberland Plateau of Tennessee. *Ecological Monographs* 49(3):255–280.

Delcourt, H.R., & Delcourt, P.A. (1985). Quaternary palynology of vegetational history of the Southeastern United States. In V.M. Bryant Jr. & R.G. Holloway (eds.), *Pollen Records of Late-Quaternary North American Sediments*, pp. 1–37. American Association of Stratigraphic Palynologists Foundation, Dallas.

Delcourt, H.R., Delcourt, P.A., & Webb, T. III. (1983). Dynamic plant ecology: The spectrum of vegetational change in space and time. *Quaternary Science Reviews* 1:153–175.

Delcourt, P.A., & Delcourt, H.R. (1977). The Tunica Hills, Louisiana–Mississippi: Late Glacial locality for spruce and deciduous forest species. *Quaternary Research* 7:218–237.

Delcourt, P.A., & Delcourt, H.R. (1979). Late Pleistocene and Holocene distributional history of the deciduous forest in the southeastern United States. *Veroff. Geobot. Inst. Eth.*, pp. 79–107.

Delcourt, P.A., & Delcourt, H.R. (1981). Vegetation maps for eastern North America: 40,000 yr BP to the present. In R.C. Romans (ed.), *Geobotany II*, pp. 123–165. Plenum, New York.

Delcourt, P.A., Delcourt, H.R., Brister, R.C., & Lackey, L.E. (1980). Quaternary vegetation history of the Mississippi Embayment. *Quaternary Research* 13:111–132.

Delcourt, P.A., Delcourt, H.R., & Webb, T. III. (1984). *Atlas of Paired Isophyte and Isopoll Maps for Important Eastern North American Tree Taxa*. Contribution Series No. 14. American Association of Stratigraphic Palynologists Foundation, Dallas.

Denton, G.H., & Hughes, J.J. (eds.). (1981). *The Last Great Ice Sheets*. Wiley, New York.

de Vernal, A., Causse, C., Hillaire-Marcell, C., Mott, R.J., & Occhietti, S. (1986). Palynostratigraphy and Th/U ages of upper Pleistocene interglacial and interstadial deposits on Cape Breton Island, Eastern Canada. *Geology* 14:554–557.

de Vernal, A., Richard, P.J.H., & Occhietti, S. (1983). Palynologie et paléoenvironnements du wisconsinien de la région de la Baie Saint-Laurent, Ile du Cap-Breton. *Géographie physique et Quaternaire* 37:307–322.

De Vries, B., & Bird, C.D. (1965). Bryophyte subfossils of a late-glacial deposit from the Missouri Coteau, Saskatchewan. *Canadian Journal of Botany* 43(8):947–953.

Dionne, J.C. (1975). Paleoclimatic significance of Late Pleistocene ice-wedge casts in southern Quebec, Canada. *Palaeogeography, Palaeoclimatology, Palaeoecology* 17:65–76.

Dort, W. Jr., Johnson, W.C., Fredlund, G.G., Rogers, R.A., Martin, L.D., Stewart, J.D., & Wells, P.V. (1985). "Evidence for an Open Conifer Woodland in the Central Great Plains During the Late Wisconsin Glacial Maximum." *Canadian Quaternary Association, Abstract* p. 41, Lethbridge.

Douglas, R.J.W. (ed.). (1971). *Geology and Economic Minerals of Canada*. Queen's Printer, Ottawa.

Ecoregions Working Group. (In press). *Ecoclimatic Regions of Canada*. Map and Report. Canada Committee on Ecological Land Classification Series 21, Ottawa.

Edelman, H.J. (1985). *Late Glacial and Holocene Vegetation Development of la Goutte Loiselot (Vosges, France)*. Drukkerij Elinkwijk BV, Utrecht.

Edlund, S.A. (1983a). Bioclimatic zonation in a high arctic region: Central Queen Elizabeth Islands. In *Current Research, Part A*, paper 83–1A, pp. 381–390. Geological Survey of Canada, Ottawa.

Edlund, S.A. (1983b). Reconnaissance vegetation studies on western Victoria Island, Canadian Arctic Archipelago. In *Current Research, Part B*, paper 83–1B, pp. 75–81. Geological Survey of Canada, Ottawa.

Edlund, S.A., & Egginton, P.A. (1984). Morphology and description of an outlier population of tree-sized willows on western Victoria Island, District of Franklin. In *Current Research, Part A*, paper 84–1A, pp. 279–285. Geological Survey of Canada, Ottawa.

Edwards, M.E., Anderson, P.M., Garfinkle, H.L., & Brubaker, L.B. (1985). Late Wisconsin and Holocene vegetational history of the Upper Koyukuk region, Brooks Range, AK. *Canadian Journal of Botany* 63:616–626.

Edwards, M.E., & Brubaker, L.C. (1984). A 23,000 year pollen record from Northern Interior Alaska. *American Quaternary Association Program and Abstracts*, p. 35.

Edwards, M.E., & Brubaker, L.C. (1986). Late Quaternary environmental history of the Fishhook Bend area, Porcupine River, Alaska. *Canadian Journal of Earth Sciences* 23:1765–1773.

Edwards, M.E., & Dunwiddie, P.W. (1985). Dendrochronological and palynological observations on *Populus balsamifera* in northern Alaska, U.S.A. *Arctic and Alpine Research* 17(3):271–278.

Edwards, M.E., & McDowell, P.F. (1984). *Quaternary Environmental History in the Southern Yukon Lowland, Northeastern Interior, Alaska*. Abstract. International Palynological Conference, Calgary.

Egler, F.E. (1977). *The Nature of Vegetation*. Aton Forest, Connecticut.

Ehrlich, J. (1934). The beech bark disease: A *Nectria* disease of *Fagus*, following *Cryptococcus Fagi* (Baer.). *Canadian Journal of Research* 10:593–691.

Engstrom, D.R., & Hansen, B.C.S. (1985). Postglacial vegetational change and soil development in southeastern Labrador as inferred from pollen and chemical stratigraphy. *Canadian Journal of Botany* 63:543–561.

Environment Canada, Atmospheric Environment Service. (1982). *Canadian Climatic Normals 1951–80*. Environment Canada, Ottawa.

Faegri, K. (1985). The importance of palynology for the understanding of the archaeological environment in northern Europe. In *Palynologie Archéologique*, Notes et monographies techniques No. 17, C.N.R.S., pp. 333–345.

Florin, M.B., & Wright, H.E. Jr. (1969). Diatom evidence for the persistence of stagnant glacial ice in Minnesota. *Geological Society of America Bulletin* 80:695–704.

Foster, D.R. (1983a). Phytosociological description of the forest vegetation of southeastern Labrador. *Canadian Journal of Botany* 62:899–906.

Foster, D.R. (1983b). The fire history of southeastern Labrador. *Canadian Journal of Botany* 61:2459–2471.

Foster, D.R. (1985). Vegetation development following fire in *Picea mariana* (black spruce) – *Pleurozium* forests of south-eastern Labrador, Canada. *Journal of Ecology* 73:517–534.

Foster, D.R., & Glaser, P.H. (1986). The raised bogs of south-eastern Labrador, Canada: Classification, distribution, vegetation and recent dynamics. *Journal of Ecology* 74:47–71.

Foster, D.R., & King, G.A. (1986). Vegetation pattern and diversity in S.E. Labrador, Canada: *Betula papyrifera* (Birch) forest development in relation to fire history and physiography. *Journal of Ecology* 74:465–483.

Fowells, H.A. (1965). *Silvics of Forest Trees of the United States*, Handbook No. 271. United States Department of Agriculture, Washington, D.C.

Fox, J.D., Zasada, J.C., Gasbarro, A.F., & Van Veldhuizen, R. (1984). Monte Carlo simulation of white spruce regeneration after logging in interior Alaska. *Canadian Journal of Forest Research* 14:617–622.

Fox, W.S. (1949). The present state of the chestnut, *Castanea dentata* (Marsh) Borkh., in Ontario. *Canadian Field Naturalist* 63(2):88–89.

Fox, W.S., & Soper, J.H. (1953). The distribution of some trees and shrubs of the Carolinian zone of southern Ontario. *Transactions of the Royal Canadian Institute* 30:3–32.

Franklin, J.F., & Dyrness, C.T. (1973). *Natural Vegetation of Oregon and Washington*. Forest Service General Technical Report PNW–8. United States Department of Agriculture, Washington, D.C.

Fredskild, B. (1967). Palaeobotanical Investigations at Sermermiut, Jakobshavn, West Greenland. *Meddelelser om Grønland* 178:1–54.

Fredskild, B. (1969). A postglacial standard pollen diagram from Peary Land, North Greenland. *Pollen et Spores* 11(3):573–583.

Fredskild, B. (1973). Studies in the vegetational history of Greenland. *Meddelelser om Grønland* 198:1–245.

Fredskild, B. (1983). The Holocene vegetational development of the Godthabsfjord area, West Greenland. *Geoscience* 10:3–28.

Fredskild, B. (1985a). The Holocene vegetational development of Tugtuligssuaq and Qeqertat, Northwest Greenland. *Geoscience* 14:1–20.

Fredskild, B. (1985b). Holocene pollen records from west Greenland. In J.T. Andrews (ed.), *Quaternary Environments: Eastern Canadian Arctic, Baffin Bay and western Greenland*, pp. 643–681. Allen and Unwin, New York.

Fremlin, G. (ed.). (1974). *The National Atlas of Canada*. Macmillan, Ottawa.

Fulton, R.J. (1984). *Quaternary Stratigraphy of Canada*. A Canadian contribution to IGCP Project 24, paper 84–10. Geological Survey of Canada, Ottawa.

Funder, S. (1978). *Holocene Stratigraphy and Vegetation History in the Scoresby Sund Area, East Greenland*. Bulletin No. 129. Geological Survey of Greenland, Copenhagen.

Funder, S. (1979). Ice-Age plant refugia in East Greenland. *Palaeogeography, Palaeoclimatology, Palaeoecology* 28:279–295.

Funder, S., Abrahamsen, N., Bennike, O., & Feyling-Hanssen, R.W. (1985). Forested Arctic: Evidence from North Greenland. *Geology* 13:542–546.

Gadd, N.R. (1971). Pleistocene geology of the central St. Lawrence Lowlands. *Geological Survey of Canada Memoir* 359:1–153.

Gadd, N.R., McDonald, B.C., & Shilts, W.W. (1972). *Deglaciation of Southern Quebec*, paper 71–47. Geological Survey of Canada, Ottawa.

Gagnon, R., & Payette, S. (1981). Fluctuations Holocènes de la limite des forêts de mélèzes, Rivière aux Feuilles, Nouveau-Québec: une analyse macrofossile en milieu tourbeaux. *Géographie physique et Quaternaire* 35(1):57–72.

Gajewski, K. (1983). *On the Interpretation of Climatic Change from the Fossil Record: Climatic Change in the Central and Eastern United States Over the Past 2,000 Years Estimated from Pollen Analysis*. Ph.D. thesis. University of Wisconsin, Madison.

Gaudreau, D.C., & Webb, T. III. (1985). Late-Quaternary pollen stratigraphy and isochrone maps for the Northeastern United States. In V.M. Bryant Jr., & R.G. Holloway (eds.), *Pollen Records of Late-Quaternary North American Sediments*, pp. 245–280. American Association of Stratigraphic Palynologists Foundation, Dallas.

George, M.F., Burke, M.J., Pellett, H.M., & Johnson, A.G. (1974). Low temperature exotherms and woody plant distribution. *Hortiscience* 9(6):519–522.

Gilbert, H., & Payette, S. (1982). Ecologie des populations d'Aulne vert (*Alnus crispa* (Ait.) Pursh) à la limite des forêts, Québec Nordique. *Géographie physique et quaternaire* 36:109–24.

Gill, D. (1973). Modification of northern alluvial habitats by river development. *The Canadian Geographer* 17:138–153.

Given, D.R., & Soper, J.H. (1981). *The Arctic–Alpine Element of the Vascular Flora at Lake Superior*, Publications in Botany No. 10. National Museum of Natural Sciences, Ottawa.

Glaser, P.H. (1981). Transport and deposition of leaves and seeds on tundra: A Late-Glacial analog. *Arctic and Alpine Research* 13(2):173–182.

Gobeil, A.R. (1938). *Dommages causés aux forêts de la Gaspésie par les insectès*, Bulletin 2. Ministère de Terres et Forêts, Service entomologique, Québec.

Gordon, A.G. (1985). Budworm! What about the forest? In D. Schmitt (ed.), *Spruce-fir Management and Spruce Budworm*, General Technical Report N.E-99, pp.3–29. U.S. Department of Agriculture, Forest Service, Broomall, Pennsylvania.

Graham, R.W. (1986). Response of mammalian communities to environmental changes during the Late Quaternary. In J. Diamond & T.J. Case (eds.), *Community Ecology*, pp. 300–313. Harper & Row, New York.

Graham, R.W., & Lundelius, E.L. Jr. (1984). Coevolutionary disequilibrium and Pleistocene extinctions. In P.S. Martin & R.G. Klein (eds.), *Quaternary Extinction*, pp. 223–249. University of Arizona Press, Tucson.

Grandtner, M.M. (1966). *La Végétation Forestière du Québec Méridional*. Les Presses de l'université Laval, Québec.

Grandtner, M.M. (1972). Aperçu de la végétation du bas Saint-Laurent, de la Gaspésie et des Iles-de-la-Madeleine. *Cahiers géographiques* 16:116–121.

Grant, D.R. (1977). Glacial style and ice limits, the Quaternary stratigraphic record, and changes of land and ocean level in the Atlantic Provinces, Canada. *Géographie physique et Quaternaire* 31:247–260.

Green, D.G. (1981). Time series and post-glacial forest ecology *Quaternary Research* 15:265–277.

Green, D.G. (1982). Fire and stability in the post-glacial forests of southwest Nova Scotia. *Journal of Biogeography* 9:29–40.

Green, D.G. (In press). Pollen evidence for the postglacial origins of Nova Scotia's forests. *Canadian Journal of Botany*.

Grimm, E.C. (1983). Chronology and dynamics of vegetation change in the prairie–woodland region of southern Minnesota, U.S.A. *New Phytologist* 93:311–350.

Grimm, E.C. (1984). Fire and other factors controlling the big woods vegetation of Minnesota in the mid-nineteenth century. *Ecological Monographs* 54:291–311.

Grüger, J. (1972). Pollen and seed studies of Wisconsinan vegetation in Illinois, U.S.A. *Geological Society of America Bulletin* 83:2715–2734.

Grüger, J. (1973). Studies on the Late Quaternary vegetation history of northeastern Kansas. *Geological Society of America Bulletin* 84:239–250.

Hadden, K.A. (1978). *Investigation of the Registration of Poplar Pollen in Lake Sediments*. M.Sc. thesis. University of Toronto, Toronto.

Hall, S.A. (1985). Bibliography of Quaternary palynology in Arizona, Colorado, New Mexico and Utah. In V.M. Bryant, Jr. & R.G. Holloway (eds.), *Pollen Records of Late-Quaternary North American Sediments*, pp. 407–426. American Association of Stratigraphic Palynologists Foundation, Dallas.

Halliday, W.E.D., & Brown, A.W.A. (1943). The distribution of some important forest trees in Canada. *Ecology* 24(3):353–373.

Hansen, B.S., & Cushing, E.J. (1973). Identification of pine pollen of late Quaternary Age from the Chuska Mountains, New Mexico. *Geological Society of America Bulletin* 84:1181–1200.

Hansen, B.S., & Engstrom, D.R. (1985). A comparison of numerical and qualitative methods of separating pollen of black and white spruce. *Canadian Journal of Botany* 63:2159–2163.

Hansen, H.P. (1943). A pollen study of two bogs on Orcas Island, of the San Juan Islands, Washington. *Bulletin of the Torrey Botanical Club* 70:236–243.

Hare, F.K. (1959). *A Photo-Reconnaissance Survey of Labrador–Ungava*, Memoir 6. Government of Canada, Geographical Branch, Mines and Technical Surveys, Ottawa.

Hare, F.K., & Hay, J.E. (1974). The climate of Canada and Alaska. In R.A. Bryson & F.K. Hare (eds.), *Climates of North America*, vol.II, *World Survey of Climatology*. Elsevier, Amsterdam.

Hare, F.K., & Ritchie, J.C. (1972). The boreal bioclimates. *Geographical Review* 62:334–361.

Hare, F.K., & Thomas, M.K. (1979). *Climate Canada*, 2d ed. University of Toronto, Toronto.

Harington, C.R., & Rice, G. (eds.). (1984). *Climatic Change in Canada 4*, Syllogeus 51. National Museums of Canada, Ottawa.

Harper, J.L. (1977). *Population Biology of Plants*. Academic Press, London.

Harrington, J.B. (1982). *A Statistical Study of Area Burned by Wildfire in Canada 1953–1980*, Environment Canada Information Report P1-X–16. Department of the Environment, Ottawa.

Harrington, J.B. (1982). *A Statistical Study of Area Burned by Wildfire in Canada 1953–1980*, Environment Canada Information Report P1-X–16. Department of the Environment, Ottawa.

Harrison, S.P., & Metcalfe, S.E. (1985). Spatial variations in lake levels since the last glacial maximum in the Americas north of the Equator. *Zeitschrift fur Gletscherkunde und Glazialgeologie* 21:1–15.

Hazell, S. (1979). *Late Quaternary vegetation and climate of Dunbar Valley, British Columbia*. M.Sc. thesis. University of Toronto, Toronto.

Hebda, R.J. (1983b). Postglacial history of grasslands of southern British Columbia and adjacent regions. In A.C. Nicholson, A. McLean, & T.E. Baker (eds.), *Grassland Ecology and Classification Symposium Proceedings*, pp. 157–191. British Columbia Ministry of Forestry, Vancouver.

Hebda, R.J., & Mathewes, R.W. (1984). Holocene history of cedar and native Indian cultures of the North American Pacific Coast. *Science* 225:711–713.

Heide, K. M. (1984). Holocene pollen stratigraphy from a lake and small hollow in north-central Wisconsin, U.S.A. *Palynology* 8:3–20.

Heinselman, M.L. (1981). Fire and succession in the conifer forests of northern North America. In D.C. West, H.H. Shugart, & D.B. Botkin (eds.), *Forest Succession: Concepts and Application*, pp. 375–397. Springer-Verlag, New York.

Held, M.E. (1983). Pattern of beech regeneration in the east-central United States. *Bulletin of the Torrey Botanical Club* 110(1):55–62.

Hemond, H.F., Niering, W.A., & Goodwin, R.H. (1983). Two decades of vegetation change in the Connecticut Arboretum natural area. *Bulletin of the Torrey Botanical Club* 110(2):184–194.

Herbert, T.D., & Fischer, A.G. (1986). Milankovitch climatic origin of mid-Cretaceous black shale rhythms in central Italy. *Nature* 321:739–744.

Heusser, C.J. (1960). *Late-Pleistocene Environments of North Pacific North America*. Special Publication No. 35. American Geographical Society, New York.

Heusser, C.J. (1969). Modern pollen spectra from the Olympic Peninsula, Washington. *Bulletin of the Torrey Botanical Club* 96:407–417.

Heusser, C.J. (1974). Quaternary vegetation, climate, and glaciation of the Hoh River Valley, Washington. *Geological Society of America Bulletin* 85:1547–1560.

Heusser, C.J. (1977). Quaternary palynology of the Pacific slope of Washington. *Quaternary Research* 8:282–306.

Heusser, C.J. (1978). Modern pollen rain of Washington. *Canadian Journal of Botany* 56:1510–1517.

Heusser, C.J. (1983). Vegetational history of the northwestern United States including Alaska. In S.C. Porter (ed.), *Late-Quaternary Environments of the United States*, vol. 1, *The Late Pleistocene*, pp. 230–239. University of Minnesota Press, Minneapolis.

Heusser, C.J. (1985). Quaternary pollen record from the Interior Pacific Northwest Coast: Aleutians to the Oregon–California boundary. In V.M. Bryant, Jr. & R.G. Holloway (eds.), *Pollen Records of Late-Quaternary North American Sediments*, pp. 141–166. American Association of Stratigraphic Palynologists Foundation, Dallas.

Heusser, C.J., Heusser, L.E., & Peteet, D.M. (1985). Late-Quaternary climatic change on the American North Pacific Coast. *Nature* 315:485–487.

Hicock, S.R., Hebda, R.J., & Armstrong, J.E. (1982). Lag of the Fraser glacial maximum in the Pacific northwest: Pollen and macrofossil evidence from western Fraser Lowland, British Columbia. *Canadian Journal of Earth Sciences* 19(12):2288–2296.

Hills, G.A. (1960). Regional site research. *The Forestry Chronicle* 36(4):401–423.

Hills, G.A., & Portelance, R. (1960). *The Reports of the Glackmeyer Subcommittee of the Northern Region*

Land-Use Planning Committee. Ontario Department of Lands and Forests, Toronto.

Hix, D.M., & Barnes, B.V. (1984). Effects of clear-cutting on the vegetation and soil of an eastern hemlock dominated ecosystem, western Upper Michigan. *Canadian Journal of Forest Research* 14:914–923.

Holland, P.G. (1981). Pleistocene refuge areas, and the re-vegetation of Nova Scotia, Canada. *Progress in Physical Geography* 5:535–562.

Holloway, R.G., & Bryant, V.M. Jr. (1985). Late-Quaternary pollen records and vegetational history of the Great Lakes Region: United States and Canada. In V.M. Bryant, Jr. & R.G. Holloway (eds.), *Pollen Records of Late-Quaternary North American Sediments*, pp. 205–244, American Association of Stratigraphic Palynologists Foundation, Dallas.

Holmgren, B. (1971). Climate and energy exchange on a sub-polar cap in summer. *Part A. Climatologie Physique*. Met. Instn. Uppsala University, Number 107.

Hooghiemstra, H. (1984). *Vegetational and Climatic History of the High Plain of Bogota, Columbia; A Continuous Record of the Last 3.5 Million Years*. Cramer, Liechtenstein.

Hopkins, D.M., Smith, P.A., & Matthews, J.V. Jr. (1981). Dated wood from Alaska and the Yukon: Implications for forest refugia in Beringia. *Quaternary Research* 15:217–249.

Horton, K.W. (1956). *The Ecology of Lodgepole Pine in Alberta*, Technical Note 45. Department of Northern Affairs and National Resources, Forestry Branch, Ottawa.

Hughes, O.L. (1972). *Surficial Geology of Northern Yukon Territory and Northwestern District of Mackenzie, Northwest Territories*, Report 69–36 and Map 1319A. Geological Survey of Canada, Ottawa.

Hughes, O.L., Harington, C.R., Janssens, J.A., Matthews, J.V., Morlan, R.E., Rutter, N.E., & Schweger, C.E. (1981). Upper Pleistocene stratigraphy, paleoecology and archaeology of the Northern Yukon Interior, Eastern Beringia. I. Bonnet Plume Basin. *Arctic* 34:329–365.

Hultén, E. (1937). *Outline of the History of Arctic and Boreal Biota During the Quaternary Period*. Bokforlagsaktiebologet Thule, Stockholm.

Huntley, B., & Birks, H.J.B. (1983). *An Atlas of Past and Present Pollen Maps for Europe: 0–13,000 Years Ago*. Cambridge University Press, New York.

Hyvärinen, H. (1985). Holocene pollen stratigraphy of Baird Inlet, east-central Ellesmere Island, arctic Canada. *Boreas* 14:19–32.

Ibe, R.A. (1983). Patterns of pollen deposition around five eastern hemlock trees (*Tsuga canadensis* (L.) Carr.). *Bulletin of the Torrey Botanical Club* 110(4):536–541.

Iversen, J. (1952–3). Origin of the flora of western Greenland in the light of pollen analysis. *Oikos* 4:85–103.

Iversen, J. (1958). The bearing of glacial and interglacial epochs on the formation and extinction of plant taxa. *Uppsala Universiteit Arssk.* 6:210–215.

Ives, J.D. (1974). Biological Refugia and the Nunatak Hypothesis. In J.D. Ives & R.G. Barry (eds.), *Arctic and Alpine Environments*, pp. 605–636. Methuen, London.

Ives, J.W. (1977). Pollen separation of three North American birches. *Arctic and Alpine Research* 9:73–80.

Jackson, S.T. (1983). *Late-glacial and Holocene Vegetational Changes in the Adirondack Mountains (New York): A Macrofossil Study*. Ph.D. thesis. Indiana University, Bloomington.

Jackson, S.T. (1986). Vegetational history along an elevational gradient in the Adirondack Mountains (New York). *Ecological Monographs*, Manuscript under review.

Jacobs, J.D., Mode, W.N., & Dowdeswell, E.K. (1985). Contemporary pollen deposition and the distribution of *Betula glandulosa* at the limit of low arctic tundra in Southern Baffin Island, N.W.T., Canada. *Arctic and Alpine Research* 17(3):279–287.

Jacobson, G.L., Jr. (1979). The palaeoecology of white pine (*Pinus strobus*) in Minnesota. *Journal of Ecology* 67:697–726.

Jacobson, G.L., Jr., & Bradshaw, R.H.W. (1981). The selection of sites for paleovegetational studies. *Quaternary Research* 16:80–96.

Jacobson, G.L., Jr., & Grimm, E.C. (1986). A numerical analysis of Holocene forest and prairie vegetation in central Minnesota. *Ecology* 67:958–966.

Janssen, C.R. (1984). Modern pollen assemblages and vegetation in the Myrtle Lake Peatland, Minnesota. *Ecological Monographs* 54(2):213–252.

Janssens, J.A., & Baker, R.G. (1984). A full-glacial bryophyte assemblage from south-eastern Iowa, U.S.A. *Journal of Bryology* 13:201–207.

Janssens, J.A., & Barnosky, C.W. (1985). Late Pleistocene and early Holocene bryophytes from Battle Ground Lake, Washington, U.S.A. *Review of Palaeobotany and Palynology* 46:97–116.

Jarvis, J.M. (1956). *An Ecological Approach to Tolerant Hardwood Silviculture*, Technical Note 43. Department of Northern Affairs and National Resources, Forestry Branch, Ottawa.

Jeffrey, W.W. (1961). *Origin and Structure of Some White Spruce Stands on the Lower Peace River*, Technical Note 103. Department of Forestry, Ottawa.

Johnson, E.A. (1981). Vegetation organization and dynamics of lichen woodland communities in the Northwest Territories, Canada. *Ecology* 62:200–215.

Johnson, W.C., Fredlund, G.G., Wells, P.V., Stewart, J.D., & Dort, W. Jr. (1986). Late Wisconsinan Biogeography of south central Nebraska: The North Cove site. *American Quaternary Association Program and Abstracts*, p. 89.

Johnson, W.H. (1986). Permafrost features in central Illinois and their environmental significance. *American Quaternary Association Program and Abstracts*, pp. 37–39.

Jordan, R. (1975). Pollen diagrams from Hamilton Inlet, central Labrador and their environmental implications for the northern maritime archaic. *Arctic Anthropology* 12(2):92–116.

Juday, G.P., & Zasada, J.C. (1984). Structure and development of an old-growth white spruce forest on an interior Alaska floodplain. In W.R. Mehan, T.R. Merrell, Jr., & T.A. Hanley (eds.), *Fish and Wildlife Relationships in Old-Growth Forests* (proceedings of a symposium held in Juneau, Alaska, April 12–15, 1982), pp. 227–234. American Institute of Fishery Research Biologists, Fairbanks.

Jurdant, M., Beaubien, J., Celair, J.L., Dionne, J.C., & Gerardin, V. (1972). *Carte écologique de la région du Saguenay-Lac-St-Jean*. Notice explicative. Rapport d'information Q-F-X-31, 3 volumes, Environnement Canada, Ottawa.

Kavanagh, K., & Kellman, M. (1986). Performance of *Tsuga canadensis* (L.) Carr. at the centre and northern edge of its range: A comparison. *Journal of Biogeography* 13:145–157.

Kearney, M.S. (1983). Modern pollen deposition in the Athabasca Valley, Jasper National Park. *Botanical Gazette* 144(3):450–459.

Kearney, M.S., & Luckman, B.H. (1983a). Postglacial vegetational history of Tonquin Pass, British Columbia. *Canadian Journal of Earth Sciences* 20(5):776–786.

Kearney, M.S., & Luckman, B.H. (1983b). Holocene timberline fluctuations in Jasper National Park, Alberta. *Science* 221:261–263.

Kelly, M., & Funder, S. (1974). *The Pollen Stratigraphy of Late Quaternary Lake Sediments of South-West Greenland*, Report 64. Geological Survey of Greenland, Copenhagen.

Kelly, P.M., Jones, P.D., Sear, C.B., Cherry, B.S.G., & Tarvakol, R.K. (1982). Variations in surface air temperatures: Part 2, Arctic regions, 1881–1980. *Monthly Weather Review* 110:71–83.

Kessell, S.R. (1979). Adaptation and dimorphism in eastern Hemlock. *Tsuga canadensis* (L.) Carr. *The American Naturalist* 113(3):333–350.

Khalil, M.A.K. (1985). Genetic variation in eastern white spruce (*Picea glauca* (Moench) Voss) populations. *Canadian Journal of Forest Research* 15(2):444–452.

King, G.A. (1984). Deglaciation and Revegetation of Western Labrador and Adjacent Quebec. *American Quaternary Association Program and Abstracts*, p. 67.

King, G.A. (1986). Vegetation dynamics near the margin of the remnant Laurentide Ice Sheet in eastern Canada. *American Quaternary Association Program and Abstracts*, p. 91.

King, J.E. (1973). Late Pleistocene palynology and biogeography of the western Missouri Ozarks. *Ecological Monographs* 43(4):539–565.

King, J.E., & Graham, R.W. (1986). Vertebrates and vegetation along the southern margin of the Laurentide Ice sheet. *American Quaternary Association Program and Abstracts*, p. 43.

Krajina, V.J. (1965). *Ecology of Western North America*, vol. I, *Biogeoclimatic Zones and Biogeocoenoses of British Columbia*. University of British Columbia, Vancouver.

Krajina, V.J. (1969). Ecology of forest trees in British Columbia. *Ecology of Western North America* 2. University of British Columbia, Vancouver.

Kuc, M. (1974). Fossil flora of the Beaufort Formation, Meighen Island, Northwest Territories. In *Report of Activities*, Paper 74–1, 193–5. Geological Survey of Canada, Ottawa.

Kupsch, W.O. (1960). Radiocarbon-dated organic sediment near Herbert, Saskatchewan. *American Journal of Science* 258:282–292.

Kutzbach, J.E., & Guetter, P.J. (1986). The influence of changing orbital parameters and surface boundary conditions on the simulated climate of the past 18,000 years. *Journal of Atmospheric Sciences* 43:1726–1759

Kutzbach, J.E., & Wright, H.E. Jr. (1986). Simulation of the climate of 18,000 years BP: Results for the North American/North Atlantic/European sector and comparison with the geologic record of North America. *Quaternary Science Reviews* 4:147–187.

Labelle, C., & Richard, P.J.H. (1981). Végétation tardiglaciaire et postglaciaire au sud-est du parc des Laurentides, Québec. *Géographie physique et Quaternaire* 35(3):345–359.

Labelle, C., & Richard, P.J.H. (1984). Histoire postglaciaire de la végétation dans la région de Mont-Saint-Pierre, Gaspésie, Québec. *Géographie physique et Quaternaire* 38(3):257–274.

Lafond, A., & Ladouceur, G. (1968). Les forêts, les climax et les régions biogéographiques du bassin de La Rivière Outaouais, Québec. *Le Naturaliste canadien* 95:317–366.

Lamb, H.F. (1980). Late Quaternary vegetational history of southeastern Labrador. *Arctic and Alpine Research* 12(2):117–135.

Lamb, H.F. (1984). Modern pollen spectra from Labrador and their use in reconstructing Holocene vegetational history. *Journal of Ecology* 72:37–59.

Lamb, H.F. (1985). Palynological evidence for postglacial change in the position of tree limit in Labrador. *Ecological Monographs* 55(2):241–258.

Lang, G. (ed.). (1985). *Swiss Lake and Mire Environments During the Last 15,000 Years*. Dissertationes Botanicae 87. IGCP. Cramer, Hirschberg.

Last, W.M., & Schweyen, T.H. (1985). Late Holocene history of Waldsea Lake, Saskatchewan, Canada. *Quaternary Research* 24:219–234.

Leopold, E.B. (1956). Pollen-size frequency in New England species of the genus *Betula*. *Grana Palynologica* 1:140–147.

Leopold, E.B., Nickmann, R., Hedges, J.I., & Ertel, J.R. (1982). Pollen and lignin records of Late Quaternary vegetation, Lake Washington. *Science* 218:1305–1307.

Lichti-Federovich, S. (1970). The pollen stratigraphy of a dated section of Late Pleistocene lake sediment from central Alberta. *Canadian Journal of Earth Science* 7(3):938–945.

Lichti-Federovich, S. (1972). *Pollen Stratigraphy of a Sediment Core from Alpen Siding, Alberta*, Report of Activities, paper 72–1B, pp. 113–115. Geological Survey of Canada, Ottawa.

Lichti-Federovich, S. (1974). *Palynology of Two Sections of Late Quaternary Sediments from the Porcupine River, Yukon Territory*, paper 74–23. Geological Survey of Canada, Ottawa.

Lichti-Federovich, S., & Ritchie, J.C. (1968). Recent pollen assemblages from the Western Interior of Canada. *Review of Palaeobotany and Palynology* 7:297–344.

Likens, G.E., & Davis, M.B. (1975). Post-glacial history of Mirror Lake and its watershed in New Hampshire, U.S.A.: An initial report. *Verh. Internat. Verein. Limnol.* 19:982–993.

Little, E.L. (1971). *Atlas of United States Trees*, Miscellaneous Publication 1146. United States Department of Agriculture Forest Service, Washington, D.C.

Liu, K.B. (1980). Pollen evidence of Late-Quaternary climatic changes in Canada: A review. Part I: Western Canada. *Ontario Geography* 15:83–101.

Liu, K.B. (1982). *Postglacial Vegetational History of Northern Ontario: A Palynological Study*. Ph.D. thesis. University of Toronto, Toronto.

Liu, K-B., & Lam N.S.N. (1985). Paleovegetational reconstruction based on modern and fossil pollen data: An application of discriminant analysis. *Annals of the Association of American Geographers* 75(1):115–130.

Livingstone, D.A. (1955). Some pollen profiles from arctic Alaska. *Ecology* 36:587–600.

Livingstone, D.A. (1968). Some interstadial and postglacial pollen diagrams from eastern Canada. *Ecological Monographs* 38:87–125.

Loucks, O.L. (1962). A Forest Classification for the Maritime Provinces. *Proceedings of the Nova Scotian Institute of Science 1959–60* 25(2):85–169. Department of Forestry, Ottawa.

Loucks, O.L. (1983). New light on the changing forest. In S.L. Flader (ed.), *The Great Lakes Forest: An Environmental Social History*, pp. 17–32. University of Minnesota Press, Minneapolis.

Löve, A. & Löve, D. (1974). Origin and evolution of the arctic and alpine floras. In J.D. Ives, & R.G. Barry (eds.), *Arctic and Alpine Environments*, pp. 571–601. Methuen, London.

Luckman, B.H., & Kearney, M.S. (1986). Reconstruction of Holocene changes in alpine vegetation and climate in the Maligne Range, Jasper National Park, Alberta. *Quaternary Research* 26:244–261.

Lutz, H.J. (1930). The vegetation of Heart's Content, a virgin forest in northwestern Pennsylvania. *Ecology* 11:1–29.

MacDonald, G.M. (1982). Late Quaternary paleoenvironments of the Morley Flats and Kananaskis Valley of southwestern Alberta. *Canadian Journal of Earth Sciences* 19(1):23–35.

MacDonald, G.M. (1983). Holocene vegetation history of the Upper Natla River area, Northwest Territories, Canada. *Arctic and Alpine Research* 15(2):169–180.

MacDonald, G.M. (1984). *Postglacial Plant Migration and Vegetation Development in the Western Canadian Boreal Forest*. Ph.D. thesis. University of Toronto, Toronto.

MacDonald, G.M., & Cwynar, L.C. (1985). A fossil pollen based reconstruction of the late Quaternary history of lodgepole pine (*Pinus contorta* ssp. *latifolia*) in the Western Interior of Canada. *Canadian Journal of Forest Research* 15(6):1039–1044.

MacDonald, G.M., & Ritchie, J.C. (1986). Modern pollen spectra from the Western Interior of Canada and the intepretation of Late Quaternary vegetation development. *New Phytologist* 103:245–268.

Mack, R.N. (1971). Pollen size variation in some western North American pines as related to fossil pollen identification. *Northwest Science* 45:257–269.

Mack, R.N., & Bryant, V.M. Jr. (1974). Modern pollen spectra from the Columbia Basin, Washington. *Northwest Science* 48(3):183–194.

Mack, R.N., Bryant, V.M. Jr., & Pell, W. (1978). Modern forest pollen spectra from eastern Washington and northern Idaho. *Botanical Gazette* 139(2):249–255.

Mack, R.N., Rutter, N.W., Bryant, V.M. Jr. & Valastro, S. (1978b). Late Quaternary pollen record from Big Meadow, Pend Oreille County, Washington. *Ecology* 59(5):956–966.

Mack, R.N., Rutter, N.W., & Valastro, S. (1979). Holocene vegetation history of the Okanogan Valley, Washington. *Quaternary Research* 12:212–225.

Mack, R.N., Rutter, N.W., Valastro, S., & Bryant, V.M. Jr. (1978a). Late Quaternary vegetation history at Waits Lake, Colville River Valley, Washington. *Botanical Gazette* 139(4):499–506.

Macpherson, J.B. (1982). Postglacial vegetational history of the eastern Avalon Peninsula, Newfoundland, and Holocene climatic change along the eastern Canadian seaboard. *Géographie physique et Quaternaire* 36:175–196.

Macpherson, J.B., & Anderson, T.W. (1985). Further evidence of late glacial climate fluctuations from Newfoundland: Pollen stratigraphy from a north coast site. In *Current Research, Part B*, paper 85–1B, pp. 383–390. Geological Survey of Canada, Ottawa.

Maini, J.S., & Horton, K.W. (1966). *Reproductive Response of Populus and Associated Pteridium to Cutting, Burning and Scarification*, Publication No. 1155. Forestry Branch, Ottawa.

Markgraf, V. (1986). Plant inertia reassessed. *The American Naturalist* 127(5):725–726.

Markgraf, V., & Lennon, T. (1986). Paleoenvironmental history of the last 13,000 years of the Eastern Powder River Basin, Wyoming, and its implications for prehistoric cultural patterns. *Plains Anthropologist* 31:1–12.

Martineau, R. (1984). *Insects Harmful to Forest Trees*. Multiscience Publications Limited, Ottawa.

Mathewes, R.W. (1973). A palynological study of postglacial vegetation changes in the University Research Forest, southwestern British Columbia. *Canadian Journal of Botany* 51(11):2085–2103.

Mathewes, R.W. (In press). The Queen Charlotte Islands refugium controversy: A paleoecological perspective. In R.J. Fulton & J.A. Heginbottom (eds.), *Quaternary of Canada and Greenland*, Geological Survey of Canada, Ottawa.

Mathewes, R.W., & Heusser, L.E. (1981). A 12,000-year palynological record of temperature and precipitation trends in southwestern British Columbia. *Canadian Journal of Botany* 59(5):707–710.

Mathewes, R.W. & Rouse, G.E. (1975). Palynology and paleoecology of postglacial sediments from the Lower Fraser River Canyon of British Columbia. *Canadian Journal of Earth Sciences* 12:745–756.

Matthews, J.V., Jr. (1974). Wisconsin environment of interior Alaska: pollen and macrofossil analysis of a 27 meter core from the Isabella Basin (Fairbanks, Alaska). *Canadian Journal of Earth Sciences* 11(6):828–841.

Matthews, J.V., Jr. (1980). *Paleoecology of John Klondike Bog, Fisherman Lake Region, Southwest District of Mackenzie*, paper 80–22. Geological Survey of Canada, Ottawa.

Matthews, J.V., Jr. (1982). East Beringia during Late Wisconsin time: A review of the biotic evidence. In D.M.

Hopkins, J.V. Matthews, Jr., C.E. Schweger, & S.B. Young (eds.), *Palaeoecology of Beringia*, pp. 127–150. Academic Press, New York.

Matthews, J.V., Jr., Mott, R.J., & Vincent J.S. (1986). Preglacial and interglacial environments of Banks Island: Pollen and macrofossils from Duck Hawk Bluffs and related sites. *Géographie physique et Quaternaire* 40:279–365.

Maxwell, J.B. (1980). *The Climate of the Canadian Arctic Islands and Adjacent Waters*, vol. 1. Ministry of the Environment, Ottawa.

Maycock, P.F. (1961). The spruce–fir forests of the Keweenaw Peninsula, northern Michigan. *Ecology* 42:357–365.

Maycock, P.F. (1963). The phytosociology of the deciduous forests of extreme southern Ontario. *Canadian Journal of Botany* 41:379–438.

Maycock, P.F. & Curtis, J.T. (1960). The phytosociology of boreal conifer–hardwood forest of the Great Lakes Region. *Ecological Monographs* 30:1–35.

McAndrews, J.H. (1966). Postglacial history of prairie, savanna, and forest in northwestern Minnesota. In T. Delevoryas (ed.), *Memoirs of the Torrey Botanical Club* 22(2):1–72.

McAndrews, J.H. (1981). Late Quaternary climate of Ontario: Temperature trends from the fossil pollen record. In W.C. Mahaney, *Quaternary Paleoclimate*, pp. 319–333. Geo Abstracts, Norwich.

McAndrews, J.H. (1982). Holocene environment of a fossil bison from Kenora, Ontario. *Ontario Archaeology* 37:41–51.

McAndrews, J.H., & Boyko-Diakonow, M. (In press). Pollen analysis of varved sediment at Crawford Lake, Ontario: Evidence of Indian and European farming. In R.J. Fulton & J.A. Heginbottom (eds.), *Quaternary of Canada and Greenland*. Geological Survey of Canada, Ottawa.

McIntosh, R.P. (1972). Forests of the Catskill Mountains, New York. *Ecological Monographs* 42(2):143–161.

McIntosh, R.P. (1985). *The Background of Ecology*. Cambridge University Press, New York.

McLennan, D.S. (1981). *Pollen transport and representation in the coast mountains of British Columbia*. M.Sc. thesis. Simon Fraser University, Vancouver.

McLennan, D.S., & Mathewes, R.W. (1984). Pollen transport and representation in the Coast Mountains of British Columbia. I. Flowering phenology and aerial deposition. *Canadian Journal of Botany* 62(10):2154–2164.

Mehringer, P.J., Jr. (1985). Late-Quaternary pollen records from the Interior Pacific Northwest and northern Great Basin of the United States. In V.M. Bryant, Jr., & R.G. Holloway (eds.), *Pollen Records of Late-Quaternary North American Sediments*, pp. 167–190. American Association of Stratigraphic Palynologists Foundation, Dallas.

Mickelson, D.M., Clayton, L., Fullerton, D.S., & Borns, H.R. Jr. (1983). The Late Wisconsin glacial record of the Laurentide Ice Sheet in the United States. In S.C. Porter (ed.), *Late-Quaternary Environments of the United States*, vol. 1, *The Late Pleistocene*, pp. 3–37. University of Minnesota Press, Minneapolis.

Mickelson, D.M., Clayton, L., & Muller, E.H. (1986). Contrasts in glacial landforms, deposits, glacier-bed conditions, and glacier dynamics along the southern edge of the Laurentide ice sheet. *American Quaternary Association Program and Abstracts*, p. 28.

Miller, N.G. (1973a). *Late-glacial and Postglacial Vegetation Change in Southwestern New York State*, Bulletin 420, New York State Museum and Science Service. New York University of the State of New York, Albany.

Miller, N.G. (1973b). Late-glacial plants and plant communities in northwestern New York State. *Journal of the Arnold Arboretum* 54(2):123–159.

Miller, N.G. (1976). Studies of North American Quaternary bryophyte subfossils. 1. A new moss assemblage from the Two Creeks Forest Bed of Wisconsin. *Farlow Herbarium of Harvard University Occasional Paper* 9:21–42.

Miller, N.G. (1980). Mosses as paleoecological indicators of late-glacial terrestrial environments: Some North American studies. *Bulletin of the Torrey Botanical Club* 107(3):373–391.

Miller, N.G. & Thompson, G.G. (1979). Boreal and western North American plants in the Late Pleistocene of Vermont. *Journal of the Arnold Arboretum* 60(2):167–218.

Millet, J., & Payette, S. (In press.). Les feux de la région du Lac à l'Eau-Claire. *Géographie physique et Quaternaire* 41.

Minore, D. (1979). *Comparative Autecological Characteristics of Northwestern Tree Species — A Literature Review*, Pacific Northwest Forest and Range Experiment Station, General Technical Report PNW-87. United States Department of Agriculture Forest Service, Washington, D.C.

Mode, W.N. (1980) *Quaternary Stratigraphy and Palynology of the Clyde Foreland, Baffin Island, N.W.T., Canada*. Ph.D. thesis. University of Colorado, Boulder.

Mooney, H.A., & Billings, W.D. (1961). Comparative physiological ecology of arctic and alpine populations of *Oxyria digyna*. *Ecological Monographs* 31:1–29.

Morgan, A.V., & Morgan, A. (1979). The Fossil coleoptera of the Two Creeks Forest Bed, Wisconsin. *Quaternary Research* 12:226–240.

Mott, R.J. (1969). *Palynological Studies in Central Saskatchewan. Contemporary Pollen Spectra from Surface Samples*, paper 69–32. Geological Survey of Canada, Ottawa.

Mott, R.J. (1973). *Palynological Studies in Central Saskatchewan: Pollen Stratigraphy from Lake Sediment Sequences*, paper 72–49. Geological Survey of Canada, Ottawa.

Mott, R.J. (1975). Palynological studies of lake sediment profiles from southwestern New Brunswick. *Canadian Journal of Earth Sciences* 12(2):273–288.

Mott, R.J. (1976). A Holocene pollen profile from the Sept-Iles area, Québec. *Naturaliste canadien* 103(5):457–467.

Mott, R.J. (1977). Late-Pleistocene and Holocene palynology in southeastern Québec. *Géographie physique et Quaternaire* 31(1–2):139–149.

Mott, R.J. (1978). *Populus* in late-Pleistocene pollen spectra. *Canadian Journal of Botany* 56(8):1021–1031.

Mott, R.J. (In press.). Late-Pleistocene paleoenvironments in Atlantic Canada. In R.J. Fulton & J.A. Heginbottom (eds.), *Quaternary of Canada and Greenland*, Geological Survey of Canada, Ottawa.

Mott, R.J., Anderson, T.W., & Matthews, J.V., Jr. (1981). Late-glacial paleoenvironments of sites bordering the Champlain Sea based on pollen and macrofossil evidence. In W.C. Mahaney (ed.), *Quaternary Paleoclimate*, pp. 129–171. Geo Abstracts, Norwich.

Mott, R.J., Anderson, T.W., & Matthews, J.V., Jr. (1982). Pollen and macrofossil study of an interglacial deposit in Nova Scotia. *Géographie physique et Quaternaire* 36:197–208.

Mott, R.J., & Christiansen, E.A. (1981). Palynological study of slough sediments from central Saskatchewan. In *Current Research, Part B*, paper 81–1B, pp. 133–136. Geological Survey of Canada, Ottawa.

Mott, R.J., & Farley-Gill, L.D. (1978). A late-Quaternary pollen profile from Woodstock, Ontario. *Canadian Journal of Earth Sciences* 15(7):1101–1111.

Mott, R.J., & Farley-Gill, L.D. (1981). *Two Late Quaternary Pollen Profiles from Gatineau Park, Quebec*, paper 80–31. Geological Survey of Canada, Ottawa.

Mott, R.J., & Grant, D.R. (1985). Pre-Late Wisconsinan paleoenvironments in Atlantic Canada. *Géographie physique et Quaternaire* 34(3):239–254.

Mott, R.J., & Jackson, L.E. Jr. (1982). An 18,000 year palynological record from the southern Alberta segment of the classical Wisconsinan "Ice-free Corridor." *Canadian Journal of Earth Sciences* 19(3):504–513.

Munroe, E. (1956). Canada as an environment for insect life. *The Canadian Entomologist* 88(7):372–476.

Murray, D.F. (1980). Balsam poplar in arctic Alaska. *Canadian Journal of Anthropology* 1:29–32.

Murray, D.F. (1981). The role of arctic refugia in the evolution of the arctic vascular flora – a Beringian perspective. In G.G.E. Scudder & J.L. Reveal (eds.), *Evolution Today*, pp. 11–20. University of British Columbia, Vancouver.

Nairn, L.D., Reeks, W.A., Webb, F.E., & Hildahl, V. (1962). History of larch sawfly outbreaks and their effect on Tamarack stands in Manitoba and Saskatchewan. *The Canadian Entomologist* 94(3):242–255.

Neilson, R.P. (1986). High-resolution climatic analysis and southwest biogeography. *Science* 232:27–34.

Nichols, G.E. (1935). The hemlock–white pine–northern hardwood region of eastern North America. *Ecology* 16:403–422.

Nichols, H. (1976). Historical aspects of the northern Canadian treeline. *Arctic* 29(1):38–47.

Nicholson, S.A., Scott, J.I., & Breisch, A.R. (1979). Structure and succession in the tree stratum at Lake George, New York. *Ecology* 60(6):1240–1254.

Nienstaedt, H., & Riemenschneider, D.E. (1985). Changes in heritability estimates with age and site in white spruce, *Picea glauca* (Moench) Voss. *Silvae Genetica* 34(1):34–41.

Oliver, C.D. (1981). Forest development in North America following major disturbances. *Forest Ecology and Management* 3:153–168.

Oliver, C.D., Adams, A.B., & Zasoski, R.J. (1985). Disturbance patterns and forest development in a recently deglaciated valley in the northwestern Cascade Range of Washington, U.S.A. *Canadian Journal of Forest Research* 15:221–232.

Olson, J.S., Stearns, F.W., & Nienstaedt, H. (1959a). *Eastern Hemlock Seeds and Seedlings: Response to Photoperiod and Temperature*, Bulletin 620. Connecticut Agricultural Experiment Station, Storrs.

Olson, J.S., Stearns, F.W., & Nienstaedt, H. (1959b). *Eastern Hemlock: Growth Cycle and Early Years*, Circular 205. Connecticut Agricultural Experiment Station, Storrs.

Ovenden, L.E. (1982). Vegetation history of a polygonal peatland, northern Yukon. *Boreas* 11:209–224.

Ovenden, L.E. (1985). *Hydroseral Histories of the Old Crow Peatlands, Northern Yukon*. Ph.D. thesis. University of Toronto, Toronto.

Overpeck, J.T. (1985). A pollen study of a late Quaternary peat bog, south-central Adirondack Mountains, New York. *Geological Society of America Bulletin* 96:145–154.

Payette, S. (1980). Fire History at the Treeline in Northern Quebec: A Paleoclimatic Tool. *Proceedings of the Fire History Workshop*, USDA, Report RM–81:126–131.

Payette, S. (1983). The forest tundra and the present tree-lines of the northern Quebec–Labrador Peninsula. In P. Morisset & S. Payette (eds.), *Tree-line Ecology*, pp. 3–24. Nordicana 47. Laval University, Quebec.

Payette, S., & Filion, L. (1985). White spruce expansion at the tree line and recent climatic change. *Canadian Journal of Forest Research* 15(1):241–251.

Payette, S., Filion, L., Gauthier, L., & Boutin, Y. (1985). Secular climate change in old-growth tree-line vegetation of northern Québec. *Nature* 315:135–138.

Payette, S., & Gagnon, R. (1979). Tree-line dynamics in Ungava Peninsula. *Holarctic Ecology* 2:239–248.

Payette, S., & Gagnon, R. (1985). Late Holocene deforestation and tree regeneration in the forest-tundra of Québec. *Nature* 313:570–572.

Peet, R.K. (1984). Twenty-six years of change in a *Pinus strobus*, *Acer saccharum* forest, Lake Itasca, Minnesota. *Bulletin of the Torrey Botanical Club* 111(1):61–68.

Pennington, W. (1980). Modern pollen samples from West Greenland and the interpretation of pollen data from the British late-glacial (Late Devensian). *New Phytologist* 84:171–201.

Peteet, D.M. (1986). Modern pollen rain and vegetational history of the Malaspina Glacier District, Alaska. *Quaternary Research* 25:100–120.

Péwé, T.L. (1983). The periglacial environment in North America during Wisconsin Time. In S.C. Porter (ed.), *Late Quaternary Environments of the United States*, vol. 1, *The Late Pleistocene*, pp. 157–189. University of Minnesota Press, Minneapolis.

Porsild, A.E., & Cody, W.J. (1980). *Vascular Plants of Continental Northwest Territories, Canada*. National Museum of Natural Sciences, Ottawa.

Porter, S.C. (ed.). (1983). *Late-Quaternary Environments of the United States*, vol. I. University of Minnesota Press, Minneapolis.

Porter, S.C., Pierce, K.L., & Hamilton, T.D. (1983). Late Wisconsin mountain glaciation in the western United States. In S.C. Porter (ed.), *Late-Quaternary Environments of the United States*, vol. I, *The Late Pleistocene*,

pp. 71–115. University of Minnesota Press, Minneapolis.
Prentice, I.C. (1983). Postglacial climatic change: Vegetation dynamics and the pollen record. *Progress in Physical Geography* 7:273–286.
Prentice, I.C. (1985). Pollen representation, source area, and basin size: Toward a unified theory of pollen analysis. *Quaternary Research* 23:76–86.
Prentice, I.C. (1986). Vegetation responses to past climatic variation. *Vegetatio* 67:131–141.
Prest, V.K. (1970). Quaternary geology of Canada. In R.J.E. Douglas (ed.), *Geology and Economic Minerals of Canada*, pp. 676–764. Department of Energy, Mines and Resources, Ottawa.
Prest, V.K. (1984). The Late Wisconsinan glacier complex. In R.J. Fulton (ed.), *Quaternary Stratigraphy of Canada*, IGCP Project 24, paper 84–10, pp. 21–36.
Radle, N.J. (1981). *Vegetation History and Lake-Level Changes at a Saline Lake in Northeastern South Dakota*. M.Sc. thesis. University of Minnesota, Minneapolis.
Railton, J.B. (1972). *Vegetational and Climatic History of Southwestern Nova Scotia in Relation to a South Mountain Ice Cap*. Ph.D. thesis. Dalhousie University, Halifax.
Rampton, V.N. (1971). Late Quaternary vegetational and climatic history of the Snag–Klutlan area, southwestern Yukon Territory, Canada. *Geological Society of America Bulletin* 82:959–978.
Rampton, V.N. (1982). *Quaternary Geology of the Yukon Coastal Plain*. Bulletin 317. Geological Survey of Canada, Ottawa.
Ramsey, G.S., & Higgins, D.G. (1982). *Canadian Forest Fire Statistics 1980*. Environment Canada, Ottawa.
Raup, H.M. (1947). The Botany of Southwestern Mackenzie. In *Sargentia*, vol. VI, pp. 1–275. Arnold Arboretum of Harvard University, Cambridge, Massachusetts.
Raup, H.M. (1981). Introduction: What is a crisis. In M.H. Nitecki, *Biotic Crises in Ecological and Evolutionary Time*, pp. 1–12. Academic Press, New York.
Raup, H.M., & Argus, G.W. (1982). The Lake Athabasca sand Dunes of northern Saskatchewan and Alberta, Canada. I. The Land and Vegetation. In *Botany* 12:1–96. National Museums of Canada, Ottawa.
Reiners, W.A., & Lang, G.E. (1979). Vegetational patterns and processes in the balsam fire zone, White Mountains, New Hampshire. *Ecology* 60:403–417.
Richard, P.J.H. (1970). Atlas pollinique des arbres et de quelques arbustes indigènes du Québec. *Naturaliste canadien* 97:1–34, 97–161, 241–306.
Richard, P.J.H. (1971). Two pollen diagrams from the Quebec City area, Canada. *Pollen et Spores* 13(4):523–559.
Richard, P.J.H. (1973a). Histoire postglaciaire comparée de la végétation dans deux localités au sud de la ville de Québec. *Naturaliste canadien* 100(6):591–603.
Richard, P.J.H. (1973b). Histoire postglaciaire comparée de la végétation dans deux localités au nord du Parc des Laurentides, Québec. *Naturaliste canadien* 100(6):577–590.
Richard, P.J.H. (1973c). Histoire postglaciaire de la végétation dans la région de Saint Raymond de Portneuf, telle que révélée par l'analyse pollinique d'une tourbière. *Naturaliste canadien* 100(6):561–575.
Richard, P.J.H. (1974). Présence de *Shepherdia canadensis* (L.) Nutt. dans la région du Parc des Laurentides, Québec, au tardiglaciaire. *Naturaliste canadien* 101(5):763–768.
Richard, P.J.H. (1975a). Contribution à l'histoire postglaciaire de la végétation dans la plaine du Saint-Laurent: Lotbinière et Princeville. *La Revue de Géographie de Montréal* 29(2):95–107.
Richard, P.J.H. (1975b). Contribution à l'histoire postglaciaire de la végétation dans les Cantons-de-l'est: étude des sites de Weedon et Albion. *Cahiers de Géographie de Québec* 19(47):267–284.
Richard, P.J.H. (1977a). *Histoire post-wisconsinienne de la végétation du Québec méridional par l'analyse pollinique. Service de la recherche* (Publications et rapports divers). Tome 1, xxiv. Tome 2, 142 pp. Direction générale des forêts, ministère des terres et forêts du Québec, Québec.
Richard, P.J.H. (1977b). Végétation tardiglaciaire au Québec méridional et implications paléoclimatiques. *Géographie physique et Quaternaire* 31:161–176.
Richard, P.J.H. (1978). Histoire tardiglaciaire et postglaciaire de la végétation au Mont Shefford, Québec. *Géographie physique et Quaternaire* 32(1):81–93.
Richard, P.J.H. (1979). Contribution à l'histoire postglaciaire de la végétation au nord-est de la Jamésie, Nouveau-Québec. *Géographie physique et Quaternaire* 33(1):93–112.
Richard, P.J.H. (1980a). Histoire postglaciaire de la végétation au sud du lac Abitibi, Ontario et Québec. *Géographie physique et Quaternaire* 34(1):77–94.
Richard, P.J.H. (1980b) L'interprétation du diagramme pollinique en termes de végétation au Québec. *Phytocoenologia* 7:127–141.
Richard, P.J.H. (1981a). *Paléophytogéographie postglaciaire en Ungava par l'analyse pollinique*. Paléo-Québec 13. Programme Tuvaaluk. Université du Québec, Montréal.
Richard, P.J.H. (1981b). Palaeoclimatic significance of the late-Pleistocene and Holocene pollen record in south-central Québec. In W.C. Mahaney (ed.), *Quaternary Paleoclimate*, pp. 335–360. Geo Abstracts, Norwich.
Richard, P.J.H. (1985). Couvert Végétal et Paléoenvironnements du Québec entre 12,000 et 8,000 ans BP l'Habitabilité dans un milieu changeant. *Recherches Amérindiennes au Québec* 15(1–2):39–56.
Richard, P.J.H. (In press). Patterns of post Wisconsin plant colonization in Quebec–Labrador. In R.J. Fulton & J.A. Heginbottom (eds.), *Quaternary of Canada and Greenland*. Geological Survey of Canada, Ottawa.
Richard, P.J.H., Larouche, A., & Bouchard, M.A. (1982). Age de la déglaciation finale et histoire postglaciaire de la végétation dans la partie centrale du Nouveau-Québec. *Géographie physique et Quaternaire* 36:63–90.
Richard, P.J.H., & Poulin, P. (1976). Un diagramme pollinique au Mont des Eboulements, région de Charlevoix, Québec. *Canadian Journal of Earth Sciences* 13(1):145–156.
Ritchie, J.C. (1964). Contributions to the Holocene paleoecology of west-central Canada. 1. The Riding Mountain Area. *Canadian Journal of Botany* 42:181–196.
Ritchie, J.C. (1969). Absolute pollen frequencies and carbon–14 age of a section of Holocene lake sediment

from the Riding Mountain area of Manitoba. *Canadian Journal of Botany* 47(9):1345–1349.

Ritchie, J.C. (1974). Modern pollen assemblages near the arctic tree line, Mackenzie Delta region, Northwest Territories. *Canadian Journal of Botany* 52:381–396.

Ritchie, J.C. (1976). The late-Quaternary vegetational history of the western interior of Canada. *Canadian Journal of Botany* 54(15):1793–1818.

Ritchie, J.C. (1977). The modern and late-Quaternary vegetation of the Campbell-Dolomite uplands near Inuvik, N.W.T., Canada. *Ecological Monographs* 47:401–423.

Ritchie, J.C. (1980). Towards a late-Quaternary palaeoecology of the ice-free corridor. *Canadian Journal of Anthropology* 1:15–28.

Ritchie, J.C. (1982). The modern and late-Quaternary vegetation of the Doll Creek area, North Yukon, Canada. *New Phytologist* 90:563–603.

Ritchie, J.C. (1983). The Paleoecology of the central and northern parts of the Glacial Lake Agassiz Basin. *Geological Association of Canada*, Special Paper 26:157–172.

Ritchie, J.C. (1984a). *Past and Present Vegetation of the Far Northwest of Canada*. University of Toronto Press, Toronto.

Ritchie, J.C. (1984b). A Holocene pollen record of boreal forest history from the Travaillant Lake area, Lower Mackenzie River Basin. *Canadian Journal of Botany* 62(7):1385–1392.

Ritchie, J.C. (1985a). Quaternary pollen records from the Western Interior and the Arctic of Canada. In V.M. Bryant, Jr. & R.G. Holloway (eds.), *Pollen Records of Late-Quaternary North American Sediments*, pp. 327–352. American Association of Stratigraphic Palynologists Foundation, Dallas.

Ritchie, J.C. (1985b). Late-Quaternary climatic and vegetational change in the Lower Mackenzie Basin, northwest Canada. *Ecology* 66(2):612–621.

Ritchie, J.C. (1986). Climate change and vegetation response. *Vegetatio* 67:65–74.

Ritchie, J.C., Cinq-Mars, J., & Cwynar, L.C. (1982). L'environnement tardiglaciaire du Yukon septentrional, Canada. *Géographie physique et quaternaire* 36:241–250.

Ritchie, J.C., Cwynar, L.C., & Spear, R.W. (1983). Evidence from northwest Canada for an early Holocene Milankovitch thermal maximum. *Nature* 305: 126–128.

Ritchie, J.C., & De Vries, B. (1964). Contributions to the Holocene paleoecology of west-central Canada: a late-glacial deposit from the Missouri Coteau. *Canadian Journal of Botany* 42:677–692.

Ritchie, J.C., & Hadden, K.A. (1975). Pollen stratigraphy of Holocene sediments from the Grand Rapids area, Manitoba, Canada. *Review of Palaeobotany and Palynology* 19:193–202.

Ritchie, J.C., Hadden, K.A., & Gajewski, K. (In press). Modern pollen spectra from lakes in Arctic Western Canada. *Canadian Journal of Botany*.

Ritchie, J.C., & Hare, F.K. (1971). Late-Quaternary vegetation and climate near the arctic tree line of northwestern North America. *Quaternary Research* 1(3):331–342.

Ritchie, J.C., & Lichti-Federovich, S. (1968). Holocene pollen assemblages from the Tiger Hills, Manitoba. *Canadian Journal of Earth Sciences* 5:873–880.

Ritchie, J.C., & MacDonald, G.M. (1986). The patterns of post-glacial spread of white spruce. *Journal of Biogeography* 13:527–540.

Ritchie, J.C., & Yarranton, G.A. (1978). The late-Quaternary history of the boreal forest of central Canada. *Journal of Ecology* 66:199–212.

Rogers, R.S. (1978). Forests dominated by hemlock (*Tsuga canadensis*): Distribution as related to site and postsettlement history. *Canadian Journal of Botany* 56(7):843–854.

Rowe, J.S. (1972). *Forest Regions of Canada*, Publication 1300. Department of Environment, Ottawa.

Rowe, J.S. (1983). Concepts of fire effects on plant individuals and species. In R.W. Wein & D.A. MacLean (eds.), *The Role of Fire in Northern Circumpolar Ecosystems*, pp. 135–154, Wiley, New York.

Rowe, J.S. (1984). Lichen woodland in northern Canada. In R. Olsen, R. Hastings, & F. Geddes (eds.), *Memorial Essays Honoring Don Gill*, pp 225–237. University of Alberta Press, Edmonton.

Ruddiman, W.F., & McIntyre, A.F. (1981). The North Atlantic Ocean during the last deglaciation. *Palaeogeography, Palaeoclimatology, Palaeoecology* 35:145–214.

Rudolph, T.D. & Yeatman, C.W. (1982). *Genetics of Jackpine*, Research Paper WO-38, pp. 1–60. U.S. Department of Agriculture, Forest Service, Washington, D.C.

Rutter, N.W. (1980). Late Pleistocene history of the western Canadian ice-free corridor. *Canadian Journal of Anthropology* 1(1):1–8.

Saarnisto, M. (1974). The deglaciation history of the Lake Superior region and its climatic implications. *Quaternary Research* 4(3):316–339.

Saarnisto, M. (1975). Stratigraphical studies on the shoreline displacement of Lake Superior. *Canadian Journal of Earth Sciences* 12(2):300–319.

Sakai, A. (1982). Freezing tolerance of shoot and flower primordia of coniferous buds by extraorgan freezing. *Plant and Cell Physiology* 23(7):1219–1227.

Sakai, A. (1983). Comparative study on freezing resistance of conifers with special reference to cold adaptation and its evolutive aspects. *Canadian Journal of Botany* 61:2323–2332.

Sakai, A., Yoshida, S., Saito, M., & Zoltai, S.C. (1979). Growth rate of spruces related to the thickness of permafrost active layer near Inuvik, northwestern Canada. *Low Temperature Science*, Series B37:19–32.

Sakai, A.K., Roberts, M.R., & Jolls, C.L. (1985). Successional changes in a mature aspen forest in northern lower Michigan: 1974–1981. *The American Midland Naturalist* 113(2):271–282.

Sakai, A.K., & Sulak, J.H. (1985). Four decades of secondary succession in two lowland permanent plots in northern lower Michigan. *The American Midland Naturalist* 113(1):146–157.

Savile, D.B.T. (1972). *Arctic Adaptations in Plants*, Monograph 6. Department of Agriculture, Ottawa.

Savoie, L., & Richard, P.J.H. (1979). Paléophytogéographie de l'episode de Saint-Narcisse dans la région de Sainte-Agathe, Québec. *Géographie physique et Quaternaire* 33(2):175–188.

Schneider, R., & Tobolski, K. (1985). Lago di Ganna – Late-glacial and Holocene environments of a lake in the southern Alps. In G. Lang (ed.), *Swiss Lake and Mire Environments During the Last 15,000 Years*, pp. 229–271. Cramer, Hirschberg.

Schweger, C. (1986). The Goldeye Lake pollen record: Could man survive the ice-free corridor? *American Quaternary Association Program and Abstracts*, p. 162.

Schweger, C.E. (In press). Paleoecology of the western Canadian ice-free corridor. In R.J. Fulton & J.A. Heginbottom (eds.), *Quaternary of Canada and Greenland*, Geological Survey of Canada, Ottawa.

Schweger, C., Habgood, T., & Hickman, M. (1981). Late-glacial–Holocene climatic changes in Alberta. In *The Impacts of Climatic Fluctuations in Alberta's Resources and Environment*, Education Reprint No. WAES–1–81. Atmospheric Environment Service, Environment Canada, Ottawa.

Schwert, D.P., Anderson, T.W., Morgan, A., Morgan, A.V., & Karrow, P.F. (1985). Changes in Late Quaternary vegetation and insect communities in southwestern Ontario. *Quaternary Research* 23:205–226.

Scott, J.T., Siccama, T.G., Johnson, A.H., & Breisch, A.R. (1984). Decline of red spruce in the Adirondacks, New York. *Bulletin of the Torrey Botanical Club* 111(4):438–444.

Shackleton, J. (1982). *Environmental Histories from Whitefish and Imuruk Lakes, Seward Peninsula, Alaska*, Report 76. Institute of Polar Studies, Ohio State University, Columbus.

Shay, C.T. (1967). Vegetation history of the southern Lake Agassiz Basin during the past 12,000 years. In W.J. Mayer-Oakes (ed.), *Life, Land and Water*, pp. 231–252. University of Manitoba Press, Winnipeg.

Short, S.K. (1978). Palynology: A Holocene environmental perspective for archaeology in Labrador–Ungava. *Arctic Anthropology* 15:9–35.

Short, S.K., Mode, W.N., & Davis, P.T. (1985). The Holocene record. In J.R. Andrews (ed.), *Quaternary Environments: Eastern Canadian Arctic, Baffin Bay and Western Greenland*, pp. 608–642. Allen and Unwin, Boston.

Short, S.K., & Nichols, H. (1977). Holocene pollen diagrams from subarctic Labrador–Ungava: Vegetational history and climatic change. *Arctic and Alpine Research* 9(3):265–290.

Shugart, H.H., Antonovsky, M.J., Jarvis, P.G., & Sandford, A.P. (In press). Climatic change and forest ecosystems. In *The WMO/ICSU/UNEP International Assessment of the Impact of an Increased Atmospheric Concentration of Carbon Dioxide on the Environment*, Chapter 9.

Shugart, H.H., Crow, T.R., & Hett, J.M. (1973). Forest succession models: A rationale and methodology for modeling forest succession over large regions. *Forest Science* 19:203–212.

Siccama, T.G., Bliss, M., & Vogelmann, H.W. (1982). Decline of red spruce in the Green Mountains of Vermont. *Bulletin of the Torrey Botanical Club* 109:163–168.

Slater, D.S. (1985). Pollen analysis of postglacial sediments from Eildun Lake, District of Mackenzie, N.W.T., Canada. *Canadian Journal of Earth Sciences* 22(5):663–674.

Smith, G.I., & Street-Perrott, F.A. (1983). Pluvial lakes of the western United States. In S.C. Porter (ed.), *Late-Quaternary Environments of the United States*, vol. 1, *The Late Pleistocene*, pp. 157–190. University of Minnesota Press, Minneapolis.

Smol, J.P. (1983). Paleophycology of a high arctic lake near Cape Herschel, Ellesmere Island. *Canadian Journal of Botany* 61:2195–2204.

Solomon, A.M. (1983). Pollen morphology and plant taxonomy of red oaks in eastern North America. *American Journal of Botany* 70(4):495–507.

Solomon, A.M., & Shugart, H.H. (1984). Integrating forest-stand simulations with paleoecological records to examine long-term forest dynamics. In G.I. Agren (ed.), *State and Change of Forest Ecosystems – Indicators in Current Research*, Report 13, pp. 333–356. University of Agriculture and Sciences, Department of Ecology & Environmental Research, Uppsala, Sweden.

Solomon, A.M., West, D.C., & Solomon J.A. (1981). Simulating the role of climate change and species immigration in forest succession. In D.C. West, H.H. Shugart & D.B. Botkin (eds.), *Forest Succession: Concepts and Application*, Chapter 11, pp. 154–177. Springer-Verlag, New York.

Sorenson, C.J., & Knox, J.C. (1974). Paleosols and paleoclimate related to late Holocene forest–tundra border migrations: Mackenzie and Keewatin, Northwest Territories. In S. Raymond & P. Schledermann (eds.), *International Conference on Prehistory and Paleoecology of Western North American Arctic and Subarctic*, pp. 187–203.

Spaulding, W.G., & Graumlich, L.J. (1986). The last pluvial climatic episodes in the deserts of southwestern North America. *Nature* 320:441–444.

Spaulding, W.G., Leopold, E.B., & Van Devender, R. (1983). Late Wisconsin paleoecology of the American southwest. In S.C. Porter (ed.), *Late-Quaternary Environments of the United States*, vol. I, *The Late Pleistocene*, pp. 259–293. University of Minnesota Press, Minneapolis.

Spear, R.W. (1981). *The History of High-Elevation Vegetation in the White Mountains of New Hampshire*. Ph.D. thesis. University of Minnesota, Minneapolis.

Spear, R.W. (1983). Paleoecological approaches to a study of treeline fluctuation in the Mackenzie Delta Region, Northwest Territories: Preliminary results. In P. Morisset & S. Payette (eds.), *Treeline Ecology*, pp. 61–72. Nordicana 47. Laval University, Quebec.

Spear, R.W., & Miller, N.G. (1976). A radiocarbon dated pollen diagram from the Allegheny Plateau of New York State. *Journal of the Arnold Arboretum* 57(3):369–403.

Sprugel, D.G. (1976). Dynamic structure of wave-regenerated *Abies balsamea* forests in the north-eastern United States. *Journal of Ecology* 64(3):889–911.

Spurr, S.H. (1964). *Forest Ecology*. Ronald Press, New York.

Stiell, W.M. (1985). Silviculture of eastern white pine. *Proceedings of the Entomological Society, Ontario* 116:95–107.

Suc, J.P. (1984). Original and evolution of the Mediterranean vegetation and climate in Europe. *Nature* 307:429–432.

Suc, J.P., & Zagwijn, W.H. (1983). Plio-Pleistocene correlations between the northwestern Mediterranean region and northwestern Europe according to recent biostratigraphic and palaeoclimatic data. *Boreas* 12:153–166.

Sugita, S., & Tsukada, M. (1982). The vegetation history in western North America. I. Mineral and Hall Lakes. *Japanese Journal of Ecology* 32:499–515.

Sukachev, V.N. (1960). The correlation between the concept forest ecosystem and forest biogeocoenose and their importance for the classification of forests. *Silva Fennica* 105:94–97.

Sutton, R.F. (1969). *Silvics of White Spruce (Picea glauca (Moench) Voss)*, Forestry Branch Publication 1250. Department of Fisheries and Forestry, Ottawa.

Swain, A.M. (1973). A history of fire and vegetation in northeastern Minnesota as recorded in lake sediments. *Quaternary Research* 3(3):383–396.

Swain, A.M. (1980). Landscape patterns and forest history in the boundary waters canoe area, Minnesota: A pollen study from Hug lake. *Ecology* 61(4):747–754.

Tauber, H. (1965). *Differential Pollen Dispersion and the Interpretation of Pollen Diagrams*, II. Series No. 89. Geological Survey of Denmark, Copenhagen.

Teller, J.T. (1985). Glacial Lake Agassiz and its influence on the Great Lakes. In P.F. Karrow & P.E. Calkin (eds.), *Quaternary Evolution of the Great Lakes*, Special Paper 30, pp. 1–16. Geological Association of Canada, Toronto.

Teller, J.T., & Clayton, L. (eds.) (1983). *Glacial Lake Agassiz*, Special Paper 26, Geological Association of Canada. University of Toronto Press, Toronto.

Teller, J.T., Thorleifson, L.M., Dredge, L.A., Hobbs, M.C., & Schreiner, B.T. (1983). Maximum extent and major features of Lake Agassiz. In J.T. Teller & L. Clayton (eds.),*Glacial Lake Agassiz*, Special Paper 26, pp. 43–45. Geological Association of Canada. University of Toronto Press, Toronto.

Terasmae, J. (1963). Three C–14 dated pollen diagrams from Newfoundland, Canada. *Advancing Frontiers of Plant Sciences* 6:149–162.

Terasmae, J. (1967). Postglacial chronology and forest history in the northern Lake Huron and Lake Superior regions. In E.J. Cushing & H.E. Wright, Jr. (eds.), *Quaternary Palaeoecology*, pp. 45–58. Yale University Press, New Haven.

Terasmae, J. (1981). Late-Wisconsin deglaciation and migration of spruce into southern Ontario, Canada. In R.C. Romans (ed.), *Geobotany II*, pp. 75–90. Plenum, New York.

Terasmae, J., & Anderson, T.W. (1970). Hypsithermal range extension of white pine (*Pinus strobus* L.) in Quebec, Canada. *Canadian Journal of Earth Sciences* 7(2):406–413.

Terasmae, J., & Hughes, O.L. (1966). Late-Wisconsinan chronology and history of vegetation in the Ogilvie Mountains, Yukon Territory, Canada. *The Palaeobotanist* 15:235–242.

Thie, J., & Ironside, G. (1976). *Ecological (Biophysical) Classification in Canada*, Ecological Land Classification Series No. 1. Environment Canada, Ottawa.

Tolonen, K., & Tolonen, M. (1984). Late-glacial vegetational succession at four coastal sites in northeastern New England: Ecological and phytogeographical aspects. *Annales Botanici Fennici* 21:59–77.

Tsay, R.C., & Taylor, I.E.P. (1978). Isoenzyme complexes as indicators of genetic diversity in white spruce, *Picea glauca*, in southern Ontario and the Yukon Territory. Formic, glutamic, and lactic dehydrogenases and cationic peroxidases. *Canadian Journal of Botany* 56:80–90.

Tsukada, M. (1982). *Pseudotsuga Menziesii* (Mirb.) Franco: Its pollen dispersal and Late Quaternary history in the Pacific northwest. *Japanese Journal of Ecology* 32:159–187.

Tsukada, M., & Sugita, S. (1982). Late Quaternary dynamics of pollen influx at Mineral Lake, Washington. *Botanical Magazine*, Tokyo, 95:401–418.

Tubbs, C.H. (1973). Allelopathic relationship between yellow birch and sugar maple seedlings. *Forest Science* 19(2):139–145.

Tucker, C.M., & McCann, B. (1980). Quaternary events on the Burin Peninsula, Newfoundland, and the islands of St. Pierre and Miquelon, France. *Canadian Journal of Earth Sciences* 17:1462–1479.

Vance, R.E. (1984). Holocene Vegetative Change in the Boreal Forest of Northeastern Alberta, Canada. *American Quaternary Association Program and Abstracts*, p. 133.

Vance, R.E. (1986). Pollen stratigraphy of Eaglenest Lake, northeastern Alberta. *Canadian Journal of Earth Sciences* 17:1462–1479.

Van Cleve, K., Dyrness, T., & Viereck, L.A. (1980). Nutrient cycling in interior Alaska floodplains and its relationship to regeneration and subsequent forest development. In M. Murray & R.M. Van Veldhuizen (eds.), *Forest Regeneration of High Latitudes*, Forest Service General Technical Report PNW–107, pp. 1–18. U.S. Department of Agriculture, Washington, D.C.

Van Cleve, K., & Viereck, L.A. (1981). Forest succession in relation to nutrient cycling in the boreal forest of Alaska. In D.C. West, H.H. Shugart, & D.B. Botkin (eds.), *Forest Succession: Concepts and Applications*, pp. 185–199. Springer Verlag, New York.

van der Hammen, T., Wijmstra, T.A., & Zagwijn, W.H. (1971). The floral record of the late Cenozoic of Europe. In K.K. Turekian (ed.), *Late Cenozoic Glacial Ages*, pp. 391–424. Yale University Press, New Haven.

Van Zant, K.L., Hallberg, G.R., & Baker, R.G. (1980). A Farmdalian pollen diagram from east-central Iowa. *Proceedings of the Iowa Academy of Science* 87(2):52–55.

Viereck, L.A. (1970). Forest succession and soil development adjacent to the Cherna River in Alaska. *Arctic and Alpine Research* 2:1–26.

Viereck, L.A., & Dyrness, C.T. (1980). *A Preliminary Classification System for the Vegetation of Alaska*, General Technical Report PNW-106. United States Department of Agriculture, Forest Service, Washington, D.C.

Viereck, L.A., Dyrness, C.T., Van Cleve, K., & Foote, M.J. (1983). Vegetation, soils, and forest productivity in selected forest types in interior Alaska. *Canadian Journal of Forest Research* 13:703–720.

Viereck, L.A., & Foote, M.J. (1970). The status of *Populus balsamifera* and *P. trichocarpa* in Alaska. *Canadian Field-Naturalist* 84:169–173.

Von Althen, F.W. (1983). *Animal Damage to Hardwood*

Regeneration and its Prevention in Plantations and Woodlots of Southern Ontario, Information Report O-X-351. Great Lakes Forestry Centre, Department of the Environment, Canadian Forestry Service, Ottawa.

Von Rudolph, E., Oswald, E.T., & Nyland, E. (1981). Chemosystematic studies in the genus *Picea* v. leaf oil terpene composition of white spruce from the Yukon Territory. *Canadian Forestry Service, Research Notes* 1:32–34.

Waitt, R.B., Jr., & Thorson, R.M. (1983). The Cordilleran Ice Sheet in Washington, Idaho, and Montana. In S.C. Porter (ed.), *Late-Quaternary Environments of the United States*, vol 1, *The Late Pleistocene*, pp. 53–70. University of Minnesota Press, Minneapolis.

Walter, H., & Lieth, H. (1967). *Klimadiagramm – Weltatlas*. Jena.

Ward, H.A., & McCormick, L.H. (1982). Eastern hemlock allelopathy. *Forest Science* 28(4):681–686.

Ward, R.T. (1961). Some aspects of the regeneration habits of the American beech. *Ecology* 42(4):828–832.

Warner, B.G., Clague, J.J., & Mathewes, R.W. (1984). Geology and paleoecology of a Mid-Wisconsin peat from the Queen Charlotte Islands, British Columbia, Canada. *Quaternary Research* 21:337–350.

Watts, W.A. (1970). The full-glacial vegetation of northwestern Georgia. *Ecology* 51(1):17–33.

Watts, W.A. (1975). A late Quaternary record of vegetation from Lake Annie, south-central Florida. *Geology* 3:344–346.

Watts, W.A. (1979). Late Quaternary vegetation of central Appalachia and the New Jersey Coastal Plain. *Ecological Monographs* 49(4):427–469.

Watts, W.A. (1983). Vegetational history of the eastern United States 25,000 to 10,000 years ago. In S.C. Porter (ed.), *Late-Quaternary Environments of the United States*, vol. 1, *The Late Pleistocene*, pp 294–310. University of Minnesota Press, Minneapolis.

Watts, W.A., & Bright, R.C. (1968). Pollen, seed and mollusk analysis of a sediment core from Pickerel Lake, northeastern South Dakota. *Geological Society of America Bulletin* 79:855–876.

Watts, W.A., & Stuiver, M. (1980). Late Wisconsin climate of northern Florida and the origin of species-rich deciduous forest. *Science* 210:325–327.

Watts, W.A., & Winter, T.C. (1966). Plant macrofossils from Kirchner Marsh, Minnesota – A paleoecological study. *Geological Society of America Bulletin* 77:1339–1360.

Webb, T. III. (1974a). A vegetational history from northern Wisconsin: Evidence from modern and fossil pollen. *The American Midland Naturalist* 92(1):12–34.

Webb, T. III. (1974b). Corresponding distributions of modern pollen and vegetation in lower Michigan. *Ecology* 55:17–28.

Webb, T. III. (1981). The past 11,000 years of vegetational change in eastern North America. *Bioscience* 31:501–506.

Webb, T. III. (1982). Temporal resolution in Holocene pollen data, *Proceedings of the Third North American Paleontological Convention* 2:569–572.

Webb, T. III. (1986). Is the vegetation in equilibrium with climate? An interpretative problem for Late-Quaternary pollen data. *Vegetatio* 67:75–91.

Webb, T. III, Cushing, E.J., & Wright, H.E. Jr. (1983). Holocene changes in the vegetation of the Midwest. In H.E. Wright Jr. (ed.), *Late-Quaternary Environments of the United States*, vol. 2, *The Holocene*, pp. 142–165. University of Minnesota Press, Minneapolis.

Webb, T. III, Howe, S.E., Bradshaw, R.H.W., & Heide, K.M. (1981). Estimating plant abundances from pollen percentages: the use of regression analysis. *Review of Palaeobotany and Palynology* 34:269–300.

Webb, T. III, & McAndrews, J.H. (1976). Corresponding patterns of contemporary pollen and vegetation in central North America. *Geological Society of America Memoir* 145:267–299.

Webb, T. III, Richard, P.J.H., and Mott, R.J. (1983). A mapped history of Holocene vegetation in southern Québec. *Syllogeus* 49:273–336.

Webb, T. III, Yeracaris, G.Y., & Richard, P.J.H. (1978). Mapped patterns of sediment samples of modern pollen from southeastern Canada and northeastern United States. *Géographie physique et Quaternaire* 32(2):163–176.

Wein, R.W., & MacLean, D.A. (eds.). (1982). *The Role of Fire in Northern Circumpolar Ecosystems*. SCOPE 18. Wiley, Toronto.

West, R.G. (1961). Late and postglacial vegetational history in Wisconsin: Particularly changes associated with the Valders readvance. *American Journal of Science* 259:766–783.

West, R.G. (1964). Interrelations of ecology and Quaternary palaeobotany. *Journal of Ecology* 52(supplement):47–57

West, R.G. (1970). Pleistocene history of the British flora. In D. Walker & R.G. West (eds.), *Studies in the Vegetational History of the British Isles*, pp. 1–11. Cambridge University Press, New York.

White, J.M., & Mathewes, R.W. (1982). Holocene vegetation and climatic change in the Peace River District, Canada. *Canadian Journal of Earth Sciences* 19(3):555–570.

White, J.M., Mathewes, R.W., & Mathews, W.H. (1979). Radiocarbon dates from Boone Lake and their relation to the "Ice-free Corridor" in the Peace River district of Alberta, Canada. *Canadian Journal of Earth Sciences* 16(9):1870–1874.

White, J.M., Mathewes, R.W., & Mathews, W.H. (1985). Late Pleistocene chronology and environment of the "Ice-Free Corridor" of northwestern Alberta. *Quaternary Research* 24:173–186.

Whitehead, D.R. (1981). Late-Pleistocene vegetational changes in northeastern North Carolina. *Ecological Monographs* 51(4):451–471.

Whitney, G.G. (1984). Fifty years of change in the arboreal vegetation of Heart's Content, an old-growth hemlock–white pine–northern hardwood stand. *Ecology* 65:403–408.

Wijmstra, T.A. (1969). Palynology of the first 30 meters of a 120 m deep section in northern Greece. *Acta Botanica Neerlandica* 18:511–527.

Williams, A.S. (1974). *Late-Glacial–Postglacial Vegetational History of the Pretty Lake Region, Northeastern Indiana*, Geological Survey Professional Paper 686B:1–

23. U.S. Geological Survey, Washington, D.C.
Wilkinson, R.C., Hanover, J.W., Wright, J.W., & Flake, R.H. (1971). Genetic variation in the monoterpene composition of white spruce. *Forest Science* 17(1):83–90.
Winkler, M.G. (1985a). *Late-Glacial and Holocene Environmental History of South-Central Wisconsin: A Study of Upland and Wetland Ecosystems*. Institute for Environmental Studies, University of Wisconsin, Madison.
Winkler, M.G. (1985b). A 12,000-year history of vegetation and climate for Cape Code, Massachusetts. *Quaternary Research* 23:301–312.
Winkler, M.G., Swain, A.M., & Kutzbach, J.E. (1986). Middle Holocene dry period in the northern midwestern United States: Lake levels and pollen stratigraphy. *Quaternary Research* 25:235–250.
Woillard, G. (1978). Grande Pile peat bog: a continuous pollen record for the last 140,000 years. *Quaternary Research* 9:1–21.
Woods, K.D. (1984). Patterns of tree replacement: canopy effects on understory pattern in hemlock–northern hardwood forests. *Vegetatio* 56:87–107.
Wright, H.E., Jr. (1968b). History of the Prairie Peninsula. *The Quaternary of Illinois*. University of Illinois College of Agriculture Special Publication 14:78–88.
Wright, H.E., Jr. (1970). Vegetational history of the central plains. In *Pleistocene and Recent Environments of the Central Great Plains*, University of Kansas Department of Geology Special Publication 3:157–172.
Wright, H.E., Jr. (1976). The dynamic nature of Holocene vegetation: a problem in paleoclimatology, biogeography, and stratigraphic nomenclature. *Quaternary Research* 6:581–596.
Wright, H.E., Jr. (1977). Quaternary vegetation – some comparisons between Europe and America. *Annual Reviews of Earth and Planetary Science* 5:123–158.
Wright, H.E., Jr. (1981). Vegetation east of the Rocky Mountains 18,000 Years Ago. *Quaternary Research* 15:113–125.
Wright, H.E., Jr. (ed.). (1983). *Late-Quaternary Environments of the United States*, vol. 2, *The Holocene*. University of Minnesota Press, Minneapolis.
Wright, H.E., Jr. (1984). Sensitivity and response time of natural systems to climatic change in the Late Quaternary. *Quaternary Science Reviews* 3:91–131.
Wright, H.E., Jr., & Heinselman, M.L. (1973). The ecological role of fire in natural conifer forests of western and northern America. *Quaternary Research* 3:316–318.
Yurtsev, B.A. (1982). Relicts of the xerophyte vegetation of Beringia in northeast Asia. In D.M. Hopkins, J.V. Matthews, C.E. Schweger & S.B. Young (eds.), *Paleoecology of Beringia*, pp. 157–178. Academic Press, New York.
Zagwijn, W.H., & Suc, J.-P. (1985). Palynostratigraphie du plio-Pleistocène d'Europe et de Méditerranée nord-occidentales: correlations chronostratigraphies, histoire de la végétation et du climat. *Paléobiologie continentale* 14(2):475–483.
Zasada, J.C., & Viereck, L.A. (1979). *White Spruce Cone and Seed Production in Interior Alaska, 1957–68*. Forest Service Research Note No. PNW–129. U.S. Department of Agriculture, Washington, D.C.
Zoltai, S.C. (1973). The range of Tamarack (*Larix laricina* (Du Roi) K. Koch) in Northern Yukon Territory. *Canadian Journal of Forest Research* 3:461–464.
Zoltai, S.C. (1975). *Southern Limit of Coniferous Trees on the Canadian Prairies*. Information Report NOR-X-128. Environment Canada, Forestry Service, Ottawa.
Zoltai, S.C., & Tarnocai, C. (1975). Perennial frozen peatlands in the western arctic and subarctic of Canada. *Canadian Journal of Earth Sciences* 12(1):28–43.

Index

K

L

M

N

O

P